0503998

W9-BNT-324

ON LINE

ROBOTS

From Science Fiction to Technological Revolution

DANIEL ICHBIAH

ROBOTS

From Science Fiction to Technological Revolution

Translated from the French by Ken Kincaid

Harry N. Abrams, Inc., Publishers

Front cover: Autonomous security robot prototype Robart. Courtesy of Spawar. *Back cover:* (left and center) Robot toys. Collection of Jean-Pierre Hartmann. Photograph by Christophe Recoura. Courtesy of FYP Éditions. (Right) Sony's android, Qrio. © Courtesy of Sony

Produced by Philippe Bultez Adams and Florence Devesa, FYP Éditions, France - robots@fypeditions.com

For Harry N. Abrams
Project Manager, English-language edition: Susan Richmond
Cover design, English-language edition: Brankica Kovrlija
Production Coordinator, English-language edition: Norman Watkins

Library of Congress Cataloging-in-Publication Data
Ichbiah, Daniel.
  [Robots. French]
  Robots : from science fiction to technological revolution / Daniel Ichbiah ; translated from the French by Ken Kincaid.
    p. cm.
  Translation of: Robots : Genèse d'un peuple artificiel.
  ISBN 0-8109-5912-7 (hardcover)
  1. Robotics. 2. Robots, Industrial. I. Title.

  TJ211.I28 2005
  629.8'92--dc22

                    2004030887

Printed and bound in Italy
10 9 8 7 6 5 4 3 2 1

Harry N. Abrams, Inc.
100 Fifth Avenue
New York, N.Y. 10011
www.abramsbooks.com

Abrams is a subsidiary of

LA MARTINIÈRE
GROUPE

31/05
BfT
37.50

Contents

Foreword

by Will Wright,
creator of the video games
Sims and *Sim City*

One of the deepest and most abiding interests of my adult life is robotics. I have been building various robot-related things since I was a teenager. So I feel honored to be given the opportunity to write the foreword to this book dedicated to robots.

I think that what really attracts me to robotics is the desire to more fully understand what it means to be human. Robotics and artificial intelligence are, fundamentally, attempts to model various aspects of ourselves. Until you have tried to build a functioning model of a human hand, it is impossible to appreciate fully what a marvelous mechanism it actually is!

A few years ago, in parallel with my video game programming activities, I started a small group of robotics enthusiasts. We would build interesting robots and take them into the street to see how average people would respond to them. One of the things that we are investigating is what happens when the lay person has an unexpected encounter with machine intelligence. It is a very enriching line of inquiry, because it enables me to explore what the social design of the intelligent machine might be. What's more, it raises primary problems that are not even considered in research laboratories. I want to see what happens when robotic devices are forced out of the laboratory into the world. What real problems do they actually encounter? Should they act like appliances? Or a pet? Or a person? There are a number of interesting robotics programs at universities around the world, yet the problem is that most of the prototypes they develop stay in their laboratories.

Over time, our definition of robots has constantly changed, and continues to constantly change. Intelligence was a clear-cut, well-understood concept until we started building thinking machines. Even when the movie *2001, a Space Odyssey* came out in 1968, people walked out of the theater commenting, "I can't believe HAL could beat that guy at chess." The assumption was that chess required a high degree of intelligence and that the natural conversation that HAL engaged in was relatively easy. We have since found that the reverse is true. Our concept of what intelligence is has been altered by this realization.

As we solve problems and robotic machines enter our lives in relatively transparent ways, we will cease to think of them as robots. It is merely a question of time. They will submerge into our environment and gradually become invisible to us. Today, I do not talk about my car's "robot codriver," but about its navigation system. It does not occur to me that it is a robot.

The primary challenge facing robotics today is that of situational awareness. We can build very capable hardware that can perform useful tasks under human control. But the hard part in automating that control is giving computers the same level of awareness that the human has. Similarly, we can attach decent cameras to a robot and send the signal into a computer. But the ability to convert the pixel data into an accurate 3D world model seems to involve substantial real-world knowledge. We need to give our robots the ability to build better models of the real world.

Ironically, though, as we build robots that are, in some sense, models of ourselves, we begin to glimpse that one of the fundamental skills that we must learn to give them is the ability to build models of their own.

THE HISTORY OF ROBOTS

What exactly is a 21st century robot? It is a very powerful computer with equally powerful software housed in a mobile body and able to act rationally on its perception of the world around it.

Interestingly, the popular imagination naturally pictures robots in human likeness. Whether tall or tiny, they are necessarily androids. It is as if they embodied humankind's tireless search to create and control a double of its own making.

Man's quest to replicate living beings goes back thousands of years, constantly incorporating developments in technology, from hydraulic engineering and clockwork to electronics. To arrive at the thinking android of the 21st century, the robot has evolved through many a stage.

Chico MacMurtrie's kinetic Fetus to Man sculpture and public clock is made up of a life-sized metal clock-jack. The human figure, or jack, are the hands of the clock, acting out a different stage in the cycle of life every hour. From a fetus in the womb at 6 o'clock, it slowly straightens up until it stands erect at 12. It then turns around to show a careworn face, and gradually bends under the weight of the years, before finally crumpling at 6 o'clock. Fetus to Man was installed on the facade of the Concorde building in the city of Lille in Northern France on December 6, 2003. It was commissioned to commemorate the city's stint as European Capital of Culture in 2004.
© Chico MacMurtrie - Photo : Jean-Pierre Duplan - Production Lille 2004

Where do robots come from?

1- *Ctesibius's water clock. Ctesibius was a gifted, ingenious Greek engineer who lived in the 3rd century BC. He repurposed the aulos, a traditional double-reed woodwind instrument that was widespread at the time, making it into a clock driven by water.*

2- *Jacques de Vaucanson was born in the city of Grenoble, France, on February 24, 1709. He built uncannily lifelike automata, two of which, The Duck and The Flute Player, made him famous.*

3- *Born in 1721, the Swiss watchmaker Pierre Jacquet-Droz built mind-boggling automata. Three are celebrated to this day: The Writer, The Musician, and The Draughtsman.*

4- *Jean-Eugène Robert, known as Robert Houdin, was born in the central French town of Blois in 1805. He created dolls and automata such as the Writer-Draughtsman.*

5- *Czech writer Karel Capek was born in 1890. It was he who coined the word "robot," which first appeared in his most famous book, RUR, published in 1921.*

Ctesibius, Vaucanson and Walter are among those ingenious engineers who, over the ages, strove to create predictable, harmonious motion in objects of their own making. Fired by their fascination with machines that appeared to move of their own accord and so looked as if they were alive, they gave them human and animal form, repertoires of gesture and set sequences of motion. For thousands of years automata relied on springs, cogs and other mechanisms, exuding a grace that even today we still admire. Then information technology entered the picture in the shape of the revolutionary microchip, which could store huge quantities of data and sequences of movement. Robots had arrived, and so sophisticated have some become that they themselves can now seek solutions to problems.

Key stages in the history of robots are:

■ **The moving masks and statues of the Ancient World.** They performed "fake miracles" designed to put fear of the gods into the faithful.

■ **Timekeeping.** From the earliest water clocks and clockwork motive power of automata to the complex inner workings of computers and androids, high-precision timekeeping has been essential to robots at every stage in their history.

■ **The automaton.** The term is derived from the Greek, *automatos*, meaning "that which moves by itself." Early automata were mechanically actuated embodiments of man's infatuation with machines that resembled living beings and could accomplish one or more tasks. Unlike its successor, the robot, the automaton performed specific, predetermined tasks and was unable to respond to external stimuli.

■ **Informatics.** Informatics is the science of automated information management. It originated with Jacquard's weaving looms and gathered growing momentum with the emergence of electronics, the invention of the transistor and the development of computers.

■ **Artificial Intelligence.** The branch of computer science that seeks to reproduce the pattern of human thought processes—analysis of a situation giving rise to decision, then to action.

■ **First-generation robots.** Highly minimalist in design, they perform specific tasks like paint-spraying or welding in an assembly plant. Although they are electronically operated, they are similar to early automata in that they are intended to perform a single task and are devoid of all ability to perceive.

■ **Second-generation robots.** They are equipped with sensors that enable them to examine their surroundings through sight or touch. They then take appropriate action.

■ **Third-generation robots.** Robots of the third generation are the culmination of thousands of years of research. Their artificial intelligence enables them to reason and act with no outside help.

Those, briefly, are the main chapters in the life story of the robot. As to the meaning of that life, etymology provides a clue. When Czech writer Karel Capek first forged the term in 1921, he knew just what he meant. He coined it from the Czech word for work, *rabota*, to mean "servant, enslaved worker." The artificial being's fate was sealed. Indeed, what else is the robot, but man's auxiliary, the embodiment of his dream of developing a race of glad and willing slaves, happy to slog away at menial chores or soar into the skies on reconnaissance missions to space as part of man's unquenchable thirst to explore the unknown?

6- *William Grey Walter was a pioneer in the field of cybernetics. Born in the USA in 1910, he was educated and worked in the UK. It was there that his research into artificial intelligence led him to create two three-wheeled, turtlelike machines with plastic shells that were powered by telephone batteries. He endowed them with two "neurons," electric circuits that amplified and transmitted signals from sensors to two motors. The turtles were drawn to light, which they detected through dedicated sensors. However, if the light was too bright they would turn away.*

Masks and statues

The origins of masks and statues that move can be traced to ancient Egyptian times. An ibis-headed mask in the likeness of the god Thot, and one of the hawk-headed Horus both seemed to be endowed with lifelike motion. Their chief characteristic is that the automated mechanisms embedded within them were used by the priest caste to exert power over both the people and the pharaohs. The miracles that priests worked to awe their contemporaries were in reality the result of their skill in operating hidden mechanisms. Temple doors opened and the statue of the god Amon lifted its arm not because the oracle spoke, but—more likely—because the flames from the holy fire caused air to expand, so compressing water and causing it run, which, in turn, set a system of ropes and pulleys in motion. A similar explanation could well be behind the milk that seeped so mysteriously from the many-breasted goddess Artemis.

The first automata that were presented as such are thought to go back to around 380 BC. A friend of the great Greek philosopher Plato, Archtyas of Tarentum, is alleged to have made a wooden pigeon. According to some accounts, it flew in loops—a motion probably caused by jets of compressed air.

TÊTES PARLANTES
Problème résolu en Mécanique qui jusqu'à ce jour avoit été regardé comme insoluble ou du moins comme très difficile
L'Accademie des Sciences a dit dans son raport que ces têtes parlantes peuvent jetter le plus grand jour sur le Mécanisme de l'Organe Vocal, et sur le mistère de la parole : Elle ajoute que cet Ouvrage est digne de son approbation par sa nouveauté par son importance ; et par son execution

In 1781 the French abbot Mical made two talking heads, which were able to produce portentous utterances like "The king brings peace to Europe".
© Private collection.

Clocks

Let us now turn our backs on seldom accredited phenomena that were either the product of trickery or the stuff of legend and turn to questions that exercised the minds of scientists of old. A fascination that exerted its hold as early as Antiquity was the regular flow of time.

From the constantly beating heart to the cycle of the four seasons that rolled on according to the position of the sun, the workings of the universe seemed to obey a rhythm. An early timepiece that emerged in response to the need to measure the passage of time at night and during the day was the clepsydra. It relied on the flow of water to keep time in an approximate manner. Records take us back to 246 BC in our search for the first great time-keeping inventor—a certain Ctesibius, who lived in the city of Alexandria. He succeeded in creating a clock so accurate that a revolution of its dial took exactly one solar year. At last a perfect match had been achieved between a man-made timepiece and a phenomenon from the natural, physical world. But the genius of Ctesibius was not to stop there.

He followed up the water clock with a hydraulic music-making machine, an early forerunner of the street organ. The strange apparatus was operated by water running through a system

of pumps, counterweights, valves and pistons that were connected to a dozen wind instruments, known as *aulos*. Hence the name given to the organ—*hydraulis*. Its renown spread swiftly to surrounding lands. During a stay in Asia Minor in 78 BC the great Roman orator and philosopher Cicero waxed ecstatic about the sounds the *hydraulis* produced, likening them to "a sweet delicacy". In Rome it became the in thing to show off at fashionable parties. On the other side of the world, in China, the same desire to bring objects to life had emerged.

Between 140 and 87 BC Emperor Woo converted a palace into an opera house complete with stage machinery actuated by acrobats, jugglers and tightrope-walkers to bring movement to a strange bestiary of animals from the natural and imagined worlds.

In the 14th century the carillon, or chime-clock, heralded the beginnings of mechanical music. It cleared the way for blade-operated musical movement, from singing birds to musical boxes. In 1865, one Charles Reuge settled in the village of Sainte-Croix, Switzerland, where he began to make musical pocket watches.
In 1953 the Reuge company resumed the production of multi-melody musical movement.
© Reuge 1865 – Lépine Pocket Watch.
Left: Musical pocket watch with automaton-driven scene, The Fountain.
Above: The same model featuring an erotic scene on the back of the watch.

From Oriental automata to European clock-jacks

In the first century of our era a Greek mathematician and engineer, Hero of Alexandria, wrote a treatise on automata in which he demystified ancient miracles by explaining that mastery of hydraulics and physics had made them possible. Most important, however, Hero spelled out the fundamentals of automatic movement, i.e., the elasticity of steam and its capacity to drive movement when heat and pressure are applied.

Arab engineers were the first to put the teachings of Hero (and Philo of Byzantium) into practice on a large scale. In 809 the Sultan Harun Al-Rasheed sent a mechanical automaton to Emperor Charlemagne as a gift. Then in the course of their eight expeditions to the East between 1096 and 1291, the European Crusaders discovered for themselves the astounding refinement of the waterclocks crafted by Al-Jazari for Harun Al-Rasheed. There were birds that opened their beaks to drop marbles on cymbals, trumpet-playing musicians, and doors that opened to reveal human figures.

To secure the constant flow of water needed to animate the automata, Al-Jazari had developed a quite ingenious system, based on the discoveries of Archimedes. His largest clock was 3.3 meters high (11 feet) and 1.35 wide (4 feet 6 inches). Whether or not he was the inspiration for the clockmakers of France when the Crusaders returned home, the late Middle Ages saw a proliferation of spires and steeples topped with clock-jacks. These were human figures made from lead or cast iron, which struck bells to mark the time of day. Their chimes would echo across cities as people went about their lives and business.

The oldest such automated clock was made in 1351 in Orvieto, Italy, while some steeples, like that of Cluny Abbey in France, were stages for the enactment of entire clockwork scenes. An angel hailed the Virgin Mary and a dove symbolizing the Holy Ghost swooped down, while God the Father blessed his creation. Such feats of precision engineering were partly intended to awe the church-going faithful who, totally ignorant of clockwork mechanics, were dumbfounded.

Clock-jacks,
Strasbourg Cathedral,
France.
© FYP 2004

By his subtle, forceful control of
1,880 neon lights on Claude Vasconi's
Lilleurope Tower in northern France,
Austrian artist Kurt Hentschläger has
made the skyscraper into a hypnotically
luminous 20-story clock-jack. Like the
Colossus of Rhodes or the Lighthouse of
Alexandria, the Lilleurope clock-jack tells
rail and road travelers that they are
approaching their destination.
© Kurt Hentschläger, Nature 04.
Courtesy of E. Valette – Production Lille 2004

Toward the golden age of the automaton

By the time of the Renaissance automata had become commonplace toys and attractions in the homes of the well-to-do. A great many of them were hydraulically operated. The gallery of Hesdin Castle in the northern French region of Picardy, where the local gentry liked to sojourn, was home to stick-wielding automata that would give guests a drubbing and-or blow white powder into their faces. All to great amusement. Around 1500 Louis XII had a mechanical lion built. The creature could walk, stop and designate the royal coat-of-arms when the king ordered it to do so. It was a fair reflection of the inventive genius of French craftsmen of the time.

Nowhere, however, did automata enjoy such pride of place as at the royal residence at the château of Saint-Germain-en-Laye, to the west of Paris. The grounds held grottoes that were packed with hydraulic machinery, which the great Florentine engineer Thomas Francini designed and made for the sole entertainment of the upper classes. In 1598 René Descartes described the automata in his book, *Treatise on Man*: "Those who enter certain of the fountains' grottoes themselves cause, and without so realizing, the very movements that take place before them, for they may enter only by treading on certain tiles, which are so arranged as to make the bathing Diana, should they approach her, hide among the reeds; should they then proceed further in her pursuit, they will cause Neptune to come forth and to brandish his trident; should they seek to take another path, they will cause a sea monster to sally forth and to vomit water in their faces, or such similar things according to the whim of the engineer who made them."

Other descriptions mention a sea where fish frolic. A clap of thunder suddenly turns into stormy waves and the scene then changes to reveal the dauphin (the king's son) descending from on high in a chariot.

It was the 18th century that was to assert itself as the golden age of the automaton. One of the greatest inventors of things mechanical was King Louis XV's protégé, Jacques de Vaucanson (1709–1792). It was said that Vaucanson had a dream—to build an artificial man. He had, however, to content himself with making a mechanical duck, which

was enough to earn him admiration. According to one account, the duck "extends its neck to take grain from a hand, swallows it and digests it". It could indeed turn the swallowed grain into a fully digested mush, which it then excreted through the usual channels. The automata he crafted included a flautist, capable of playing eleven different tunes.

Vaucanson's creations—and those of his disciples—enchanted the whole of Europe and were exported as far as the USA.

The brief life of Vaucanson's duck

"It is not known exactly what became of Vaucanson's three automata after he had parted with them. They departed, it is thought, for foreign lands and the duck, found in an attic in Berlin, was purchased by one Georges Tiets who took four years to restore it to its original state. This piece, exhibited in Paris in 1844, required some repair and was entrusted to the care of Robert-Houdin. It is well established that, even in Robert-Houdin's time, Vaucanson's automata prompted the emulation of copyists, and we believe that the duck described by the celebrated conjurer was precisely one such replica. We are of the opinion that the deception denounced by Robert-Houdin was considerably too crude and little worthy of the great inventor's mechanical genius."

© *From* Secrets of Conjuring and Magic, *by Robert-Houdin, 1868.*

Page left
Above: Portrait of Jacques de Vaucanson.
© *Courtesy of the Grenoble Museum of Automata*

Below: Painting of Jacques de Vaucanson and his famous mechanical duck.
© *Courtesy of Ville de Blois*

Page right
Above: Two pictures of Vaucanson's duck.
Courtesy of the Grenoble Museum of Automata

Below: **Anas Mechanicas Arcana, 1997, a work by Frédéric Vitoni in tribute to Jacques de Vaucanson and exhibited by the Grenoble Museum of Automata.**
© *F. Vitoni, Grenoble Museum of Automata*

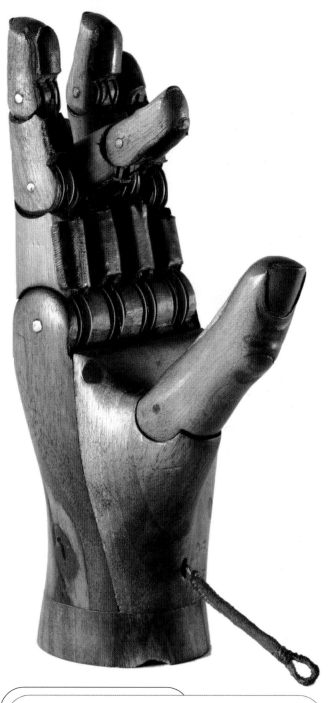

The first programmable machine

The needs of the French weaving industry of the 18th century led to the invention of the first randomly programmable machine. At the time the Chinese style with its silk outfits and complex patterns of embroidery was all the rage. The cloth weavers of the French city of Lyons were at their wits' end: the manufacture of figured fabric involved operating a huge number of warp threads with needles attached. Local weaver Basile Bouchon got to thinking about the problem. As luck would have it, his father was an organ-maker. It occurred to Bouchon that it should be possible to adapt organ technology to looms by using strips of perforated card to guide the needles. The principle was straightforward: if a needle went through a hole, it would go through the cloth. If there was no hole, it would be driven back. The process was developed in 1729 and automated by Vaucanson in 1745. Some decades later Joseph-Marie Jacquard, a weaver, came up with the idea of making the punched pattern cards separately from the actual loom. Jacquard unveiled his weaving loom in 1801—the first machine to process information automatically and to work on the principle of separate programs. It sold in the thousands and cleared the way for computers and robots able to operate by themselves.

Coll. Vivier Merle (Cl. Gontier)

Artificial limbs

Walnut and boxwood prosthetic hand from the 20th century.
The first artificial limbs date back to 400 BC. Herodotus recounts in Calliope, book nine of his Persian Wars, how Hegesistratus cut off his own foot to escape from the Spartans, then procured himself a wooden replacement.

© *Private collection – Michell and Jean-Pierre Hartmann – Photo: C. Recoura*

THE HISTORY OF ROBOTS · 19

The hands of the famous automaton The Writer *built by Pierre Jacquet-Droz (1721–1790).*
© Neuchâtel Museum of Art and History, Switzerland

The global automaton craze

At about the same time, Swiss watchmaker Pierre Jacquet-Droz and his son, Henri-Louis, designed and built three very human-looking automata. One could draw with all the finesse of an artist. Another—made in 1773 in the likeness of a young woman—could play the harpsichord while looking from her hands to the musical score. She would then rise to take a bow at the end of her piece. The third of these android dolls was crafted the following year and called *The Scribe*. It could hold a quill and write down a given text of up to 40 characters in an elegant hand. So lifelike and near-perfect were Jacquet-Droz's automata that

voices were raised in Spain accusing him of witchcraft and demanding his arrest.

Other renowned automaton engineers contemporary with Jacquet-Droz include the Austro-Hungarian Baron Von Kempelen. He created a chess player that inspired as much awe as controversy as to whether a human being was hidden inside it.

Watches in the 18th century generally tended to be miniature reproductions of the complex workings of large clocks. Tiny male and female clock-jack couples dubbed "Martin and Martine" appeared on dials. Such was the popularity of these clockwork gems and their mechanical figurines that they soon ceased to be luxury items. The growing market prompted

fierce competition among watchmakers to produce the most original miniature scenes.

Towards the mid-19th century the automaton vogue assumed a global scale. In the Japanese city of Osaka, Takeda Omishojo's mechanical puppet theater won such resounding popularity that the genre was named after him— *Takeda Karakuri* (*karakuri* being the term for the automated wooden dolls). At the same time, far-away Europe was being swept by the craze for conjuring automata, able to perform tricks worthy of the master magician Robert-Houdin (1805–1871), who revolutionized magic by introducing hydraulic and clockwork mechanisms. Darlings of bourgeois society, these mechanical magicians could levitate, perform sleight-of-hand, swallow balls and tell the future.

A decisive step towards the coming of the robot was made in 1854 when Professor George Boole of Queen's College, Cork, in Ireland, set out the system of logic that bears his name. Boolean algebra uses only two values, 0 and 1. The sheer simplicity of the basis of Boole's system had opened the way to formal representations of logical thought and thus to the feasibility of a machine endowed with the ability to think. The computer was coming.

As the 20th century dawned the public's infatuation with automata waned and the advertising industry stepped into the breach. In 1909 the Parisian department store Au Bon Marché staged an animated window-dressing scene in which automata reenacted the first successful expedition to the North Pole, led by Rear Admiral Robert E. Peary. Other automata tapped on the windows to attract the attention of passersby.

It was about this time that the first devices that could be likened to the robots of the future came into being. One was the electric dog built in 1915 by Hammond and Miessner. It was drawn to light, which it detected through a sensor. What set it apart from mere automata was precisely that an outside stimulus (light) caused it to respond and behave in a certain way. Spanish inventor L. Gondola Torres built numerous machines that operated automatically—the word is of his coinage. One was a chess player able to make the final three moves of a game.

Songbird in a Cage, *made around 1910 by Blaise Bontems, a clockmaker from eastern France. He took the art of mechanical birds to near perfection, faithfully reproducing the warbling of the real thing.*

© Private collection Frédéric Marchand
Photo : Christophe Recoura

AUTOMATES, PRESTIDIGITATION.
DIMANCHE PROCHAIN
DEUX SEANCES
M. ROBERT-HOUDIN SE CHARGE DES SOIREES PARTICULIERES.

Le Génie des Roses (Automate).

Le Chasseur et la Colonne au gant.

The "Fantastic Evenings" of Robert-Houdin

1- The Genie of the Roses
2- The Hunter and the Glove-holder
3- The Patissier of Palais-Royal – engraving
4- The Fantastical Orange Tree – engraving
5- The Chess Player – painting by
E. Partemelot

<u>Above:</u> Poster advertising a performance on January 14, 1852, of one of Robert-Houdin's "Fantastic Evenings."

Interview
Philippe Sayous

Director of a movie production company, Philippe Sayous has made several originally styled documentaries, two of which explore the history of automata, **Les Automates de Neuilly-sur-Seine** *and* **Les Androïdes Jacquet-Droz.** *His Paris-based company markets automata and musical boxes and also makes them, using period production methods. He is a globally acknowledged authority in the field of automata and runs the website automates-anciens.com.*

Philippe Sayous,
director of TIL Productions

In the documentary you made about Jacquet-Droz you say that android fever swept the 18th century. How do you explain the infatuation?

It was a very scientifically minded age. People were asking questions about the nature of man. Was it possible to create an artificial being that could usefully perform difficult tasks in his stead? Could artificial limbs be made?

Why then did they build a young girl playing the harpsichord, and a writer?

Partly as a tribute to the arts, but primarily because drawing, writing and music were the most difficult tasks to set a machine.

Vaucanson was the first great creator of android automata. He made them with the purpose of furthering science. How do his automata reflect his interest in science?

Vaucanson had studied medicine and he constantly sought advice from doctors and surgeons when making his androids. Each automaton was built with a specific aim in mind. His flute player, for instance, was a study in how a human breathes. As for the duck, it was all about digestion.

How did an automaton like the flute player work mechanically?

The central mechanism was a revolving cylinder studded with short pegs that drove rods which, in turn, activated the automaton's limb movements.

How did Vaucanson's duck work?

The duck could waddle in water, quack, flap its wings, move its neck. It craned its neck forwards to take the grain it was offered. Three mechanisms mimicked the basic digestive processes of breaking food down, reducing it to a mush and excreting it.

Did Jacquet-Droz's organ grinder work with basic mechanics similar to those of the street organ?

Yes, it also used a revolving cylinder with studs. Some of them were in fact quite long, so as to draw out notes.

Would you say that Jacquet-Droz's automata were never bettered?

Some were even better and that's what my next film will be about. Round about the middle of the 19th century there was a mechanical engineer, Stèvenard, who was also a conjurer and ran a miniature automata theater in Paris. Some were as complex as Jacquet-Droz's androids but on a smaller scale. You needed an ultra-powerful magnifying glass to be able to see the links in the chains he used. One automaton was an illusionist. It would take some cups and place them on a table. Then, when it removed them, there was an egg. The egg would open and out would come a hummingbird. It flapped its wings, then withdrew into the

The Little Marquess, a three-movement android by Christine and Laurent Duchaussois
© TIL Production – P. Sayous
Photo: Christophe Recoura

advertised as holding "100,000 curiosities". One whole room was devoted to mechanical curiosities. They included androids, some of which were Jacquet-Droz's models. Yet this museum, too, burned down.

As you showed in one of your documentaries, the research carried out by Vaucanson and Jacquet-Droz led to the first articulated prosthetic limbs being made. Can you tell us about that?
We came across a letter from a baroness to Jacquet-Droz and his son thanking them and describing the artificial hand that she had had grafted. The contemporary French artist Monestier has designed a similar hand, which he created as an *objet d'art*. Somebody fitted with it can thread a needle and fold paper.

Automata remained popular throughout the 19th century. How do you account for the wane in enthusiasm after that?
After the First World War interest fell away. It should be said that automata were very expensive to build. If one that could make only five or six movements was to be made nowadays in accordance with the rules of the art, it would cost about of $22,500.

egg, which would close and vanish again. The trick would last about ten minutes. Sadly, we don't know very much about Stèvenard.

How do you account for the fact that so many famous automata have disappeared?

We have no idea what became of some of them. Others were quite simply destroyed. Vaucanson's duck is alleged to have been destroyed in a fire at the museum of St. Petersburg. Similarly there was the Barnum Museum in the US, created by the impresario and showman of the same name and

Jean-Pierre Hartmann's automata and imaginary toys

1- Falconry and Flying Carpet, *brass and bronze automaton. The carpet appears to float, the falconer raises his arm and the bird flaps its wings.*

2- The Caravan, *seven-movement alloy and bronze automaton.*

3- Automaton driven by a musical box.

4- The Wasp and the Butterfly, *bronze and alloy automaton.*

5- Black Icarus, *bronze and alloy automaton. Both wings are articulated by two hinges, which allows the automaton to reproduce the wing motion of a real bird. Icarus is holding a burning torch.*

6- The Bathyscape, *bronze automaton with six lights. The figure pitches, the porthole opens, and the propellers rotate.*

7- Thomas Edison, *a domestic-surveillance robot. An infrared sensor enables it to detect any human presence. It stamps its feet, extends its head-lamp, talks, and can dial into a telephone line.*

A computer for brainpower

The 1940s ushered in a major new development that saw electronics take over from the slow, unwieldy cogs and levers of mechanical systems. In a study published in 1913, the great Danish physicist Niels Bohr had described how electrons could move from one atom to another at lightning speeds of thousands of kilometers per second. Bohr's findings soon gave rise to ideas for producing circuits that would make use of such incredible speed of motion. In 1937 Alan M. Turing published his milestone paper, *On Computable Numbers*, in which he set out the broad principles of a machine that would perform computation at electronic speeds and would thus be capable of processing enormous amounts of data in Boolean binary code (0 and 1). Turing believed that such a machine would be able to work from symbols to solve any problem that was put to it and could therefore rival human brainpower. Turing was the driving force behind the first computer, which was built in 1943. The British put its sheer calculating power to use during the World War II. It played a decisive part in cracking Enigma, the code the Nazis had developed to exchange messages. Computers were well and truly on their way. They would play a key role in the development of the thinking machine, or robot. The first computer worthy of the name was commissioned in February 1946 at the University of Pennsylvania. Now regarded as an IT dinosaur it bore the name of ENIAC. It took up 140 square meters (1,506.94 feet) and bristled with 19,000 vacuum tubes and hundreds of cables and relay switches. Nevertheless, Harvard University continued to make use of it for over fifteen years. The mid-20th century saw the advent of a crucial new invention that would replace bulky, unreliable vacuum tubes—the transistor. The work of William Shockley and John Bardeen, the transistor was an electronic component able to amplify and switch electric current. It could not have come more opportunely, just at a time when people were beginning to think of ways of making computers smaller. In the 1950s several transistors were implanted in a piece of silicon, a chemical element whose natural form is silica, or sand: the integrated circuit was coming.

The first remote-controlled arm

During the Second World War Americans and German were involved in a race against time to develop the first atomic weapon. It was the search for the atom bomb that gave rise to the first remote-controlled arms (limbs, not weapons). Because radio-active materials were too hazardous to be handled by humans, machines had to do it. These machines were claws driven by a system of rods and pulleys and controlled by a human operator working at a safe distance behind a protective glass screen.

In 1954 a scientist from eastern France by the name of Raymond Goertz came up with idea of fitting articulated arms with electric motors. The advantage of the system was that an operator could be hundreds of meters distant, well away from any hazardous materials, while the arm's claws responded to commands transmitted to it through electric wires. The concept of the robotic arm was nigh.

Artificial Intelligence

The advent of the computer tantalizingly pointed to a future where Turing's "thinking machines" would be a reality. One morning in September 1956 Herbert Simon, professor of computer science at Carnegie Mellon University in Pittsburgh was strolling in the park. Suddenly it came to him that it must be possible to program a computer to simulate a logical thought process. He spent the rest of the weekend with another computer scientist, Alan Newell, working on his insight. The result was a program dubbed Logical Theorist, which was able to prove basic math theorems by itself. Although the concept of artificial intelligence had now been born, a long time would pass and many hopes would be dashed before it came of age. Two other pioneers in artificial intelligence were John McCarthy and Irvin Minsky. The ideas they expounded aroused the interest of the scientific community. Knowledge engineers designed expert systems, i.e., sets of rules that computers can follow to simulate human activity, while one Norbert Weiner, a mathematics professor at the Massachusetts Institute of Technology (MIT), defined a new discipline that he called cybernetics. One of its declared aims was to study the foundations underlying communication and automatic control in man-made and living systems.

Wiener's theories, which he set out in his book *Cybernetics or Control and Communication in the Animal*, examined how to enable a robot to perform a given task. They influenced a Missouri-born British scientist, W. Grey Walter, who set about putting Wiener's theories into practice. Grey built and programmed two walking tortoises, whom he called Elmer and Elsie. They wandered about his apartment in Bristol by themselves, using a system of sensors that included photo-electric cells to get their bearings. The work of Walter and others drew wide interest and was publicized by lengthy articles in publications like *Scientific American*.

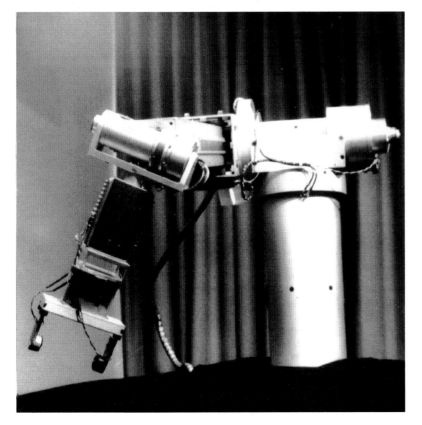

The robotic arm built in 1969 by Victor Schelman at the Stanford Research Institute in California was powered exclusively by electricity. The motive force of all its predecessors was hydraulics or compressed air.

Robot generation I

The advent of sensors and the emergence of artificial intelligence seemingly spelled the emancipation of the robot. That day would come, but not quite yet. For one thing, computers were still too bulky and costly for their components to be built into robots on a large scale. And sensors were not even thought necessary.

Functionally speaking, the first robots to come to the market were similar to the automata of yore, while their design was so unflattering that Vaucanson would have turned up his nose.

A cocktail party in 1956 brought together George Devol and Joseph F. Engelberger, who agreed to start working on a robot together. In 1958 Engelberger, working from his garage, created a company, prompted by his acute commercial acumen, which told him that the robotics market would one day be big business. The first robot that Devol made was called Unimate (from Universal Automation).

It was intended for the prosperous automobile industry, in accordance with Engelberger's intuition that the time had come to relieve men of burdensome, uncomfortable and dangerous toil. The automakers themselves could not but look approvingly on labor that was never tired, grumbled or demanded pay rises. George Devol obtained the first US patent for a robot in 1961— number 2,998,237. It was not long before Unimate took its place on a General Motors assembly line in New Jersey. Hooked up to high-pressure die-casting machines, the 1.5-ton articulated arm could handle pieces of metal weighing up to 150 kilograms (330 pounds). Engelberger and Devol's company, Unimation, eventually sold 8,500 Unimates, a crude kind of robot that performed only one task and was unable to perceive or respond to external stimuli. Companies similar to Unimation flourished. One was AMF, which produced its first industrial robot in 1963, while a labor shortage in Norway led to a Tralfa robot being enlisted to paint wheelbarrows.

These articulated hands and arms, built to perform repetitive tasks like welding and spray-painting automobile bodies, were the first examples of robotics being put to practical use.

Artificial intelligence, however, had still not proved itself, even after Marvin Minsky published his milestone work *Steps Towards Artificial Intelligence*. Controversy broke out between artificial intelligence devotees like Herbert Simon, who upheld that by 1985 machines would be able to do anything humans could do, and skeptical researchers like Dreyfus and Ryle. They argued that Minsky's vision would never come true. Nevertheless, the idea that a computer would one day think for itself exerted its hold on movie director Stanley Kubrick. His 1996 film *2001, A Space Odyssey* portrayed a spaceship computer, HAL, who actually believes himself superior to the human astronauts. It should, of course, be said that robots had already became part of the cast in real-life space exploration. The moon landing vessel *Surveyor* was equipped with an automated robotic arm.

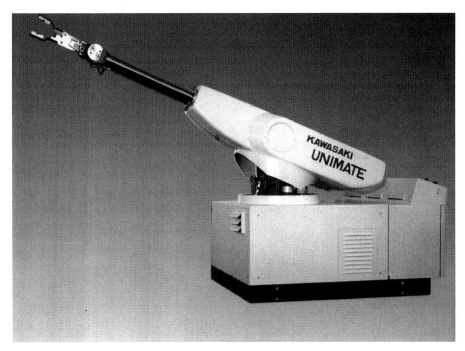

In 1966 Kawasaki bought a license to manufacture industrial robots from the US company Unimation. Japan's first industrial robot, the Kawasaki Unimate, was a welder in automobile plants.
© Courtesy of Kawasaki Heavy Industries Ltd.

Robot generation II

The visions painted by sci-fi novelists like Isaac Asimov had shaped expectations of robots that were rather more sophisticated than the drudgery of Unimation arms welding away in a General Motors plant. The cybernetics concept forged by Wiener had outlined the direction research should take—enable automata to gather information from and about its surroundings and act accordingly. It was a challenge taken up in 1968 by a team at the Artificial Intelligence Center (AIC) at Stanford University,

California. The AIC's first autonomous, mobile robot, Shakey, was no beauty. Its head was a camera that swiveled and tipped, while its body was a chunky computer squatting on a three-wheeled platform. Shakey the Robot could, indeed, move by himself, but at a maddeningly slow pace—he had to spend an hour thinking before undertaking the slightest movement. He was dubbed "Shakey" because he … well, shook, when he moved. Still, what mattered was that, despite his ponderous appearance, he could sense his environment and navigate his own course thanks to his sensors—

a TV camera, a range finder for measuring distances and bumpers for feeling physical obstacles. Shakey could carry objects into the next room. He had broken new ground. Only two years later, in 1970, General Motors was using an industrial application of robotic vision in its foundry in Ontario, Canada.

Generation III: robots with AI

It was in the mid-1970s that the microcomputer entered the picture, its multitasking capability providing the robot with what had been its missing link. If a robot could endlessly repeat a single task, then why should it not be able to perform all sorts of jobs? It was just a question of modifying its software, or enabling it to make the modifications itself. A robot whose intelligence is easily programmable would be a very different machine from its immediate predecessor, the automaton.

What distinguishes third-generation robots is that they can perform tasks by taking appropriate action themselves without humans having to intervene. The first intelligent humanoid robot was born at Waseda University, Tokyo, in 1973. Its name was Wabot-1. Although it bore only a remote resemblance to a human being, it could see, grasp, walk stiffly on two legs, and even speak a few words of Japanese. Its mental capacity was compared to that of a baby, but then again, humanoid robots were still in their infancy. The Hitachi corporation followed suit in 1974 with the Hi-T-Hand robot, which used force sensing to slide pins into holes. By the end of the

1970s, roboticist Hans Moravec had unveiled the first robots to operate autonomously outdoors. Now that they were endowed with the ability to perceive and a smattering of intelligence, robots could start diversifying into new activities—patient care, exploration, safety and warfare. A prime advantage was that, whether they were autonomous or remote-controlled, they could go where no man could or would go—to the rims of volcanic craters, the ocean bed or contaminated areas. In 1982 the Ontario-based company Pedesco used a robot to clean up a nuclear-production site after a fuel spill, while in 1984 Robot Defense Systems produced Prowler, the first in a long line of military robots designed to intervene on the battlefield. In order to

improve the efficiency of robot programming, dedicated languages were developed. One example was Robot Basic, marketed by the US company Intelledex.

However it was Japan that was asserting itself in the mid-1980s as the global robotics giant: its 150 manufacturers accounted for 60 percent of world output. Artificial intelligence had also finally made its mark. The theories that Marvin Minsky and Seymour Papert had expounded were now being applied to multi-agent robot systems. The proponents of multiagent systems believe that robots can be made to work together in order to solve complex problems. In 1985 Professor Rodney Brooks of MIT began work on behavior-based robotics, making small robots that learned empirically from their environment rather than being programmed to recognize it. Brooks's ambition is to have colonies of robots working together, rather like ants.

The age
of the robot

In the 1990s robots proliferated. They drew on a wide range of disciplines and skills, from sculpture and molding to electronic sensing and artificial intelligence. So commonplace have they become on automobile-industry assembly lines that most of the jobs done by people focus on servicing and repairing them.

In the medical field Robodoc assisted in a hip-replacement operation in 1992, while out in space the robotic rover *Sojourner*, explored the soil of Mars as part of the 1997 Pathfinder mission, so helping to revive waning interest in space travel. It was at about the same time that Tokyo University's H7 and Honda's P3 paved the way for the first robots that were truly humanoid in that they could walk like humans. And when Sony marketed its robotic dog Aibo to such overwhelming demand in 1999, it became clear that on the eve of the third millennium, the age of the robot had finally come. A telling sign of these robotic times came in 2003 when the humanoid robot Asimo made an official state visit to the Czech Republic as ambassador of peace and goodwill between robots and humans.

The robot has wooed and won over industry, made its mark in hospitals and secured its place in space travel. It is now poised to embark on a new phase in its development by stepping into everyday life and, in particular, into the home. From home helper who does the housework to baby-sitter who reads bedtime stories to the children; from bus driver whose paramount concern is the highway code and passenger safety, to museum guide always keen to hold visitors' interest. The number of career paths opening up to *Homo sapiens*'s artificial assistant has multiplied. The 21st century will see the robot entering most walks of human life.

ROBOTS IN FICTION

From Frankenstein's monster to R.U.R.'s robots in revolt and Maria of Metropolis, the earliest stories told of man-made beings usually described them as degenerate creatures, obsessed by the desire to supersede their creator, seize control of civilization and ruthlessly enslave human beings. Entire generations were weaned on a perception that is tough on robots. It took the works of Isaac Asimov and amenable artificial beings like R2-D2 to temper that harsh image. But the damage had been done. And even in later years, merciless machines like the Terminator and the android spy who brings Alien on-board the Nostromo act as timely reminders that robots have no minds of their own and merely do the evil they are programmed to do.

Are robots fated to be our foes? Or could they become our friends, displaying such moving magnanimity as the replicant in Blade Runner or the emotional sensitivity of Bicentennial Man?

Writers and moviemakers should be lauded for being the first to raise such questions and exerting a far-reaching influence on the way in which roboticists and the general public perceive our electronic, programmed progeny.

Underground Robot,
clay animation film
by Webster Colcord.
© Courtesy of WebsterColcord.com

Robots in their own write

Sumerian, Egyptian, Chinese and some African mythologies allude to human techniques of creating lifelike artificial beings. In most cultures, the power to create a human doppelgänger raises a fundamental question. Can humankind make creatures in its own likeness without dire consequences? Can we, sullied by original sin and made to labor and sweat for our bread "till [we] return unto the ground", defy God's bidding by giving birth to a mechanical slave who would relieve us of strenuous and menial chores? No, we cannot, has been the sentiment of writers and filmmakers alike. Whether affected by the biblical story of work as the price paid for Adam and Eve's lapse in the Garden of Eden, or by myths like that of Prometheus, described on the next page, they believe that it would be unnatural to assume the role of the creator of life. Inventing an automaton invariably turns out to be fatal to the inventor.

Yet through the centuries there have been authors who have entertained kinder notions of automated servants, even extolling their beauty and suggesting they might have a soul. In the eighth century BC, Book XVIII of Homer's *Iliad* describes the smith-god Hephaestus, a son of Zeus and Aphrodite. He built a set of three-legged tables with golden wheels ranged around his smithy and ferried the wares that he made there to Mount Olympus by themselves. To assist him in his craft, Hephaestus also made two "golden handmaids...who worked for him and were like real young women, with sense and reason, voice also, and strength, and all the learning of the immortals; these busied themselves," narrates Homer.

Twenty-eight centuries later, the science-fiction writer Isaac Asimov propounded an insight of genius. Man holds the trump card, he wrote in 1940. Robots should be programmed to respect humans and to protect them if they are in danger.

Asimov notwithstanding, the consensus that continued to prevail was that robots could never be anything more than machines, even when "enlivened" by sophisticated programs. The dividing line between man and his manufacture was awareness, the faculty that human beings have of knowing or thinking about why they act. Devoid of this quintessentially human trait, robots could only follow the dictates of their software. There have always been, and still are, works that explore the development of self-awareness among androids, but they are works of fantasy fiction rather than probes into the future. Because it would never be self-aware, the robot was easily manipulated prey for the dastardly, not to mention the shortcomings inherent in the simplicity of the automaton. It would only take some bug to throw a spanner into the works of the world's most powerful computer and it would do what its creators had never planned that it should.

Friend or foe? The fears and hopes prompted by the robot have their roots in mythology and seem to spring principally from two legends—those of Prometheus and Pygmalion.

Prometheus:
creation kills the creator

The idea of servants who relieve their masters of thankless tasks is an ancient human dream. Indeed it antedates humanity, as revealed by inscriptions on tablets found around 3,000 BC in Mesopotamia, the birthplace of writing. They describe the birth of man, created to serve the gods. The account goes like this: "One day the worker gods felt they were being exploited and revolted. The god of craftwork suggested creating puppets endowed with life that would be put to work in place of the gods. Thus were humans born. So that they might not revolt, they were created much inferior to gods."

In this light, man is considered as nothing more than an imperfect automaton, subject to the whims and fancies of the gods. The myth of Prometheus echoes that vision, adding punishment for whomsoever sought to create a semblance of life.

Ancient Greek Creation myths tell how Prometheus the Titan fashioned the first man from a piece of clay mixed with water. Zeus, however, took his revenge.

Prometheus the Titan created the first man from a piece of clay mixed with water. In doing so, he braved the power of the god supreme, Zeus. He then stole a spark of fire from the sun, which he gave to man, who went forth and multiplied. Later, Prometheus added insult to injury by tricking Zeus. Asked to arbiter a dispute over which parts of a sacrificial bull should be offered to the gods and which given to men, he dismembered a bull and cut it into pieces. From its hide he made two bags. He filled one with meat and marrow, concealing it beneath the stomach, the least prized part of the animal. Into the other bag he put bones, which he covered with fat. Prometheus asked Zeus to choose which bag he wanted. Zeus fell for the deception and chose the bones and fat. Prometheus was thus able to give humans an anatomy as rich and varied as the bull's. Furious at being hoodwinked, Zeus exacted cruel revenge, first on Prometheus then on the race of mortals, by ensuring that Pandora (fashioned from clay) would open her box, so releasing evil, suffering and sickness into human existence.

On the orders of Zeus, Hephaestus chained Prometheus to a pillar in the Caucasian mountains where an eagle tore out his liver all day long, all year round. Left, a sculpture from the 18th century by Nicolas-Sébastien Adam depicts Prometheus in chains.
© Paris - Musée du Louvre

Pygmalion: the sublimated woman

The story of Pygmalion told by the Latin poet Ovid in *Metamorphosis* around 20 BC casts the story of the man-made being in a happier light. Pygmalion, king of Cyprus and sculptor, despaired of finding his ideal. He fashioned a statue of perfect beauty, whom he named Galatea. He fell in love with his work and beseeched the goddess Aphrodite to breathe life into her. She became his queen and bore him children. The theme has given rise to many subsequent works, the best known being George Bernard Shaw's play *Pygmalion* in 1913 and George Cukor's musical movie *My Fair Lady* in 1964.

Broadly speaking, Prometheus and Pygmalion reflect the two dominant outcomes of stories recounting man's attempts to create beings in his own like-ness. On one hand, they ulti-mately destroy their creator, who pays the price of his effrontery through being punished by a power greater than his. On the other hand, their sheer near-per-fection begs for that last, ulti-mate ingredient to be bestowed upon them—the breath of life.

Extracts from a superb copy of the Paul Wegener film, Der Golem, 1920.
© *Munich Film Museum*

From the Golem to Romanticism

In the centuries that elapsed between ancient and modern times, artificial beings vanished from the imaginary inventory of literature. True, in 1495 Leonardo da Vinci sketched the blueprint of a sort of mechanical knight that moved its head and jaw, swung its arms, stood up unaided and was not unlike a humanoid robot. Yet literature remained unmoved by the Renaissance scientist's flight of fancy.

Central Europe, however, told different stories—that of the Golem, for instance. Born from the bowels of the earth, the name of this androgynous giant was derived from the Hebrew which, paradoxically, meant "shapeless mass, body without soul". The first mentions of the Golem are to be found in Chapter 139 in the Old Testament Book of Psalms and in the Kabbala. It then resurfaced in folk tales told and passed on in Central Europe's Jewish communities. In the 17th century it entered the stories of lay and non-Jewish writers. This body without a soul was depicted now as a man, now as a woman, and some-times as a sheer destructive force.

Frankenstein, a movie directed by James Whale, 1931.
© *Collection Cahiers du Cinéma - D. Rabourdin*

Frankenstein:
of science born

In the early 19th century, Romanticism's penchant for the dark side led writers to venture into new fictional territory. E.T.A. Hoffman wrote a novel entitled *The Sandman* in 1816. It tells the story of Nathanael's doomed love for Olympia. He realizes his love cannot be requited on finding out that she is an automaton made by the dastardly Dr. Coppelius. He refuses to believe that she is not flesh and blood until Coppelius pulls out her eyes to prove she is nothing but an animated dummy. Grief drives Nathanael into madness. The novel was to inspire Léo Delibes's ballet *Coppélia*, which ran at the Paris Opera from 1870 to 1961.

In the same year that Hoffman published *The Sandman*, the Romantic poet Lord Byron was staying in the Swiss mountains with some friends. They included his doctor, John William Polidori, the mercurial poet Percy Shelley and his wife, Mary. They began making up stories in an impromptu contest to see who could tell the most frightening one. Polidori's attempt led to his writing *The Vampyre*, a short story about the adventures of an aristocratic vampire. But it was Mary Shelley's *Frankenstein* that

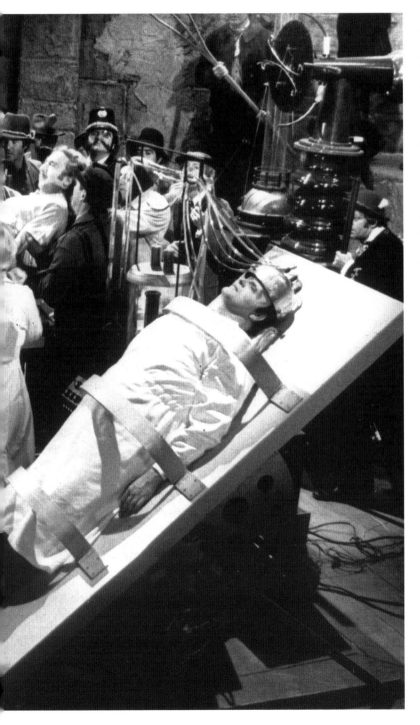

would survive the test of time. Eventually published in 1818, Shelley's work ushered in a new kind of literature, with a plot that was at once realistic and scientific.

It tells the tale of a creature created by Doctor Frankenstein. Though it resembles a human being, it is in fact assembled from the spoils of Frankenstein's grave-robbing and has been stitched and cobbled together. Appalled by its hideousness, the doctor abandons his creature. So monstrous is it that it is spurned by all. Its grief turns to hate, which drives it to crime.

What is new about the novel *Frankenstein* is that the artificial being is not created by a god, magic, or some supernatural agency, but by science. Significantly, even in the work of the profoundly emancipated Mary Shelley, the vanity of seeking to create life cannot go unpunished. The moral of her story is closer to Prometheus than to Pygmalion and prefigured the robot literature that was to emerge over 100 years later.

Baron Frankenstein robs graves in search of a body that he can use in order to create and give life to an artificial being. Scene from Frankenstein, *directed by James Whale in 1931, starring Boris Karloff and Colin Clive.*
© Collection Cahiers du Cinéma - D. Rabourdin

Does automation enslave or ennoble man?

Industrialization and mechanization spread fear and confusion throughout the 19th century and the early 20th century. Confined to soul-destroying tasks, the working class came to loathe the machines that enslaved them. Men and machines were pitted against each other in what seemed at the time an eternal opposition, despite the voices occasionally raised to argue that ultimately society would be changed for the better.

Nineteenth-century writers were divided in their perception of automata. Some depicted the beauty of the robot, others its urge to destroy. In *Steam Man of the Prairies* (1865), Edward S. Ellis explored America's fascination with the mechanization of industry, while the following year French novelist Villiers de l'Isle-Adam wrote *Eve Future*, in which Hadaly, a clone of the beautiful Alicia, enchants all those who lay eyes upon her.

In 1909 Italian poet and writer Filippo Tommaso Marinetti, founder of the Futurist movement, wrote a play called *Elettricità sessuale* (Sexual Electricity) with characters who were machines in human like-

ness. Marinetti professed faith in the beauty of machines and the power of technological evolution. Italian artists followed in his wake, such as Paladini with *Mechanical Aesthetics* in 1923, Filia, Curtoni and Caligaris, who wrote *The Mechanical Idol* in 1926. Italian Futurist fiction posited that the machine's inner and outer excellence made it an ideal replacement for deeply flawed humanity. The Futurists' strange infatuation with machines would influence later schools of thought and utopias, where the superiority of computers and robots was ineluctable. It was a notion that prominent roboticists Hans Moravec, Kevin Warwick and Hugo de Garis would make their own.

Karel Capek invents the robot

Czech writer Karel Capek coined the term "robot," and embodied the mechanical being in his play *R.U.R.*, which stood for Rossum's Universal Robots. Rossum is an inventor who has produced androids in his workshops on an island. Their capacity for work is incomparably greater than man's. Significantly, the word "robot" is derived from the Czech word "rabota", meaning "worker, serf". The implication is that this artificial slave is designed to relieve humans of dirty work and menial chores.

Industrialists are much taken with Rossum's robots, which they employ widely in their plants as shopfloor workers, typists, etc. However, a newcomer to the island brings trouble. Her name is Helen and she tries to incite the downtrodden, exploited robots to stand up for their rights. She fails because the robots are not programmed for violent action. When the boss' son falls in love with her, she persuades him to ask an engineer to further perfect the robots. As a result, they learn to reason for themselves, even though they do not have feelings. They rise up against their human oppressors and begin to kill them.

One day a boat arrives on the island. It turns out that it is carrying pamphlets written by robots urging their peers on the island to kill all humans. The robots on the island take over the factory and kill all humans with the exception of one, who is spared because he is the only human still working with his hands. The rebel leader, Radius, sums things up in these terms: "The power of man has been annihilated. A new world is born. The time of the robot has come." By the end of the play two robots have fallen in love. They are renamed Adam and Eve and sent out to avoid the sins that destroyed their predecessors.

Written in 1920 and first performed in 1921 at the Garrick Theatre in New York, *R.U.R.* heavily influenced all robot stories that followed in its wake. In addition to its denunciation of the production line and exploitation of workers, which leads them to revolt, its plot laid the foundations for a deep mistrust of the robot's potential deviance. Capek's play was, once again, a moral tale in which man gets his comeuppance for seeking to ape his own creator.

Above:
Karel and Josef Capek at the Hôtel des Américains in Paris, 1911.

Right:
Karel Capek and Professor Vocadlo in Surbiton, UK, 1924.

© Courtesy of Kristina Vánová, Památník Karla Capka

Capek also propounded a number of other beliefs that would surface in books and movies. Some examples are:

- Robots are more efficient and servile than humans. As Domin, a character from *R.U.R.* puts it, "A robot can replace two-and-a-half workmen."

- The perfectibility of androids will eventually lead them to compete with their maker, expressed in the statement "The human machine was terribly imperfect. Sooner or later, it had to be removed."

Karel Capek lampooned as a robot by his brother Josef, a well-known artist and, some say, the inventor of the word robot.
© Courtesy of Kristina Vánová, Památník Karla Capka

Hysterical and destructive

Capek had set the tone. *R.U.R.* spawned the warmongering, murderous robots that peopled the popular literature of the 1920s and 1930s. So mediocre was the bulk of the stories written that few have come down to us. Pulp fiction gave pride of place to the bloodletting robot. *The Jameson Satellite*, a short story by Neil R. Jones, published in 1931 in the magazine *Amazing Stories*, ushered in a long series of tales featuring Zorome, a robot with a brain poached from a deep-frozen astronaut.

Karel Capek distanced himself from the tacky mythology that he had engendered. "I recoil in horror from any idea that metal contraptions could ever replace human beings and awaken something like life, love and rebellion. Such a grim outlook is nothing but an overestimation of the power of machines and a grave insult to life," he said in 1935. But it was too late and writers unfailingly portrayed robots as the cold-blooded, soulless enemy of man.

Interview
Stéphane Nicot

Stéphane Nicot is a renowned French authority on science fiction. He is editor of the review Galaxies, *and co-authored the* Dictionnaire de la science-fiction *at* Livre de Poche *(Dictionary of Science Fiction) with Denis Guiot and Alain Laurie. He has contributed to numerous anthologies and organizes science-fiction events and festivals.*

The literature of the 1920s and 1930s paints robots as man's enemy. What are a robot's main traits in the fiction of that time?
They're vaguely humanoid tin-can men. Actually they're pretty anthropomorphic, but not life-like. They're the menacing contraptions of the worst pulp fiction. In American folk imagery of the time they go together with other threatening figures like the cruel Indian and the hostile alien from outer space.

Apparently there are no great science fiction works of that time featuring robots. How do you account for that?
You shouldn't forget that in the1920s and 1930s English-language science fiction was almost exclusively short stories.

The sci-fi novel didn't come into the picture until after World War II, which some people ascribe to the paper shortage limiting the production of pulp. Most novels that addressed the robot theme were European, particularly French.

What do you think was the first great science fiction novel about robots?
There's no question, it was *I, Robot* by Isaac Asimov, even if it was fixed up from individual short stories, but that was common practice in the golden age of American science fiction. With the Three Laws of Robotics, he firmly instated robots as one of English-language science fiction's major themes. The Laws of Robotics were

in fact formulated in 1941 by John W. Campbell, editor of the great science-fiction review *Astounding Stories*, as Asimov frankly admits in his autobiography, *I, Asimov*.

Since the 1970s it would seem that robots no longer hold much attraction for science fiction writers. Is that true?
Yes, the robot has become banal, a consumer item. Not exactly conducive to fantasy. So science fiction has shunned it because it's an aging theme with nothing novel left.

Interview
Christian Denisart

Swiss stage director Christian Denisart was originally a musician before turning his interest towards theatre in 1998 and writing some twenty musical scores. In 2001 he founded his own company that will be performing a traveling play entitled Robots in 2005.

A man, a woman, and three machines make up the intriguing cast of Robots, a tragic dumb show, where eight musicians are the narrators.
© Courtesy of Christian Denisart - Photo: Daniel Balmat

As an artist when did you start getting interested in robots?
I belong to the countdown generation, by which I mean the generation that grew up looking ahead to the year 2000. The result was that I spent part of my youth dreaming of the future. I was born the year before man set foot on the moon and I saw the coming of the PC. There were two main areas of science about which we used to fantasize and which then became true. Space travel and robotics. As a composer I recently worked with a troupe from Geneva that creates plays on scientific themes. The one I worked on was about matter and antimatter and was shown 100 meters below ground in the CERN particle accelerator. I had the opportunity to talk with researchers and I realized we spoke the same language. I was scared they would go on

about cogs and codes, but we just talked about ordinary things and about dreams and poetry. They're creators, too, who work in a very abstract world. At about the same time I saw a video from a university in Japan, which showed a robot in a cage, which swung its arms and moved around like a monkey. What struck me was how smooth and easy his movements were and, what's more, it was poetic. I didn't know just how far robotics had developed and I thought, "It's time you got interested in robots."

With hindsight don't you think it is amazing that so many powerful ideas were explored in *R.U.R.* by Karel Capek?
What I find amazing is that as well as being a precursor in robotics, the play is now emerging as a precursor in the area of cloning. Capek called them robots because the word derives from forced labor; they do the work man doesn't want to do. But if you read the play today you realize that the robots aren't mechanical but biological. The play has proved visionary in one field and looks like it will be again in another

field. What's more, ideas that recur in robotic mythology, like emotional ties between machines, were already in the play.

What do you think are the salient features of robots in literature and the movies?

Most robot stories are based on the assumption of underlying artificial intelligence. Robots are seldom described as perfectly engineered machines, and usually become aware of their sorry condition and revolt. That pattern is more a metaphor for slavery than a real exploration of the nature of machines. In other words, many works build from a starting point that doesn't really interest me. I think the image of degenerate, out-of-control science is a phony one. It's up to us to shoulder responsibility for the problems of science.

You say that artists and scientists share a dream. Could you elaborate on what you mean?

Science is more and more difficult to explain to the public at large, and more and more costly. At the same time researchers want recognition for their work and art is one of the nicest ways of giving them that recognition. From our point of view science is wondrous and gives us a lot of inspiration. New technologies influence us; look at the synthesizer in music and video in the movies. Artists and scientists can be mutually useful.

In what way is your play, *Robots*, a different experience from other work that you have staged?

When a new technology that can be used in art emerges, there are always pioneers ready to explore new ground. The advent of the synthesizer threw up bands like Kraftwerk. Then a second generation comes along and incorporates the new technology into a more mainstream context. As far as we're concerned, we wanted to see if the robotic medium had its place on the traditional stage. We wanted a 19th-century-type stage set, so we could see how the robots would fit in and whether the audience would forget the technological novelty and concentrate on the story.

For your play you secured backing from university researchers and industry. What appealed to them so much?

Some researchers make robots for exactly the same reason as man walked on the moon—because it's a dream come true. Only afterwards do they say "Hey, that's useful, too." Humankind tries to make humanoids because they are our artificial mirrors. Then comes the question, What can they do? Receptionist? Museum guide? It is more annoying than constructive

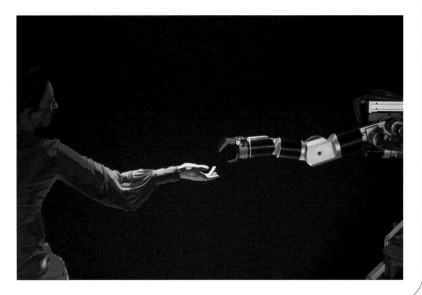

to ask that question for the moment, so roboticists focus on ways of making it pleasurable to be with a robot. What researchers are primarily interested in are man–machine interfaces. Devices are more and more difficult to use, you have to pore over a 60-page manual to work out how your video recorder works. Our play provides an answer, or at least suggests answers, to problems of how to communicate with machines. Take the shape of a robot, for example. Should it look humanoid or cubic? Our play raises questions of that order. In art schools a robot's design also raises questions related to shape and form. Even if a robot has no eyes, spectators will still talk about its head, or face, and find ways to "anthropomorphize" it. Then there's the question of materials and textures. Robots are always made out of new alloys. But why not wood or fabric? The man–machine relationship is very important, of course. In our early rehearsals we saw that actors could instill an enormous amount of life into a machine. A robot with a glass of wine is weird, but when the actor asked it for some wine and it poured him a glass, it was the beginning of a relationship and the robot came alive, as it were.

The Autonomous Systems Lab of the Swiss engineering school L'Ecole polytechnique fédérale de Lausanne (EPFL) helped to build the robots. The chief technological challenge is to ensure that the autonomous robots operate and interact with the actors and the stage set for three hours. That is the job of Professor Siegwart and his team. Together with BlueBotics, an offshoot of Autonomous Systems Lab, they installed the world's biggest autonomous and interactive robot show at Switzerland's national exhibition, Expo.02.

© Courtesy of Christian Denisart - Photo: Daniel Balmat

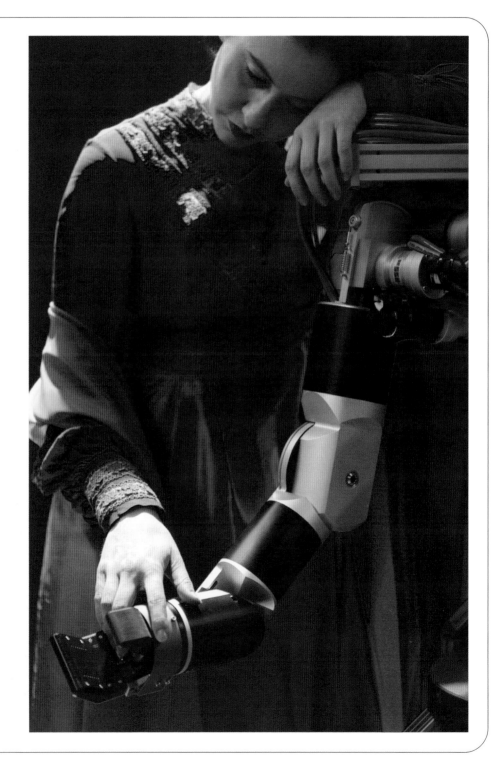

Through your play, _Robots_, you hope to change people's behavior towards robots. What is the message that you'd like to get across here?

The bottom line is that robots are no more than sophisticated machines; what matters is the way we perceive them. If you give a coffee percolator eyes and wheels, you'll grow fond of it even though it's still nothing but a percolator. When we see a robot going somewhere, we wonder what it's thinking about and ascribe it an attitude. In fact we remind the audience every now and then that what they're watching is no better than a high-class coffee percolator. Everything is in the way we look at robots. We work on a hot dance between a "robota" and a male actor. There are spectators who'll want to be up there in the actor's place, even though the bottom line is that the actor is dancing alone.

Why is your play without words?

Robots can't speak properly. When you see them being demonstrated at trade fairs, they go up and down stairs, dance a little, and it's very impressive. But as soon as your hear their voice, the spell's broken, they're like toys. The reason is that language is the

product of thought and intelligence and robots have neither. So we chose to use music as the narrative language. It gives the play a Chaplinesque feel.

Special effects for the stage are far more limited than the movies. How do you make up for that?

We use stage effects, like lighting and music and the stage presence of the actors, to compensate for the lack of special effects. The main difference is that the robot is there before your very eyes. If I could put on _The Three Little Pigs_ with three real, little,-trained porkers, people would say, incredible! Maybe they'll say the same when they see the robot. I

think that's how we'll be able to impress the audience, with the robots. They'll be right up there on the stage. There's a very different interaction compared to the movies.

Recent movies like _A.I._ and _Bicentennial Man_ have sought to soften up audiences and make them emotional about robots that are portrayed as having a spark of life. What's your view of such an approach?

In those movies robots are pondering the meaning of their existence. It's an interesting metaphor but robots weren't necessary. I think it would be better to say, You have this inanimate object and it affects you as if it were alive. We imagine it's alive and we fantasize about it. In theater we can use all the machines' poetry, as if we were to work with life-size puppets. François Junod built the body of our dancing robots and he has given "her" a powerful poetic charge.

The heyday of science fiction stories about robots was the 1950s. Do you think the time has come for new fictional forms featuring robots?

In the 1950s that stuff was pure science fiction. The tin-can robot exerted as much fascination as Tintin on the moon. But once man had set foot on the moon, it was no longer possible to believe in those stories. Reality had outstripped fiction. For decades now people have been thinking about robots' shapes, size, reactions, behavior. In fact, they existed in our imaginations before people started building them.

Robots are becoming consumer items. Do you think that they still exert enough fascination to be reinstated in works of fiction?
Could Asimo or Aibo inspire stories? If they did, they would be stories that mixed fact with fiction. We are entering the realist school of robotic fiction.

The three machines were designed by the Lausanne Cantonal Art School (ECAL), whose reputation has crossed borders through design shows in Milan, Berlin and Venice. The challenge was to turn a machine into a true stage character.
© ECAL,
Luc Bergeron/Stephan Küster

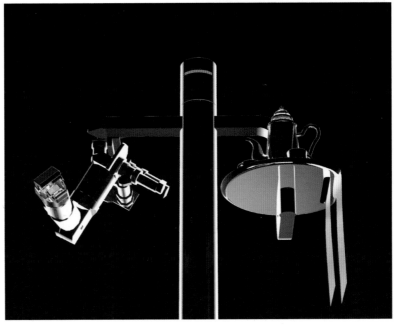

Isaac Asimov creates ethical robots (1938)

It was just beginning to look as if there was not much science-fictional mileage left in robots, when an inspirational scientist came to their rescue in 1938. Weary of the dismally penned stories that unfailingly churned out robot invasions and robot attacks, the Russian-born biologist Isaac Asimov published the first of his many novels and short stories built on the immutable ethics of his Three Laws of Robotics. Designed to instill in his android robots the knowledge that they must never harm humans, the three laws were:

1. A robot may not injure a human being or, through inaction, allow a human being to come to harm.

2. A robot must obey orders given it by human beings, except where such orders would conflict with the First Law.

3. A robot must protect its own existence as long as such protection does not conflict with the First or Second Law.

Asimov predicates the existence of robots on their always being our helpers and servants, simply because that is what we want. Man could create a morally flawless artificial being whose behavior was perfect. It would protect another's life before even thinking of defending itself.

By leveraging the dramatic potential of his laws, Asimov wrote an impressive corpus of work, which is considered to be the Golden Age of robotic mythology. They were published as short stories throughout the 1940s before being compiled and released in novel form in 1950. Asimov explores the subtleties of robotic intelligence by forcing his robots to solve all sorts of dramatic situations within the constraints of the laws that compel them to respect and protect human life.

With hindsight Asimov's robotic tales seem to build on foundations that are utopian and way too optimistic. In the story *Robbie*, Asimov describes a scene that is set in 1996, where a robot that has been programmed with a strong maternal instinct performs an incredible feat. Just as a giant, runaway bulldozer is about to crush a little girl lost in a great factory, good old Robbie leaps to the rescue. The duty to save human life overcomes Robbie's instinct of self-preservation

Isaac Asimov.

and little Gloria is snatched from the jaws of death.

Asimov's vision should not be considered flippantly, however. His work was behind the creation of the very first robotics companies. When Joseph Engelberg met George Devol in 1956 it was to talk about Asimov, and their passionate discussion about the master's writings was to lead ultimately to the first robot manufacturing enterprise, Unimation.

The notion of the well-intentioned robot was very much an idea of the times in the mid-20th century. In 1939 another writer, Eando Binder, also painted a friendly picture in his series of stories entitled *I, Robot*,[1] in which the robot, Adam Link, recounts his adventures. In his 1949 novel *The Humanoids* Jack Williamson follows in Asimov's footsteps with robots whose motto is "to serve, obey and guard men from danger." Yet he

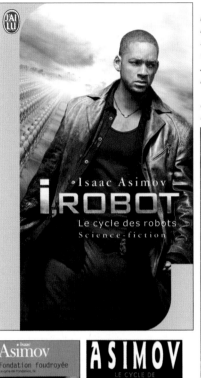

The American author Isaac Asimov (1920–1992) is the most popular author in the history of science fiction. His career spanned some 60 years and his work was read and enjoyed worldwide.
A survey by French literary magazine Lire revealed that Foundation was one of the 100 best-loved books in France.
© Courtesy of: J'AI LU, Folio, Pocket SF, Omnibus, Denoël Présence du futur

also explores the dangers inherent in too strict an application of the motto. The robots are so benevolent that they do all they can to look after and preserve the lives of interplanetary settlers whom Williamson portrays as riddled with weakness and faults. As a result, the settlement in space withers away.

In Lester Del Rey's *Instinct* (1951), robots re-create the extinct human race. In their efforts to bring their original creator back to life they discover that what separates humans from robots is instinct. In 1958 Edmund Cooper revisited a founding myth when he wrote *Pygmalion 2113*, in which a refriger-

ated man is roused from hibernation by android robots. The female android who looks after him falls in love with him, unaware that he is fomenting revolt against the ruling robots.

(1) The title that Isaac Asimov wanted to give his first collection of stories was Mind and Iron. His publisher preferred *I, Robot*, which was the title of an earlier short story by Eando Binder.

Philip K. Dick invents the perfect clone

In the 1960s Western culture was brimming with new ideas, while a sense of freedom released creativity in all the arts. Science fiction writing sought to squeeze the good robot out of the straitjacket into which Asimov had laced it. It went down new, darker, stranger paths, like the breakthrough Lester Del Rey story *Helen O'Loy* (1966), about a female robot whose master has programmed her to behave like a model wife.

Another master of science fiction appeared at this time—Philip K. Dick, affectionately known as PKD. His work is in a realistic vein and features characters with whom readers can identify. Robots appear in his work, but, not bound by any Asimovian law, they are strange, disturbing entities. In his 1964 novel *The Penultimate Truth*, millions of people are crammed into underground shelters waiting for World War III to finish. Fed on lies from the robot rulers above, little do they know that it has been over for decades. In *Do Androids Dream of Electric Sheep*[2] (1968), subsequently adapted for the screen as *Blade Runner*, PKD describes androids so sophisticated that they blend totally with humans. The story reaches almost unbearable pitches of suspense, as in the scene between the bounty hunter Rick Deckard and a highly intelligent opera singer, who is, in fact, an android. Sensing that Deckard suspects her, she turns the tables on him by talking him into doubting his own nature. After all, how does he know he is not an android programmed to believe he is a human. The idea of a robot that has reached such degrees of sophistication that it thinks of itself as human is a recurrent theme in PKD's work.

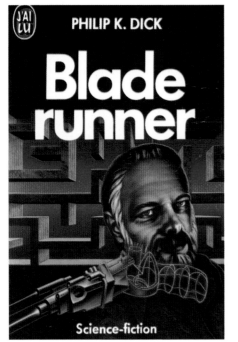

The familiar robot

The fact is that robots have been gradually disappearing from science fiction literature. In the last decade they have been practically driven off the written page as the robot myth has been undermined by science. It has proved incapable of living up to works of imagination and producing androids with even a hint of the seductivenss of Maria from the 1926 movie *Metropolis*. The prosaic drabness of industrial robotic arms first took the gloss off android glamor, while artificial intelligence has not fared much better, failing to live up to the grand dreams of visionaries.

For generations science fiction prompted aficionados to embrace scientific careers and fed the dreams of researchers. It also fostered a sense of anticipation that robots do not look like satisfying for decades to come. What's more, space travel captured young minds when man landed on the moon in 1969. Astronauts were made of flesh and blood, with their oh-so-human fears and courage. They were the new heroes.

In the late 1960s and early 1970s, while America was still vibrant with the energy of the counter-culture born from the hippie movement and student protests, writers were reinventing plots that reflected the pervasive sense of emancipation at the time. The robot was no longer either all good or all bad and authors took off on flights of fancy. In *Demon Seed* (1973), Dean R. Koontz broke new ground with an unprecedented kind of hybrid. An intelligent machine that feels cooped up in its body, takes a woman hostage in her house and seeks to impregnate her to ensure its posterity[3].

Cliffort D. Simak's 1972 novel *A Choice of Gods* was one of his last. It describes a world devoid of humans where robots operate of their own free will. It is a sign of the times that the plot treatment is comical. In *The Reproductive System* (1968), John T. Sladek's scathing humor takes as its premise that people are normal but that robots are so crazy they revolt. More irreverently, Robert Silverberg replaces the pope by a robot in his story *Good News from the Vatican* (1971).

The final debunking of the robot myth comes from Robert Sheckley's hilarious stories published under the title *The Robot Who Looked like Me* (1972). The first story is about a businessman who does not have time to seduce the woman with whom he is very much in love. He has an android built in his own image and sends it regularly to dine with his beloved. Everything goes smoothly. The narrator spends his free time watching videos of the cozy evenings and, seeing that the robot is behaving as intended, he tells it to ask for his girlfriend's hand in marriage. It is then that the android disappears without a word. The narrator rushes to his fiancée's house and confesses that he has been courting her by proxy. She then tells him that she, too, has been so busy that she had done exactly the same things. The lovers at the candlelit dinners were both androids!

---

(2) The motion picture *Blade Runner* coins the word "replicant" to denote the androids. PKD does not, however, use the term in *Do Androids Dream of Electric Sheep*.

(3) In 1977 *Demon Seed* was made into a motion picture with Julie Christie in the role of the hostage Susan.

Cyberpunk versus robot

For the fashionable science-fiction writers of the 1980s, robots were out. In *Tik-Tok*, written by John T. Sladek in 1983, a domestic robot with a talent for painting turns to crime after mistreatment blows its Asimovian moral circuits. In his *Fourth Law of Robotics* (1989), in which robots try to glean the secret of human emotion, Harry Harrison propounds an extra robotic law to add to Asimov's three: "A Robot may reproduce whenever such reproduction does not contradict the First, Second or Third Laws." But robots that enjoy being bad or having fun merely bucked the trend: futuristic writers were looking elsewhere for inspiration.

William Gibson and Bruce Sterling blazed a trail by introducing the cyberpunks who, ten years ahead of their time, explore the deviant, overblown abuses of information systems and networks. The heroes of these stories are hackers who infiltrate the computers of multinational corporations.

The 1990s ushered in literature that focused more on the potential of nanotechnology than robotics. Neil Stephenson wrote the now-famous *The Diamond Age*, while Paul Anderson penned the deeply unsettling *Nanodreams*.

It was not until 2003 that the first major novel to feature robots in a long time was published. Titled *Robota* it was written jointly by science fiction author Orson Scott Card and prodigious movie-set designer David Chiang (he was artistic director for the second *Star Wars* trilogy). *Robota*'s plot involves robots from outer space landing on Earth to warn its inhabitants that they are under threat from a great danger. The only hope of salvation is to teach the earthlings how to build highly advanced robots to save them.

The bottom line is that robots no longer belong to science fiction and maybe it is better that way. The third millennium has ushered Aibo and other personal robots into our everyday lives and driven their fictional counterparts out of futuristic books and films into the present. Ultimately, just as they are becoming ordinary parts of life, so they may reappear in realistic work as part of the backdrop of ordinary urban life. Science fiction traditionally draws on the present to build scenarios for the future. Robots could, therefore, be integrated into works that explore the future of Earth. How might robots help save the planet, improve public health, reduce growing poverty, develop new medicines? The great challenges facing the 21st century will be the stuff of many a story, with robots merely routine tools that are part and parcel of the social landscape.

Robots in film and television

Motion pictures have, more than any other art form, made robots part of popular culture, while exploring in depth how they might evolve. The power of the image gave them shape and substance, painting them as monstrous, ridiculous, fascinating or fearsome. However, they have had a rollercoaster history on both big and small screens.

Robots first appeared in film in the early 20th century—antedating literature—when the motion picture industry was in its infancy. The feat is all the more impressive when the rudimentary nature of special effects are taken into account. Moviemakers sensed early on that robots offered potential for producing high drama and suspense in silent moving pictures. The political atmosphere of the early 20th century was conducive to fantasies of fear and loathing. A world war was decimating cities and claiming millions of young lives, while the great Russian empire had collapsed and the Bolsheviks were intent on exporting their economic system to Europe and the US. The very "unhumanness" of the robot and the deviance associated with it epitomized the mass anguish sweeping the world.

A 1916 German movie by Otto Rippert, called *Homunculus*, featured the first android in the short history of cinema. Desperately lonely, it hopelessly sought ways of concealing that it was a robot. Four years later, the horror movie *The Golem*, by Paul Wegener and Carl Bœse, portrayed a clay robot based on East European folklore. In 1922 robots appeared in an Italian film by Cretinetti, aka André Deed, called *L'Uomo meccanico*, or *Mechanical Man*. The unscrupulously intriguing Mado diverts a science project to her own evil end of creating a malevolent, destructive machine. Another character saves the day by creating a good robot to combat its evil counterpart.

Scene from Homunculus, *by Otto Rippert. This black-and-white science-fantasy movie in six episodes tackles the theme of Nazism in the expressionist register in which many German films of the time bathed. Homunculus is an artificial being steeped in hate and violence, who dreams of dictatorship. The fiend is borrowed from Scandinavian folklore, particularly the* Kalevala *from Finland, which tells tales of monsters similar to the Golem and Frankenstein.*

Extract from André Deed's film, L'Uomo meccanico, *or Mechanical Man, 1922.*

Metropolis and the perfidious beauty

The turning point in popular perceptions of the robot came in 1926 with the perfidious, seductive android that appears in the Fritz Lang film *Metropolis*. The movie is a remake of the slave revolt led by Spartacus, only this time the rebel leader is the sexy but evil robot Maria.

In her statuesque, chrome-plated bodyshell, she is flamboyant and magnificent, as visually captivating as she is supremely intelligent. The superbly designed robot remains to this day one of the finest in the history of motion pictures. Her metal shell is not an end in itself, for clad in a human skin, Maria is the exact replica of the young woman on whom she is modeled.

It was on a visit to the US in 1924 that, surrounded by the skyscrapers of New York, German director Fritz Lang had his vision of Metropolis, the futuristic city. The movie is set in 2026. Freder, son of the master of Metropolis, Fredersen, enjoys the luxury lifestyle of the privileged in the upper quarters of the city with its gardens and luxury mansions. One day he is struck by the beauty of a young woman, Maria, whom he glimpses and tries to seek out. He fol-

lows her underground only to discover a great industrial city where workers on assembly lines slave away at a clockwork tempo and live in degrading conditions.

Fredersen has asked the scientist Rotwang to create a robot called Futura to replace the workers. At the same time, the government thinks that Maria is dangerous because she champions workers' rights. She urges them to resist and Fredersen asks Rotwang to capture her and give Futura her features. The Maria look-alike now begins to charm the workers, creating divisive rivalry among them. They do not realize that they are being hoodwinked by an android, who is driving them to their own destruction. Luckily, the real Maria escapes and manages to save the day. Futura is burned at the stake. Her artificial skin melts, revealing her true metal structure beneath.

Metropolis *by Fritz Lang, 1926.*
© Courtesy of MK2

Freder brokers talks between Fredersen and the workers' representative. The first motion picture to depict a robot in such spectacular fashion, *Metropolis* remains a cult movie to this day. Superbly directed and filmed, it enjoyed substantial resources for the time, with shooting taking place over 310 days and employing 36,000 extras. In 1984 musician Giorgio Moroder bought the rights to the film and produced a color version, complete with theme music by himself, Freddy Mercury and other vogue artists of the time. Its success gave *Metropolis* a fresh lease on life and popularity among a younger generation of moviegoers.

Maria, from the film Metropolis, *by Fritz Lang, 1926.*
© *Courtesy of MK2*

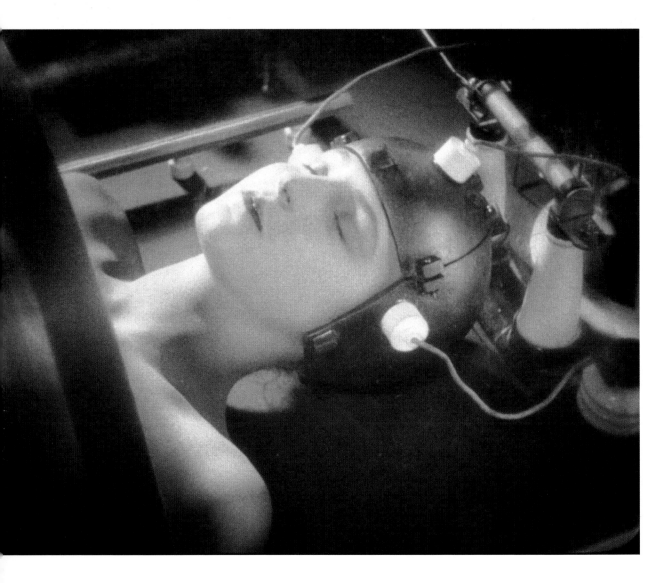

The man-hater

As World War II loomed, the world was plunged into a growing malaise and sense of upheaval. The cinema reflected this. While the barbarous, armed might of Germany raised the specter of war and threatened to unleash its soldiers on the world, there was a proliferation of B movies in which robots were hell-bent on destruction.

James Whale's 1931 *Frankenstein*, loosely drawn from Mary Shelley's novel of the same name, resurrected the myth of the creator destroyed by its creation. Russian cinema mirrored the drabness of the oppressive Soviet political system that had grown on the foundations of deviant Marxist theory. In *Gibel sensatsii* (Loss of Sensation, 1935), Aleksandr Andrievski tells the story of a people of mechanical workers manipulated by the bourgeoisie against the interests of flesh-and-blood proletarians.

On the Western front, meanwhile, B movies painted the robot in an evil light. There was a steady stream of trash movies—a godsend to penniless producers who were able to give spectators cheap thrills by getting actors to don sheets of metal. *Doctor Satan's Robot* (1940) by William Witney and

EXHIBITORS' CAMPAIGN

CARL LAEMMLE PRESENTS

FRANKENSTEIN

THE MAN WHO MADE A MONSTER!

A UNIVERSAL PICTURE with
COLIN CLIVE — MAE CLARKE — BORIS KARLOFF — JOHN BOLES

John English was one example of a long line of turkeys in which a mad hack scientist makes an invincible robot.

In the running for the worst robot film of all time was Phil Tucker's *Robot Monster* (1953). The felonious fiend in question had, almost literally, to be seen to be believed. Spectators put on special 3D glasses to see how closely Ro-Man from Mars resembled a skin-diving gorilla and how he laid waste to the world, sparing only a family of six. He cannot bring himself to kill them because he fancies one of the daughters, Alice. *Robot Monster* is regularly cited in worst-movie anthologies and is typical of films made on a shoe-

string. It was shot in just four days and its director pasted in much old archive footage. Interviewed by the Medved brothers for their book *The 50 Worst Films of All Time*, Tucker explained how all the robot costumes he found were too expensive. He then remembered George Barrows, who specialized in ape parts and was only asking $40 a day. He had his own ape outfit and when Tucker asked him to try on a skin-diving mask, he felt that the garb worked.

Invaders from Mars (1953), by William Cameron Menzies, borrowed the belabored theme of a robot that assumes the likeness of another character in the film to pass itself off as a human. Menzies was a respected production designer and designed the sets for *Gone with the Wind*. Behind the camera he worked hard on the movie's visuals, but the story line was an object lesson in the paranoia spawned by the hysterical anticommunism of the time and the belief that the enemies of the nation could be the guy next door.

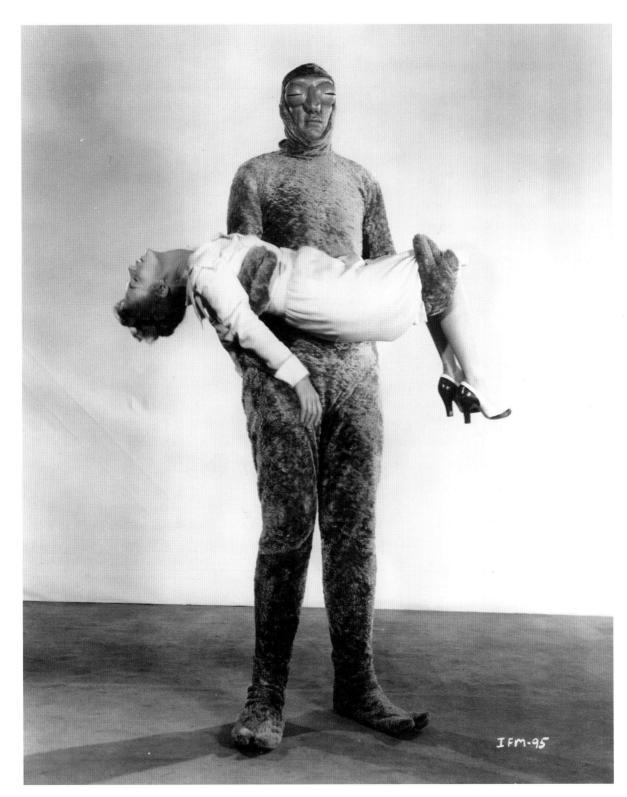

IFM-95

Much-loved people lovers

Ridiculed by its screen portrayals as a hysterical, homicidal fiend bent on destruction, the robot had a lot of ground to make up. Hidebound by story lines that made it hideous and terrifying, it mirrored the fears of a world over which hung the threat of global catastrophe.

The science-fiction literature of the 1940s had gone some way to refurbishing the robot's tarnished image, with the work of Asimov throwing a kinder light on androids, which obey the three laws of robotics. Movies of the early 1950s gradually caught on, depicting a new kind of altruistic, protective robot that would take devotion to self-sacrificial lengths. Some of these people-loving bots won a following among children and two made the transition to the small screen.

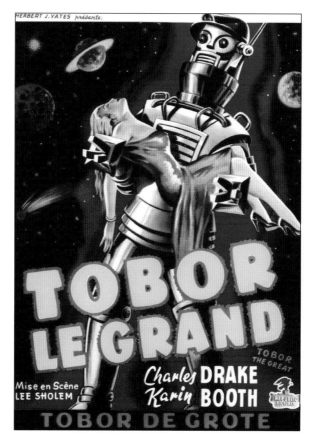

Tobor the Great, filmed in black-and-white in 1954 by Lee Sholem.
© Republic Pictures Corporation

Invaders from Mars
by William Cameron Menzies.
© 1953, 20th Century-Fox - Les Cahiers du Cinéma

Tobor

Tobor the Great, filmed in black-and-white in 1954 by Lee Sholem, was typical of the gradual change of attitudes toward the robot in the movies. Tobor actually began its career on television and was still a robot that was easily led into bad ways. But on graduating to the big screen, its transformation was complete as it became a benevolent bot that saved lives. The robot that was to become the indefatigable,

invincible Tobor the Great first appeared in a TV series, *Captain Video and his Video Rangers*, that ran from 1949 to 1955. Emblazoned on his chest was his trademark "I-Robot" sign.

He was a well-intentioned android, but led astray by the power-hungry Atar. She hopes to use the deadly ray gun he has in his chest to conquer the universe. Captain Video manages to overcome Tobor in the nick of time.

In *Tobor the Great*, Tobor is kind and enthusiastic and fond of kids. Doctor Harrison and Professor Nordstrom create Tobor to replace an astronaut and they are about to reveal this to the press when Nordstrom and his son are kidnapped by a spy. The faithful robot tracks them down. Compared to other science-fiction movies of the time, *Tobor the Great* is well crafted, relatively credible, and conveys the image of a peace-loving robot.

The Day the Earth Stood Still
by Robert Wise.
© *Collection Cahiers du Cinéma*

The Day the Earth Stood Still

The Day the Earth Stood Still (1951) by Robert Wise remains a robot cult movie. Adapted for the screen from the Harry Bates novel *Farewell to the Master*, it warns against the dangers of nuclear technology and the devastating effects of the atom bomb. The bombing of Hiroshima and the immense dangers inherent in the possession of such weaponry have traumatized the American people. Wise wanted to hammer home the message that nuclear testing had to stop.

In the movie the robot, Gort, will do anything to protect its master, while remaining profoundly respectful of the sanctity of all human life. Its behavior is inspired by Asimov's Laws of Robotics.

The story begins when a flying saucer arrives in Washington and out steps Gort, accompanied by an extraterrestrial, Klaatu, who bears a message of peace to all the peoples of the earth. But hardly has he stepped outside his craft when a panic-stricken soldier shoots and wounds him. Gort has to use force to ensure that peace prevails, so he destroys all the military's weapons.

Klaatu finds shelter in the home of a widow, Helen, but is denounced by the man who hopes to marry her, and assassinated by the US army. Gort retrieves his body and brings him back to life. Klaatu then delivers an address to all the scientists of the world and warns them that unless the human race renounces nuclear weaponry and agrees to live in peace, the earth and all life on it will be destroyed.

Gort embodies the quiet, strong android dear to Asimov with its loyalty, the wisdom it nurses within its metal shell, its devotion to its master's life and its respect for all forms of life.

The actor who played the part of Gort was an usher at the Chinese Theatre on Hollywood Boulevard who measured over seven feet tall. He was clad in a rubber jumpsuit that he could only wear for thirty minutes at a time.

Popular Robby

Forbidden Planet by Fred MacLeod Wilcox was released in 1956. One character in the movie was the benevolent robot Robby, whose good nature and sense of humor endeared it to the public at large.

When Commander Adams and his crew land on the remote planet Altair-4, they are welcomed by Robby, a robot of superior intelligence built by the strange Doctor Morbius who has a weakness for insane experiments.

Although Robby played only a supporting role in *Forbidden Planet* he was an instant hit, especially among children, and stores were soon stocking Robby toys. He was to appear in later pictures such as *The Invisible Boy* (1957) by Hermann Hoffman, where he has to fight against the power of a super-powerful, wicked computer that tries to turn him against humans. When the computer orders Robby to kill a boy called Jimmy, the robot destroys it in strict application of Asimov's laws. Robby would also feature in *Earth Girls are Easy*, *Hollywood Boulevard*, *Gremlins* and in TV series such as *The Addams Family*.

Right:
Poster and scene from MacLeod Wilcox's **Forbidden Planet** *with the famous robot Robby.*
© *Warner Home Video*

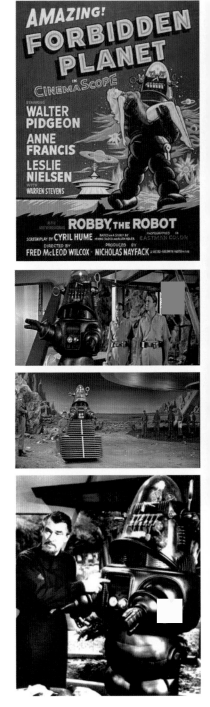

Above:
Forbidden Planet, *1956,*
by Fred MacLeod Wilcox
© Collection Cahiers du Cinéma

Right:
Robby also appeared in
Hermann Hoffman's The
Invisible Boy, *1957.*
© Warner Home Video

B9, the faithful companion

From 1963 to 1965 an American TV series etched strange and fantastical tale of extraterrestrials on the popular imagination. Produced by Leslie Stevens and Ben Brady, *The Outer Limits* offered fascinating story lines that ran the science fiction gamut and were watched in homes across the US. Robots were, of course, a regular feature, but the TV robot that was to make its mark was the faithful B9 from *Lost in Space* produced by Irwin Allen.

The Robinson family was flying across space to the star Alpha Centauri, where it had been ordered to found a settlement. To stop the mission, Smith the spy sabotaged the robot accompanying the family. As a result the vessel *Jupiter II* is "lost in space"[4] and week after week adventures befall the family, usually in the shape of evil aliens. When B9 sensed danger it would flail its arms and go into a spin, uttering the cry "Danger! Danger! Will Robinson!", which was to become part of household lore. Such was B9's impact that a model of the robot today stands in the conference hall at SRI International, one of the US's most important centers of robotics and artificial intelligence research.

The same year, the first episode of *Doctor Who* was broadcast in the UK. It would run for another 26 years, clocking up 685 episodes. The series took its name from its main character, Doctor Who, an eccentric time-traveling scientist. He was regularly pitted against the Daleks, murderous robots that resembled pepperpots studded with flashing lights. The cult series spawned its first feature film, *Doctor Who and the Daleks*, directed by Gordon Flemyng, in 1965, followed by a sequel the next year, *Daleks— Invasion Earth 2150 A.D.*

France at the time was under the spell of the *Nouvelle Vague*. One of its leading lights was director Jean-Luc Godard, who, in an unusual departure from his experimental, social œuvre, made a science fiction film, *Alphaville* (1965). Eddie Constantine plays the role of secret agent Lemmy Caution who, passing himself off as a reporter, gets into Alphaville to track down its ruler, the dictatorial computer Alpha 60. Its subjects, the Alphabètes, behave like robots.

---

[4] The series was made into a motion picture in 1998 by Stephen Hopkins, starring Gary Oldman and William Hurt. A DVD and new series are slated for 2005.

2001: A Space Odyssey

2001: A Space Odyssey *was made in 1968 by Stanley Kubrick from Arthur C. Clarke's story,* The Sentinel. *The film was made with support from NASA, IBM, and Vickers-Armstrong. The scale models were built with Cape Kennedy engineers as consultants. Kubrick often used real machines. The computer's name, HAL 9000, is IBM shifted one letter up in the alphabet.*

© Collection Cahiers du Cinéma

Interview
Philippe Druillet

*Internationally renowned
for his comic strips* Lone Sloane
and Salambô, *Druillet is a
multifaceted artist who paints,
sculpts and makes movies.
In 1974 he founded* Métal
Hurlant, *a French publishing
house that specialized in
science fiction comics and has
now acquired a cult status.
He nurses a passion for robots
and owns several hundred.*

**You have long collected robots.
What drives your passion
for them?**
I've got two collections. One is
mostly 1970s Japanese robots
made from metal. The other
consists of a second generation
of robots with a more plastic feel
and look, and special effects like
flashing lights and throbbing.
I'm interested in them for esthetic
reasons rather than as a collector.
I'm involved in science fiction,
so I thought they were beautiful
objects. Their metallic look was
like the armored look I've always
drawn. Toys are a creative,
inventive field and I believe they
mirror society. Robots are creative
in their use of color, their
mechanisms, and the little screens
on their chests. The Japanese
have worked wonders. Some have
invented new shapes that recall

Philippe Druillet.
© FYP

the automata of the 19th century.
Robots also have an African,
totemic side to them. There are
African terra-cotta sculptures
from the 13th century that are
monolithic in the same way as
Japanese robots, even though
they are not made of metal.
They give off a sense of raw
power. The anthropomorphism
reminds me of medieval knights.

**What pieces are you most proud
of in your collection?**
The Japanese robots from the
1950s and 1960s. They're worth a
fortune, as it happens. In *Star
Wars* Lucas used a Japanese robot
from that period as the model for
his warriors. I've owned between
400 and 500 robots, but what
I liked was to see them as a
whole. When you acquire so
many objects, they become a
force, an absolute work of art, a
multicolored tableau that's totally
magical. I find them esthetically
pleasing, but I like to work them

and set them against each other.
Some shoot sparks and others
have guns that go up and down,
but I always prefer the ones that
are musical and colorful.

**When did your fascination with
robots begin?**
It started when I was very young,
when I started to read American
science fiction like Asimov and
Van Vogt. Robots were also part
of my movie culture, with
Forbidden Planet and *The Day the
Earth Stood Still*. As someone
who draws and designs, I was
drawn to robots because of their
medieval, armored side. The
robotic laws remind me of Moses'
Ten Commandments.

**In what way have robotic
concepts influenced your own
creative work?**
They've never influenced it.
I focus much more on the human
body, in the Terminator style. For
me Sloane is a sort of Bioman,
somebody who's had implants
to prolong his life, enhance his
vigor, his speed, his vision of
things. I consider the robot from
an aesthetic point of view, with
its medieval appeal, which is
evocative, romantic, and
nostalgic. My work is apart from
all that. There's only *Six Voyages*,
where I drew a robot that was
40 meters tall. But the first
Transformers were based on

my drawings. A Japanese delegation presented me with them at the *Métal Hurlant* offices in 1974.

Which do you think are the greatest robots in fiction?
There are only two that I really love: Robby from *Forbidden Planet* and Gort from *The Day the Earth Stood Still*. I read interviews with Lucas who said that he could never forget *Forbidden Planet*. The great well of light in *Star Wars* comes from *Forbidden Planet*. When you see that at sixteen years of age, it has quite an effect. Then there were all the B films with their cardboard creatures. Let's not forget the robot from *Metropolis*. That was the first time there had been a woman robot, the design was sublime. I've got to mention Frankenstein, even if it wasn't a robot because it was made from flesh and blood. But it still carries the idea of the freak who is rejected because of his appearance, who only wants company and to be loved. Like *King Kong* and *Freaks*, which is one of the greatest of love films. What I find extraordinary is the theme of the outcast. I think it started with *The Golem*.

The robot was long an essential part of science fiction that looked to the future. Now it's a reality
with Sony's pet-animal robots and Honda's android. What are your feelings?
That's the neurotic side of the Japanese. They're so perfectionist that they can build a cat or a dog that becomes a companion. The science fiction writers of the 1950s never suspected there would be anything like a pet robot, which is emotional and toylike and appeals to the need to possess and to shun reality. There's probably nothing more beautiful in the world than a cat. But with Aibo, its owner has everything a cat can offer without the hassle of taking it out or feeding it, unless it's with synthetic food. Owners are in control of the robot pet, but a cat scratches. It's fascinating, there are the pluses of having a pet without the drawbacks. The need for love has been met by an electronic system, which for me is proof of deep neurosis. Maybe, in the future, that kind of emotional transfer will focus on artificial men or women as a way of satisfying fantasies.

In what directions could science fiction authors now take the robot?
Nowhere, there are none left. Robots remind me of old spacecraft from the 1950s, which smell of oil that has just been changed—and that's why I like them. They've lost their mythical power. Science fiction writers have explored everything remotely to do with robots. The 30,000 or 40,000 works written between 1930 and 1970 circumscribed everything that was possible and imaginable.

Would you like the idea of having an android at home. What tasks would you set it?
I'd love one. I'd design it myself. But I know that eventually I'd feel like telling it to get out, back to the junk heap.

No future without robots

Robots gradually became stock characters devoid of panache in TV series, used simply to create a futuristic feel. Science fiction motion pictures were considered pure entertainment or second-rate stuff, with no artistic value compared to great works like Coppola's *The Godfather* or the spellbinding extravagance of Fellini and the Italian masters.

Robots appear in George Lucas's first movie, *THX 1138* (1971), where a police state entrusts law and order to white-clad, sexless humans under constant surveillance from CCTV cameras and android cops. Although a little loosely structured, *THX 1138* offered a vision of society that heralded the work of the Wachowski brothers thirty years later in *Matrix*. *Silent Running* (1972) by Douglas Trumbull focused on environmental issues dear to the counter-culture. Asked to destroy his great greenhouse in space, Lowell Freeman refuses and, aided by three robots, he determines to save his plants even at the expense of his three accomplices.

One well-crafted movie is *Westworld* (1973) by the writer Michael Crichton who, after pub-lishing his first book in 1969, and having others adapted for the big screen, decided to try his hand at directing. *Westworld* skillfully mixes three genres—science fiction, Westerns and comedy.

In a futuristic Disneyworld one of the attractions is a Wild West town. Robots dressed as cowboys entertain visitors (one is played by Yul Brynner, with two steel balls in lieu of eyes) in mock shoot-outs. Every evening the robots shot dead by visitors are regenerated. Meanwhile they are evolving and learning. Suddenly they start shooting back and trying to kill all the visitors to Delos Park.

A sequel was released in 1977—*Futureworld* by Richard T. Helffron. Long closed after the robots' revolt, the theme park reopens. But two snooping reporters get wind of a plot to replace heads of state by robots...which takes them to the theme park.

Woody Allen, too, used robots in his 1973 sci-fi satirical comedy, *Sleeper*. He plays the part of a health-food store owner in the 1970s who is cryogenically frozen when he dies on the operating table. He wakes up in 2173, when the US has become a police state where robots do the dirty work and menial service jobs.

Star Wars robots' saving grace

Fifty years had passed since Fritz Lang fascinated movie-house goers with Maria in *Metropolis*. Since then the robot had ceased to fire imaginations. Appearing in countless numbers of trashy movies, it had never been given the aura or depth of treatment it had enjoyed in Asimov's work. Robots had fallen from their rusty pedestal. Then, in 1977, they were suddenly back on the silver screen, enjoying unexpected pride of place in *Star Wars* by George Lucas.

Star Wars marked a turning point in motion picture science fiction, thanks to a brilliant sceenplay, consummate directing, and special effects like nothing that had ever been seen before. Lucas also broke new ground by incorporating characters into the script from a wide range of cultures and periods—from medieval heroes and comic strips to the *Odyssey* and, so Asimov claimed, from his *Foundation* series of novels. The melting pot gelled, and on release the movie asserted itself as a classic with its undeniable qualities. *Star Wars* broke box office records and spawned an enormous merchandising trade with spin-offs that included the two robots C3PO and R2D2. Although part of the supporting cast, they won popular acclaim.

Clumsy but cute

R2D2 is designed to repair computers and spacecraft. It is on the cumbersome side and does not enjoy moving. It talks through funny electronic squeaks and buzzes, while its hemispherical head spins. Its comical side swiftly made it the darling of the public.

C3PO and R2D2 form the best-known robotic duet in the Star Wars films by George Lucas.

R2D2 is assisted by a translator robot, C3PO, an android that Lucas, in a stroke of genius, decided to portray as a snobbish, uptight butler who is as cowardly as he is brilliant—he masters six million languages. C3PO is as human as can be, despite his golden metal body.

R2D2 plays a crucial role in the story line when Princess Leia entrusts him with a message for Obi Wan Kenobi just before she is captured by Dark Vader. After being landed on the planet Tatooine, R2D2 soon finds himself on Luke Skywalker's farm. Once C3PO has translated the message entrusted to R2D2, the saga can begin.

In the year of its release *Star Wars* revived public passion for both science fiction movies and robots. Two sequels would follow: *The Empire Strikes Back* and *The Return of the Jedi*. A second trilogy was released in 1998.

Above:
Born on August 24, 1934, in Birmingham in the UK, Kenny Baker played the part of the robot R2D2 in three movies of the first Star Wars trilogy.
© Courtesy of Jean-Régis de Vansay

Right:
Even though R2D2 is not anthropomorphic and can only squeak and squeal, the actor inside is a real person who gives the robot personality.
© 2004 Lucas film Ltd & TM All Right Reserved - Used under authorization - Courtesy of Fox Pathe Europa

Measuring
5 feet 6 inches
(1.7 meters) and
clad in a golden,
metal armor,
C3PO can speak
more than six
million languages.
The peerless
interpreter
constantly
reprimands those
around it and acts
like a robotic
diplomat.

Alien and the evil android

Ridley Scott's 1979 *Alien* is a landmark in science fiction film. It creates a unique atmosphere, where unbearable suspense in a closed-in, claustrophobic space mixes with elements of terrifying gore. In one scene, which has gone down in the annals of horror movies, a robot tears its way out of the body of a crew member. One of *Alien*'s great merits is to take science fiction in a new direction, away from mere futuristic technology. Its roots lie more in the eerie tales of H.P. Lovecraft.

A spaceship, the *Nostromo*, is on its way back to Earth, manned by a crew of five men and two women. It receives an SOS message from a small planet, where the crew members alight to find some mysterious eggs. Lieutenant Ripley, played by Sigourney Weaver, has remained alone on board the craft and realizes too late that the message was warning them of danger.

When the crew return to the ship, they unwittingly bring with them an egg, which mutates into the Alien and kills all the crew members one by one. Ripley takes refuge in a shuttle and manages to eject the creature into outer space.

Ripley has noted the increasingly strange behavior of Ash, the crew's scientist. When she realizes that he is trying to retrieve the Alien, they skirmish verbally, then physically fight. She strikes him with a fire extinguisher, tearing off his head to reveal the android that had secretly joined the crew. It had been sent by the company that had chartered the *Nostromo* in order to bring the Alien back to earth. Ridley Scott draws on the science fiction and horror tradition of the artificial being using human likeness to its own wicked ends.

Alien was an enormous box-office success, raking in $80 million in the US and worldwide, and followed by three sequels. In *Aliens*, directed by James Cameron in 1986, Ripley has the support of an android called Bishop, which this time turns out to be a reliable ally. The third sequel, *Alien, Resurrection* (1997) features Call, Ripley's android companion, played by Winona Rider, who gives her character a deeply sensitive personality.

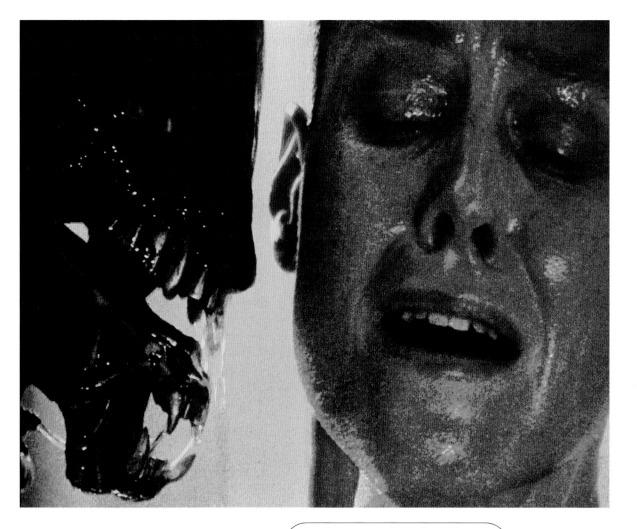

Alien *is the remake of a*
1950s science fiction movie,
Terror from Beyond Space,
featuring androids that are
so sophisticated it is
impossible to tell them apart
from humans.

Ridley Scott: can't tell them apart at all

In the very early 1980s, at a time when literature had shunned robots, the movies continued to offer them openings. Some were major roles, but not in the traditional metal mold. What is noteworthy is that they played important roles precisely because they were androids that could not be distinguished from humans. The two most remarkable androids of this type were both the work of the same movie director, Ridley Scott.

The obvious charm of the replicants

In 1982, three years after *Alien*, Ridley Scott released the now legendary *Blade Runner*, a screen adaptation of the Philip K. Dick novel, *Do Androids Dream of Electric Sheep?* The film moves spectators to sympathize with refined, sensitive artificial beings, one of which is played by the beautiful Daryl Hannah. The futuristic Los Angeles, in which the film is set, was visually flabbergasting and *Blade Runner* became a milestone in science fiction film, paving the way for the cyberpunk genre.

In 1982 Ridley Scott's **Blade Runner** *was released, adapted for the screen from the Philip K. Dick novel,* **Do Androids Dream of Electric Sheep***? Left: Batty, the leader of the rogue replicants, played by Rutger Hauer, with Daryl Hannah as Pris, also a replicant.*
© Collection Cahiers du Cinéma

Bounty hunter and former policeman Rick Deckard, played by Harrison Ford, is hired to kill four replicants, very human-looking androids, that have turned dangerous. Tyrell, who makes replicants, first asks him to find out if his personal assistant, Rachel, is one, or whether she is human. Deckard works out that she is android, which does not stop him from falling for her. They form a couple throughout the rest of the movie.

Deckard manages to kill three of the replicants, including the gorgeous Pris (Daryl Hannah), before he is overpowered by the powerful Batty. Oddly, Batty, who is dying, spares Deckard's life, prompting critical questions about the importance of life. Deckard then runs away with Rachel.

Blade Runner is a cult film among cult films. Yet it was not an overnight success. Indeed, it was only once the video was released that it became a contemporary classic that broke new ground for its sets and lighting.

Terminator
the ruthless robot

Another director left his mark on cinema of the 1980s with a powerful, uncompromising portrayal of an android whose grim determination never wavers. He was James Cameron and the movie was *Terminator*. Released in 1984, it was the first film he had ever directed from start to finish, but it showed what a demanding filmmaker he was. Excellent as the robot bent on accomplishing its mission come what may, Arnold Schwarzenegger would never be better.

In 2029, when robots have risen up in revolt against the human race, one of them, the terrifying Terminator, travels back in time to 1984 and, one by one, kills all the women who bear the name of Sarah Connor. A waitress with that name manages to escape and discovers that she is protected by a man who has also come from the future. His name is Kyle Reese and he explains to Sarah that the Terminator is a killing machine sent by the robots waging war in the future against humans. Terminator's objective is to prevent the birth of a boy named John to Sarah Connor, for it is he who will lead humans to victory against the robots.

One of the strongest themes in the movie is that machines can have no moral code other than that programmed into them. Even when it has been sliced in two, Terminator still tries to accomplish its mission—parts from the robot continue to try to kill. Only destroying every last one of its chips can stop it from piecing itself together and starting again.

Success was instant. The film grossed nearly two billion dollars worldwide and yielded two sequels. *Terminator 2*, also directed by Cameron in 1990, makes full leverage of morphing technology to drive home the same message—the robot does not know what it is to be despondent or discouraged and, as long as its chips function, so does it. Killer robots can be smashed to smithereens until they are nothing but

minute balls of mercury spilled on the ground. Yet each ball still remembers what was programmed into it. One by one they reassemble and the newly reconstituted robot rushes to the attack. Jonathan Mostow directed *Terminator 3* in 2003.

Robocop
the robotic man

In *Robocop* (1987) by Paul Verhoeven, the hero is a human being who has been rebuilt by means of electronic implants. He has thus become a cyborg, further blurring the ever-more-tenuous boundary between humans and their man-made creations.

In an attempt to stem the tide of violence swamping Detroit, a private security conglomerate undertakes research into robot-

The robots in Terminator *can behave like humans.*
A neuronal network enables Terminator-type robots to learn from experience.
© Gaumont Columbia Tristar Home Video

ics. Shortly afterward, a policeman by the name of Murphy is
badly beaten and left for dead by
a gang of killers led by Clarence
Bodicker. He undergoes major
surgery and his limbs and damaged organs are replaced by
electronic implants. He becomes
a cyborg by the name of Robocop. When his memory returns,
Murphy sets out to find his
attackers and learn how the conglomerate employing him might
be involved in Bodicker's gang.
Two sequels have followed: Irwin
Kershner's *Robocop 2* (1990)
and *Robocop 3* by Fred Dekker
in 1992.

*Robocop is not, like
Terminator, an entirely
automated system.
It/he is a cyborg,
a part biological,
part electronic,
cybernetic hybrid.*
© 1987, Orion Pictures
Corp. - Collection
Cahiers du Cinéma

Robots seeking love

After a long time in the wilderness, the robot was now back in motion pictures in a big way as the third millennium loomed. Two films sought to draft it into a new kind of plot that addressed the issue of robots' emotional needs.

Bicentennial Man (1999), directed by Chris Columbus, was the first film to be adapted from an Isaac Asimov short story, *The Positronic Man*, written in 1976. A domestic robot of above-average capabilities has performed its household duties for 200 years with the ambition of becoming a human being in its own right. A roboticist, Rupert, helps it to realize its dream through a human-organ implant. It emerges from the operation with a comical, soft-centered look, rendered by Robin Williams. As Williams said: "Chris Columbus has managed to touch on all the questions raised by the appearance of robots: master–slave relationships, intolerance, humanity. The robot I play has something special, it's inquisitive, it's easily fascinated." Making the robot was a technological feat in itself and involved Robin Williams laboring in an outfit made up of 250 separate parts and weighing 40 kilos (88 pounds).

David, the first robot with feelings, and Gigolo Joe, an adult robot programmed to satisfy your every desire. The script for A.I. was loosely adapted from Supertoys Last All Summer Long, a Brian Aldiss short story published in 1969 in a special issue of Harper's Bazaar. Following the death of Stanley Kubrick, it was Steven Spielberg who made the movie, stressing androids' human side.
© Photo: David James - 2001 Warner Bros & Dreamworks LLC - Collection Cahiers du Cinéma

Steven Spielberg, too, tackled the question of the emotional life of the highly sophisticated robots of the future. *A.I. Artificial Intelligence* (2001) is set in the middle of the 21st century, at a time when overpopulation and global warming have compelled the government to regulate the birth rate. Couples may have only one natural child, but may also adopt androids. David is one, programmed to feel love exclusively for its human parents. The Swintons adopt him after their daughter has been cryogenically preserved following a serious illness. The mother, Monica, however, is incapable of showing the least affection for the little android and when her daughter has been cured, she asks him to leave home. The unhappy David now trudges off to look for the human within.

A.I. was originally an idea that came to Stanley Kubrick in the mid-1960s, when, after reading the Brian Aldiss story, *Supertoys Last All Summer Long*, he began thinking about a film in which the characters would be robots. He had great trouble in finding a screen adaptation that satisfied him, and the script he finally chose in 1982 was mostly his own work. He asked a number of comic-strip artists to draw him a robot world and also approached graphic artists at ILM, George Lucas's special effects company. When Kubrick was working on *The Shining* and Spielberg on *The Raiders of the Lost Ark*, the two directors got together and talked long about the artificial-intelligence project. It was finally filed away, principally because Kubrick felt the technology could not live up to his vision of the film.

In 1993 *Jurassic Park* struck a chord with Kubrick when he saw that computer-generated images were powerful enough to produce the effects he would need for his film on artificial intelligence. He planned to start shooting after *Eyes Wide Shut*.

Kubrick's death in March 2000 sounded, so it seemed, the death knell of his artificial-intelligence movie. It was to turn out, however, that he had given Spielberg a copy of the script.

The first and third acts struck Spielberg as fabulous, but the second, he felt, needed to be developed before he could think about getting down to the movie. The final result was a multifaceted work, which has come in for criticism for its contention that robots can never develop awareness.

Two thousand and four was the year of *I, Robot*, directed by Alex Proyas and adapted from an Asimov novel. Robots go peaceably about their tasks in 2035. Humans have grown accustomed to having full confidence in them, for the Three Laws of Robotics have been programmed into them. Yet Detective Del Spooner, played by Will Smith, is investigating the murder of a roboticist and suspects first one robot, then all robots.

The film was released at a time when robots had gradually started to crop up in routine walks of life, particularly the home, through the robotic pet Aibo, and the spread of autonomous vacuum cleaners. Reflecting that trend, *I, Robot* has the merit of setting the action in a world where robots are ordinary denizens of ordinary society—a reality that, though still on the horizon, is on its way. As Rob Enderie of the market research institute the Enderie Group, says: "The success of the film *I, Robot*

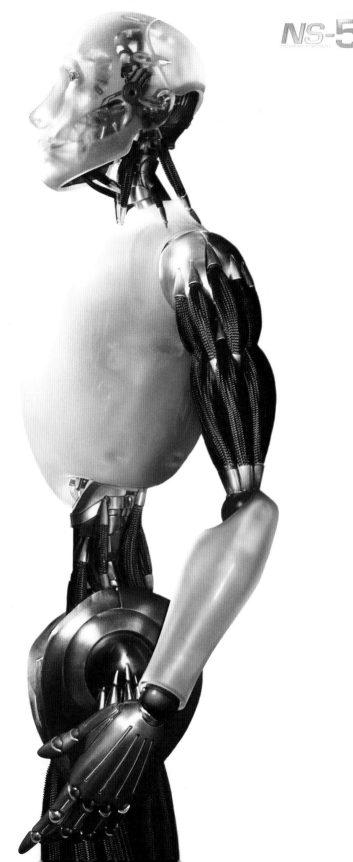

has prompted increased demand for robots. Once that demand has been clearly demonstrated, it won't be long before someone thinks of a way to meet it."

What roles will robots play in movies yet to come? The amazing advances in computer-generated imaging now make it easy to create them on the screen. When it comes to subject matter, however, the motion picture industry looks as if it has explored every avenue opened by 50 years of science fiction literature. The chances are, though, that robots will increasingly appear in film of all types, as they have in literature, as they enter the home. They are no longer a byword for future shock. Accordingly, they will take their place in movie genres where once they would never have been seen. Sooner or later, there will be a remake of *Love Story*. Will the robot be a shoulder to cry on, or will it cry on a shoulder?

The robot NS-5, or Sonny, suspected of murdering its creator.
© 2004 Twentieth Century Fox Film Corporation

Robots from *mangas* and cartoons

From autumn 1944 to the spring of 1945, US B29 bombers pounded Tokyo mercilessly. The Empire of the Sun, which had once thought itself invincible, reeled under the final atomic onslaught on the cities of Hiroshima and Nagasaki. Defeat stared it in the face. It surrendered.

At the last meeting of the War Research Committee in August 1945, attendees were informed of the Potsdam Declaration, which brought official confirmation of Japan's defeat. The sight of its devastated cities annihilated any lingering hopes survivors might have placed in age-old myths of the might of Japan. The country was now occupied by US troops under the command of General MacArthur.

To help Japan rebuild its industrial base, the American Committee for Economic Reconstruction decided to place special emphasis on technological development, and the US channeled millions of dollars into the effort. A new page in Japanese history had been turned and corporations like Sony, Honda and Matsushita were to play a central role in transforming the country into an economic world power. There would also be a transformation in its forms of artistic expression. Through its *manga*s and cartoons, Japan would sublimate its frustration at being constitutionally denied the right to warfare by creating all-powerful heroes: robots.

The rise of mangas

While the US exported its comics and pop music into Japanese streets, schools, homes and the Tokyo metro, a homegrown comic-strip culture developed, called *manga*. The beginnings of *manga* comics can be traced back to the 1930s and 1940s, and even beyond to medieval scrolls of the 13th century, known as *ukiyo-e*. After World War II, however, *manga*s took on a new lease on life and began to create their own heroes, the best known of whom remains *Tetsuwan Atom*, or Astro Boy.

Although *manga*s broached wide-ranging themes, the trauma of recent defeat and destruction lay at their core. It manifested itself in two apparently opposite but complementary ways that would also emerge in Japanese robotics design. Some mirrored what is known as the *kawai* sensibility, defined by its cute, child-like graphics. *Kawai* appealed to large groups of Japanese youth, who took refuge in the sweetness and light of its lyrical, escapist world.

Other *manga*s developed an imagery whose ingredients were high technology and apocalyptic visions, replete with attacks by aliens from outer space and the destruction of life on Earth as we know it. They sought, too, to restore the pride of a wounded people by forging a multiplicity of allegorical stories about haughty, warlike peoples.

One pioneering practitioner of the *manga* medium was Osamu Tezuka. After a string of successes *like New Treasure Island* and *King Leo*, Tezuka turned his talents and imagination to the story of a robot. He was to create the first hero of what would become a *manga* genre, progressing from comic strip, to television series, to movie—a pattern that was to establish itself.

Astro Boy, the robot in the home

Astro Boy put in his first appearance in a *manga* comic in 1951 in the guise of a diminutive robot. To all intents and purposes he resembled a little boy, yet he was endowed with special powers. Rockets in his boots could blast him into orbit, while his fingers were ray guns.

A grieving rocket scientist from the space agency builds Astro Boy to make up for the loss of his son, and deliberately makes his creation superhumanly powerful and intelligent—he can speak dozens of languages.

Eventually, however, he tires of the artificial surrogate, realizing that the little robot will never replace his true son. He sells Astro Boy to a circus boss who exploits his special powers. It is then that a professor begins calling for the release of Astro Boy and other robots who, he argues, are entitled to their rights. Great adventures now await the little boy robot, who is as endearing as he is super-powerful.

The importance of Astro Boy resides in part in his depiction as a robot who lives the traditional Japanese family lifestyle and goes to school like other children. What's more, as his adventures unfold and he asserts his place in the world of humans, the underlying implication that a robot might have feelings grows more explicit. The message Astro Boy seeks to get across is that robots and humans could live together and get on naturally. Overnight he became a part of every Japanese family.

Astro Boy was the first robot to be made famous by comic books. His popularity extended beyond Japan, and the boy android was adopted internationally. His success also exceeded its original medium, and in 1963 Tezuka adapted his *manga* stories for television. Astro Boy was now also the hero of an *anime*, or cartoon series. Tezuka, who had founded a studio, Mushi Productions, showed creative ingenuity as he forged a minimalist *anime* style of no more than four to five frames per second. The Astro Boy cartoon series was broadcast weekly in black-and-white episodes that were 25 minutes long. Success was swift.

Meanwhile, across the Pacific, a similar attempt to make a robot part of the family had not encountered popularity on a comparable scale. In the wake of their wildly successful TV cartoon of a "modern Stone Age family", *The Flintstones*, Bill Hanna and Joe Barbera decided to use a similar recipe, transferring an all-American family to a far-distant future. They came up with *The Jetsons*, of which the first episode was broadcast in 1962. The cartoon teemed with robots, like the comical, housekeeping Rosie the Robot. But *The Jetsons* never really caught on and before long it was discontinued[7].

Astro Boy's international popularity, though, went from strength to strength, with the Japanese International Exhibition at Osaka in 1970 giving him a further boost. Millions of visitors made the acquaintance of the humane, humanist little boy bot, and countries queued up for the rights to broadcast his adventures. The very American Walt Disney went so far as to propose Tezuka a partnership.

Astro Boy attained legendary status and so deeply embedded in the Japanese psyche did he become that even now it is not uncommon for the country's roboticists to refer to him as their touchstone[8].

---

(7) *The Jetsons* was revived in 1984 to a much improved reception and ran for four years. In 1990 the TV series was adapted as a motion picture.

(8) A second Astro Boy cartoon series, this time in color, began in 1980. Produced by Noboru Ishiguro, it grew to cult status. A third series was broadcast in 2003, produced by Konaka Kazuya.

Doraemon, Popular as Mickey Mouse

Astro Boy has inspired subsequent *robotto anime* (robot cartoons), which also have their roots in *kawai* culture. They cast robots in a friendly light as the socially integrated companions of ordinary people. The best known of the *kawai* robots is Doraemon, who first appeared in 1969, the work of Fujimoto Hiroshi and Motoo Abiko. He is a chubby, round, blue-colored robot tomcat from the 22nd century. His pockets are crammed with all sorts of ultrapowerful gadgets and devices that he brought with him from the future.

Doraemon has been sent into the past to watch over a little boy called Nobita and to stop him from getting up to pranks that could steer his family's future off course. He has a typically sweet *kawai* appearance. He is small, pudgy and zealously smart, always using his intelligence for the well-being of Nobita to ensure his destiny stays on track. He might be a consummate cat, but he still has weak spots, like his terror of mice. The mixture of techno-bot and comic failing was precisely what endeared him to so many people in a nation still trying to find its bearings.

More than 100 million copies of Doraemon comic books have been sold, and 23 full-length feature films have recounted his adventures. In 1979 the Asahi TV channel broadcast the first episode in a cartoon series of 1,800 episodes. Doraemon is as famous and well-loved in Japan as Mickey Mouse.

The influence of kawai anime on Japanese robotics

The *manga*s and *anime* that have grown up around the *kawai* sensibility are an integral part of Japanese culture. They draw a wide and devoted following from among young people and heavily influence fashion, from dress codes to accessories and hairstyles. *Manga*s and *anime* are a homegrown version of Western pop culture. Many of them regularly feature robots.

Escaflowne, by Shoji Kawamori, was premiered in 1996. It is a cartoon, or *anime*, which tells the story of a robot that changes into a dragon to combat the bad Zaibach. He is aided and abetted by a young girl called Hitomi.

Among more recent productions one of the most impressive is the *Chobits* series of *manga*s and *anime*. The first story appeared in 2002, the work of Morio Asaka from Studio Clamp, which was created by four female artists. With its highly refined graphic artistry, *Chobits* raises numerous questions about the nature of life and how real real is.

In *Chobits* computers have been replaced by very human-looking cyborgs called Persocons. One day a young guy whose name is Hideki finds a dysfunctional Persocon, Chii, who has been thrown out on the garbage heap. He soon discovers that she behaves very strangely indeed. Could she be a *chobit*, one of those very special computers that can work without an OS and show feelings?

So deeply do *kawai* robots permeate Japanese society that many of their traits are mirrored in commercial devices now on the market. They help to understand why the personal robots that have been produced since 1999, from Sony's Aibo to NEC's PaPeRo, look so cute (some even have cuddly fur), even though, to a Western eye, they resemble toys. They also account for the readiness of a company like Sony to invest millions of dollars in creating a miniature android that can sing, dance, and entertain, and has absolutely no practical use. Plainly, the robots that are increasingly entering Japanese homes and keeping the elderly company are embodiments of Astro Boy and Doraemon.

Mechas: the dark side of robotics

There is another, darker, side to *manga* comics, represented by *mecha*s. Over the years *mecha* stories have come to account for three-quarters of the total *manga* output.

Mechas are giant robots controlled by humans or aliens who clash in spectacular combat. Some are reminiscent of the ancient samurai, who once symbolized the power and art of the Japanese warrior tradition. Others are more like gigantic, monstrously deformed, post-nuclear insects—for the Hiroshima syndrome is always at the back of the mind.

The *mecha* craze made its 1980 debut with the cartoon series *Robotech*, also known as *Macross*. (Robotech is the name of the technology used by the aliens in the series, and Macross is the island on which their spacecraft crashes.) Produced by Katsuhito Akiyama, the *Robotech/ Macross* TV series was such a huge success that it led to full-length animated feature films. It has developed a cult following throughout Japan and is particularly popular among boys.

Ten years after an extraterrestrial spaceship crashed on the island of Macross, Earth has come under attack from giant warrior invaders. Unable to resist the onslaught, the government of Macross manages to reactivate the craft and beam it across space towards Pluto. A long journey across the solar system now begins. Despite the name, Robotech, robots are not the main characters in the series. Theirs is a supporting role in the service of heroes.

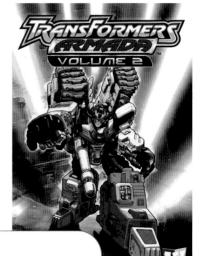

Transformers Armada

The debut appearance of shape-changing transformable robots came in 1980 with the Japanese Diaclone toys. In 1984 the Hasbro company acquired the rights to these transformable toys with a view to making a cartoon series, which would be produced and directed by the Marvel Production and Sunbow Production studios. In 2002 Aeon Inc. reworked and updated Transformers, releasing the new series under the name Transformers Armada.

© Courtesy of Miam.com/TF1 vidéo.

Cubix

Connor and his father move to the robot city of Bubbleville. Connor meets Abby, Mong, Chip and Héla, and is allowed to join their club when he repairs Cubix. The friends set out in pursuit of Sewwix, the robot poet, who is responsible for the power cuts plaguing Bubbleville.

© Courtesy of Miam.com/TF1 vidéo.

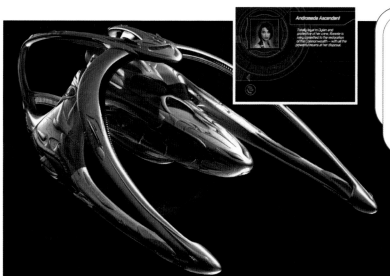

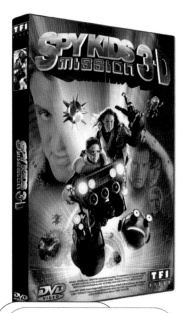

Andromeda

Science fiction TV series produced by Robert Hewitt Wolfe and Gene Reddenberry, who created Star Trek. Andromeda *is a space vessel with its own individual personality and able to assume human form.*

© Gaumont Columbia Tristar Home Video - Courtesy of Élysées édition

Spy Kids, Mission 3D

The best-known story from the Spy Kids *saga by Robert Rodriguez, which distinguishes itself from the first two parts by its use of 3D. Using a high-resolution video camera developed by James Cameron, Robert Rodriguez designed new equipment and accessories to refine and simplify the entire 3D process. He also supervised the digital creation and design of the computer-generated characters, creatures and vehicles.*

© Courtesy of Miam.com/TF1 vidéo.

Vision of Escaflowne

This heroic fantasy saga was adapted by Katsu Aki from an anime and made into a shonen manga *(comic book especially for boys).*

Chobits

The story of Chobits is set in a futuristic Japan, peopled with robots and androids. Hideki is a schoolboy who finds an android, Chii, in a garbage can. This manga *comic book was highly popular.*

Chobits : © Clamp/Kodansha ltd - Pika Édition
Vision d'Escaflowne : © Katsu Aki/ Kadokawa Shoten Publishing Co - Pika Édition

France discovers the mecha's through Goldorak

Strangely enough, France's introduction to the world of mechas came via a cartoon series, Goldorak, that had relatively little success in Japan. The first episode of Goldorak, created and produced by Go Nagai and Kazuo Komatsubara, was broadcast in 1978 and soon became a very popular fixture.

Goldorak is a gigantic robot with a horned head. It is controlled by Actarus, an extraterrestrial prince who has found refuge after the forces of Vega have destroyed his planet. To prevent the same fate befalling Earth, Actarus does all he can to stave off the Vegans. He is assisted by Goldorak, who plays a decisive part in overcoming the aliens from Vega. Goldorak has a companion by the name of Alcor, who also operates a giant robot, Mazinger.

The characters created by Go Nagai are a reflection of the infatuation with raw power that struck a chord with a cross section of Japanese youth at the time.

In the wake of Goldorak, French mecha enthusiasts began to feel the appeal of robots, and Albert Barrillé created and produced the series Il était une fois l'espace (Once upon a Time in Space). In the 1982 episode, La revanche des humanoïdes (Venge-ance of the Humanoids), Pierrot, Psi, and their robot, Métro, crash on a hostile planet inhabited by humanoids that are preparing to take over the universe.

In 1987, René Laloux produced Gandahar, adapted from a series of novels by Jean-Pierre Andrevon, first published in 1969. Gandahar is the saga of attempts by an army of destructive robots to invade a planet where women play a dominant role.

Mechas go global

The 1980s saw the irresistible rise and rise of mechas. The straightforward story lines and spectacular combats between giant robots against the backdrop of a post-apocalyptic world—the action is often set in the immediate aftermath of a cataclysmic disaster—were ingredients in a recipe that proved increasingly popular.

The titanic gladiatorlike clashes also caught on in the US when, in 1984, the US family entertainment company Hasbro began importing toy Transformers, the terrible robotic giants that could transform themselves into fighting machines. In a move to promote sales, Hasbro supervised the development of an animated TV series featuring the Transformers. Record-breaking sales ensued—$100 million in the year they were first mar-keted—and an international mecha craze took hold.

A few years later, in 1987, the release of a new anime feature film made a powerful contribution to popularizing "Japanimation" on a global scale. Akira, directed by Katsuhiro Otomo, was the screen adaptation of a manga comic book that ran to thirteen volumes. Its grim, post-nuclear atmosphere is similar to the feel that prevails in mechas. It tells the story of a gang of young motorbike rebels and their adventures in a giant, nightmarish crater peopled with mutants, which is, in fact, Tokyo after the apocalypse.

There is a similar feel and theme to Patlabor (1988), created and produced by Studio Headgear. Released almost simultaneously as a manga comic book and an anime cartoon, Patlabor recounts the struggles of a police force made up of heavily armed, titanic robots (the Patlabors) in a futuristic Tokyo whose survival is under threat. Ordinary citizens can only look on as battle-crazed mechanical monsters clash with each other. Patlabor was later adapted as a movie by Mamoru Oshii, who would go on to make the celebrated Ghost in the Shell. Armitage III (1995) by Taro Maki and Hiroyuki Ochi, from a character created by Chiaki Onaka, is another classic of its kind. It consists of a quadrilogy of anime, in which the action is set on Mars in a future when people have settled there. The Martians are paranoiac in their suspicion of robots and gradually eliminate them one by one. Which is where Armitage steps in. She is a third-generation android with an extremely human face and body—she has a penchant for tight hot pants—and a sharp intelligence that she uses to find out why the Martians are acting as they do.

Neon Genesis Evangelion by Anno Hideaki was initially a television series whose first episode was screened in October 1995. Later adapted for the big screen as a feature-length cartoon, it is hard-core apocalyptic *mecha*.

An explosion has wiped out half of the earth's population and it has been invaded by the Angels, extraterrestrials hell-bent on laying waste to all. Only children of 14, born at a special time, have the power to resist them. They fight back by using their brainwaves to control experimental tentacular robots, the Evangelions. As the series unfolds it emerges that aliens are not so very different from the Evangelions. They, too, are robots controlled by human beings: Once again technology is a powerful, deadly tool in the hands of the forces of evil.

The Ghost in the Shell *phenomenon*

The *anime* cartoon *Ghost in the Shell* by Mamoru Oshii was released in 1995 and, to widespread acclaim, soon asserted itself as a landmark work of Japanimation. James Cameron, who directed *Terminator* and *Titanic*, described it as "visionary".

The film is a serious exploration of the nature and direction of human evolution, addresses the issue of computer piracy in a networked world, and casts a young bionic woman as the heroine. She is the agent Kusanagi, half woman, half machine, whose mission is to track down the Puppet Master, a virtual being and perfect secret agent. Kusanagi finally catches up with her protean quarry. The Puppet Master, who was born from an artificial-intelligence project, attempts to merge with Kusanagi.

Although its vision of technological evolution is not so unremittingly bleak as most *mecha*s, *Ghost in the Shell* raises philosophical questions about dystopias where technology goes unchecked, where the Internet blurs borders, humans are augmented, and reality is virtuality.

In 2003 the creators of *The Matrix*, the Wachowski brothers, came up with an animated robot film initiative, *Animatrix*. It consisted of nine short-footage Japan-ese animated films, each of which took a facet of the *Matrix* world for its theme. Some treated periods prior to *The Matrix Reloaded*, while others delved into how and why the *Matrix* originated. One cartoon, *Matriculated*, by Peter Chung, features a group of rebels who reprogram a robot to serve the human cause. Some great masters of animation, like Square, the creator of *Final Fantasy*, took part in the project.

Kawai *or* mecha?

Whether steeped in the cute, soft-focus *kawai* genre or the muscular mayhem of *mecha*s, Japanese *anime* cartoons, together with *manga* comics before them, have established the robot as a recurrent character and theme in one of Japan's most vigorous art forms.

The omnipresence of robots in the *anime* and *manga* genres does indeed seem to have played a decisive part both in their widespread acceptance in Japan and in the undisputed lead the country has built up over the rest of the world in the design, research, development, and production of robots.

It is a remarkable fact that the two fields in which Japan stands out from the crowd on the world stage are *manga* comics and *anime* cartoons, on the one hand, and robots on the other.

Underground robot

An error in its coordinates causes a robot from the future to crash-land on Earth, where it has to face and do battle with numerous monsters. That, in a nutshell, is the story line of the short animated film, Underground Robot, which features a robot superbly designed by Webster Colcord, who hopes to make an animated feature.

© Courtesy of WebsterColcord.com

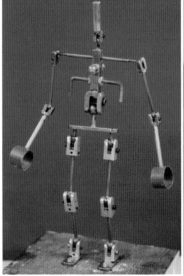

Interview
Tanino Liberatore

Tanino Liberatore has designed covers for magazines and albums such as Frank Zappa's Man of Utopia *as well as stage and movie sets. He is also the author of the three volumes of the adventures of Ranx, a broken-nosed robot that is unlike any before or since.*

© Courtesy of Tanino Liberatore

How did you get the idea for the Ranx character?
He was born in the mind of Stefano Tamburini in 1977. Stefano and I drew the first episodes in black and white for the magazine *Cannibale*. Ranx was an old character of his whose nose he'd cut off. Then, when Tamburini created the magazine *Frigidaire*, which was less underground, he asked me to flesh out Ranx with color. I put my heart and soul into it, I was so scared it was going to be terrible. It was the first time I'd done a comic book in color. My friends didn't much like what I'd done, but luckily everyone else thought it was fantastic. Initially it was only supposed to be an easy comic 16-pager. But meantime there was the big comics festival at Lucca in Italy. Some people from the French comics publishers *L'Écho des Savanes*, bought *Ranx*. So we went on drawing. The robot itself and

its environment reflected what was happening in Italy at the time with the student movements and police repression. The robot was supposed to help the delinquent students procure illegal substances. Things were very tough in Italy at the time. The original idea was to regulate Ranx's aggressiveness, but it got stuck at the maximum after the character who was supposed to control it got killed.

What would you say distinguishes Ranx from other robots usually found in comic books and works of fiction?
Apart from the fact that he is made out of photocopier spare parts, the fact that he is more human than humans. He is manipulated by a girl, who's the negative element in the story and he does everything she tells him to do. At the time, Tamburini and I had relationships with women

who showed that we weren't the strong men. The women were the manipulators, and Ranx reflects that.

Why did you choose to give him a humanoid look?
Tamburini and I designed him like that from the outset for fun. We didn't want to do what everybody else was doing all the time. We felt it gave him more impact. Another thing was that it made it easier for him to help the students; he didn't draw attention to himself. Just as an aside, I made Ranx an album cover with Frank Zappa in the image of Ranx. His nose is cut off and he's got the same glasses. Zappa got in touch with me when he saw the comic and said, " I want a Frank Xerox."

How did Alain Chabat[1] come to collaborate on the third volume of *Ranx*?
Chabat was a fan right from the start. He used to have a program on the French radio station Radio Monte Carlo, and he interviewed me twice about the first books. They were the best interviews I ever had. Afterwards we became friends and we worked together. Then Tamburini died and it was obvious that Chabat was the only person who really felt the spirit of the series. At the time, he was working in

the film *Gazon maudit*[2] and he didn't really have the time to focus on comics. But he helped out with script and dialogues. What can I say about the end result? It had been fifteen years since the second book, so it took Ranx and his readership time to get reacquainted.

What is the message you want to get across through the adventures of Ranx the robot?
There was never a preconceived message. We lived at a certain time and we mirrored what we saw every day in the street. At the time, Tamburini and I were very young, we set the action and our characters in the concrete jungle, in an atmosphere where our friends were alcoholic and into drugs. I think we were the only two who never took anything.

What do you think are the best robots you have seen in comics?
If you're asking about actual robots, I couldn't really say, but the best overall science fiction graphics in comic books that I've seen are without question Moebius[3]. He has revolutionized everything and others have drawn inspiration from what he did. Ninety-five percent of graphic artists are influenced by him.

What are the robots that have most impressed you in the movies?

The Terminator, because it's such an obvious copy of Ranx. A journalist told me that Cameron was familiar with my comic books. There are things that he lifted wholesale, like the mechanical parts in the arms. The difference is that he had a much bigger budget than I did. One thing's for sure, *Ranx* came out before *Terminator* and even before *Blade Runner*. I liked the first *Robocop*, too, for the directing, the plot and its overall feel.

Could there be a fourth *Ranx* book?
I don't think so. But, as they say— never say never. To make it interesting I'd have to portray him as old and alcoholic.

(1) Alain Chabat is a French comic actor and movie director.
(2) *Gazon maudit* is a ménage à trois movie with a twist. Chabat plays a philandering macho husband whose wife has an affair with a gay woman who gives her the attention Chabat's character is giving other women.
(3) This Moebius is a French graphic artist who creates science fiction comic books. Not to be confused with the themed science cartoonist Moebius (in reality American scientist Professor Ed Edelsack and British cartoonist Leon).

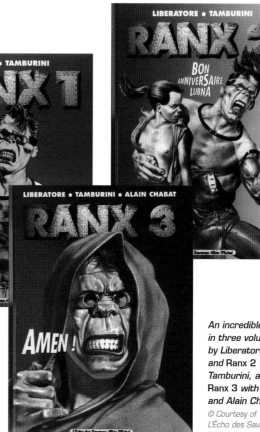

An incredible saga in three volumes by Liberatore: Ranx 1 *and* Ranx 2 *with Tamburini, and* Ranx 3 *with Tamburini and Alain Chabat.*
© Courtesy of L'Écho des Savanes/ Albin Michel

Animatronics

From the terrifying shark in Spielberg's *Jaws* to the killer ape in *King Kong*, who wreaks havoc among the skyscrapers, and the beast in the *Brotherhood of the Wolf*, motion pictures have created artificial animals inspired by automata. The technology behind them has a name: animatronics. It is the result of a clever combination of factors that are similar to robotics and include building the animal's body, controlling its articulated limbs and remote controlling it.

We would be less awestruck at the sight of the *Tyrannosaurus rex* in *Jurassic Park* if we could see the skillful handlers who control the giant dinosaurs from handheld units, moving them around at their beck and call. Just as awesome, though, is the sheer complexity of the machinery that actuates and drives the *T. rex*, for all the world as if it had come back to life.

When animatronics are good, they are very good and it's impossible to see the slightest trace of any trickery. In *Jurassic Park III*, so consummate is the mastery of the special effects that there is no way of knowing whether the giant reptiles we see on the screen are the result of computer-generated digital effects or

filming an animatronic. The word animatronics is coined from "animation" and "electronic". It denotes the technologies used to make artificial animals move and to create the illusion that some imaginary monster or extinct dinosaur is alive and well and bearing down on us. As naturalistically as an animatronic may move, however, it has to look like the real thing.

The first filmmaker to use a process akin to animatronics is thought to have been the Frenchman Georges Méliès in 1912. In his movie *La Conquête des pôles*, a giant Abominable Snowman swallows Arctic explorers on an expedition to the North Pole. Its movements are remote controlled by a system of cables connected to pulleys hidden under the cardboard skin of the snow monster.

His technique caught on and in 1924 the German director Fritz Lang used it in his movie *Die Nibelungen* to animate the dragon. Ten operators inside the monster's body pulled wires and pulleys to make it move in on its prey.

In the 1930s hydraulic mechanisms were used to move large creatures, with valves regulating the flow of fluids, which drove levers, which created motion. Walt Disney soon showed interest in mechanical processes for

the attractions in his amusement parks.

In June 1963, the Tiki birds became a new feature in Disneyland. They were driven by audio-animatronics, a technology whereby taped sounds caused a metal surface to vibrate, triggering an electric mechanism that created a particular movement or caused a bird's beak to open. Later a combination of compressed air and hydraulic systems were used to a more sophisticated effect.

In the 1960s another technique, that of push–pull cables simplified the manipulation of robots while ensuring their movements remained smooth. They functioned in a way that was similar to brakes and gear changers on bicycles. An operator pulled a lever to actuate a wire that slid through a sheath to relay the movement. The downside of push–pull was that wide-angle shots caught operators in the act of pulling and pushing.

The robot B9

1- The robot from the film Lost in Space, *animated by Matthew Denton.*
© Courtesy of Matt Denton

2- Puppeteer Todd Jones (right), watches the robot moving overhead in two monitors on the control unit. Todd Jones was equipped with an exoskeleton fitted with sensors on all its joints to control the robot's movements.
© Courtesy of Matt Denton

The robot's skeleton, complete with electronics and wiring.
© Courtesy of Phil Ross

T. rex by Nimba Creations

It took eight months to make this life-size Tyrannosaurus rex and a further four weeks to install it in a theme park. The sheer quality and skill of animatronics engineers Siobhan Hall and Tom Lauten impressed Richard Taylor of the New Zealand-based Weta Workshop, which designed the special effects for Peter Jackson's Lord of the Rings *trilogy. As a result Weta asked them to come back and work again for the remake of* King Kong *by Peter Jackson. Most animatronic animals made for motion pictures have a very short life span. Those that are made for theme parks and other public places, however, must be rugged enough to withstand changing weather conditions. That was the challenge that Nimba Creations successfully rose to with its awe-inspiring T. rex.*

© *Courtesy of Nimbacreations.com*

Interview
Siobhan Hall

Siobhan Hall and Tom Lauten are special effects artists who founded their company, Nimba Creations, in 1999. Their most ambitious animatronic project to date is the creation of a full-size Tyrannosaurus rex *powered by hydraulics and pneumatics and using the same digital technology to control dinosaurs as in* Jurassic Park.

Siobhan Hall works on a Bigfoot animatronics face for a television program to be broadcast in 2006. The mask uses servomotors and wire actuators for facial expressions.
© Courtesy of Nimbacreations.com

How do you perceive the relationship between robotics and animatronics?

This is an interesting question; a lot of people confuse the two. Robots that are constructed for scientific purposes are very different from those made for the entertainment industry. An animatronic on a film set only has to appear as if it can talk or move; it's an illusion that is helped by camera angles, lighting and editing.

But haven't you ever had to find solutions to the same kind of problems as robotics engineers, like problems with the way a biped walks?

We have never had to make an animatronic that has to walk around on its own—thank goodness! Often creatures you see in movies are realized on

screen in a number of different ways, including separate animatronics for different sections such as separate arms or legs, puppeteers performing in costumes, miniature photography and of course, digital compositions. We have to consider how to bring any of our creations to life on screen, but we have the luxury of trying to find the easiest way to do it rather than saying, "OK, let's make a full-size dinosaur walk around the set!" This is actually the basis for all special effects, creating the illusion that the audience is seeing something that doesn't really exist.

Do you have much contact with robot laboratories?

We don't tend to deal with 'robot laboratories', but we have been fortunate to enjoy good relationships with some of the top people in the movie-effects industry. You may think that each different company would be very guarded about discussing their methods and procedures, but every true professional that we have met has been good enough to chat with us about many of their projects, including animatronics. During our *T. rex* build, Henson's Creature Shop were good enough to recommend a material to us that proved to be invaluable, and it's that kind of openness that keeps the FX industry advancing as well as it does.

Is the expression of emotions the hardest part to realize?

More than the actual mechanisms it often comes down to performance, which is why professional puppeteers are so well respected. Making joints that move in many directions isn't an unknown science, so building it isn't what gives it the emotion. It's making sure that the performance isn't 'robotic' looking that gives it a realistic look.

What impact has digital imaging technology had on your field?

Well, there is no denying that digital technology is moving at an amazing speed. We have a digital suite that is capable of 2D and 3D animations and editing. We have also implemented a kit that allows us to scan our sculptures and models into computers in order to either animate them or to scale them up to infinitely larger sizes by carving out the forms with a digitally controlled polystyrene milling machine. The 3D scanner we have is also so portable that we can scan objects away from the workshop, on set or on location, and we can even scan actors' faces and bodies without harming them. Still I think we are at a stage where it is being used a little more than it needs to be. The *Lord of the Rings* trilogy is a perfect example of the way a filmmaker can use all the resources at hand to create stunning movies; where you really have to concentrate to work out what's digital and what isn't, mainly because every discipline of special effects was used perfectly from miniature building to animatronics. Hopefully more people will go back to producing good models, puppets and animatronics to make their films, rather than automatically going for digital effects.

Apart from the movie business, what are the potentials of animatronics?

It's been a lot of fun applying the level of quality we would use for films and TV shows to theme park animatronics—it really surprises people! There is still a preconception that animatronics that aren't in movies are generally like old shop window manikins with one or two basic movements on a badly made model. So to produce realistic, full sized, fully animated dinosaurs that sit in cages roaring at passersby has been a real thrill.

Tom Lauten's animatronics costumes and prosthetics

These costumes and prosthetics were designed by Nimba Creations project director Tom Lauten for short films and image inserts in video games for Games Workshop Plc. Some of his creations are exhibited at Games Workshop head office, in Nottingham, UK.

© Courtesy of Tom Lauten. Designs: © Games Workshop Plc.
www.games-workshop.com

Christian Hofmann's bestiary

Founded in 1878, the German company Christian Hofmann GmbH builds animatronic people and animals for a very wide range of theme and amusement parks.

© Courtesy of www.hofmann-figuren.de

Animatronic animals all over the screen

In 1975 Joe Alves and Roy Arbogast took animatronics through a quantum leap with their work on Steven Spielberg's *Jaws*. The three men met on the set of the TV series *Night Gallery* in the early 1970s. For Spielberg's movie, Alves designed a maquette for a giant shark that would be truly frightening. Bob Mathey, who was in charge of

This hyperrealist artificial face was designed by David Hanson. Like robotics, animatronics brings together art, science and technology to endow a robot's face with eerily realistic expressions. Sculptor and robot designer Hanson realizes his creations for theme parks like Disneyland as well as for Universal Studios and MTV. He also works with NASA/JPL and MIT.

© *Courtesy of David Hanson*

mechanical FX, now built three resin-and-fiberglass sharks. Each one was eight meters (26 feet) long, weighed three tons and required up to thirteen operators. So powerful were they that they generated forward thrust comparable to that of an airplane. The success of *Jaws* spawned a vogue for disaster movies and gave animatronics a new lease on life.

Technology now became easier to handle, with push–pull wires connected to motors that increased the range of possible movements. A handler radio controlling an animatronic could actuate eight articulations at any one time via a computer that stored an inventory of stock expressions and motions. Spielberg's endearing E.T. made his movement, gestures and faces with just such a system in the 1982 movie.

In the 1980s Stan Winston's wizardry made him the grand master of animatronics. He began his career at Disney as assistant makeup artist in the late 1960s, which led him into television. His work on articulated faces for a series called *Heartbeeps*, which featured robots, earned him a reputation as the robot man. It was in 1980 that Winston made the acquaintance of a budding movie director named James Cameron,

who had a script with the title *Terminator*. He told Winston about his plans for a picture that would feature a character who was half human, half machine. What he also wanted was that the robot's skin should wear and tear gradually throughout the film to show glimpses of the metal body beneath. Eventually the entire metal skeleton would be revealed. Ever the perfectionist, Cameron did not want the robot to be an actor clad in a metal shell. For the whole point of the Terminator was that it was a robot inside a human skin—not the other way round.

Winston thought the solution lay in animatronics and he got down to designing a full-sized puppet modeled on the body of the actor Arnold Schwarzenegger. On its 1984 release *Terminator* made not only Cameron's name, but Winston's too.

The next film that Cameron directed was *Aliens*, two years later. As demanding as ever, he drove Winston to create the largest animatronic puppet ever created for a film. It was the alien queen, a monster that brought him his first Oscar.

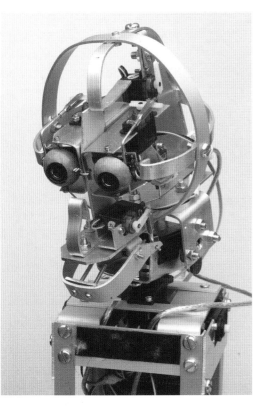

Flint, the anthropoid robot

Built by Bob Mottram in 2003 for research into interactive, sociable robotics, Flint is similar to a two-year-old chimpanzee in size and appearance. It can automatically remember the faces of people it has become familiar with and can understand and respond to expressions and gestures. Flint is part of a project whose aim is to develop an interactive being that is more robotic than animatronic.

© Courtesy of Bob Mottram

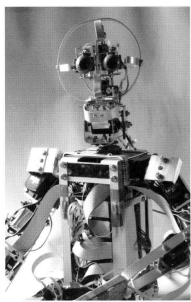

Digital imagery, a threat to animatronics

In the 1990s animatronics found itself contending with a serious competitor—computer-generated images (CGI). Steven Spielberg revealed the potential of digital animation, but not at the expense of animatronics: digital dinosaurs roamed through Jurassic Park side by side with full-size models realized by Stan Winston, one of which was a 14-meter long Tyrannosaurus rex, whose remote-controlled pupils could dilate.

Animatronics offers numerous advantages. First, it is often less costly than CGI. What's more, actors are much more comfortable with a robot on the set. "The motion picture industry still needs, as much as ever, figures and characters that face the camera and with which actors can interact. It helps to make a move credible," says George York who heads YFX Studio, a company specializing in animatronics.

Animatronic techniques came into their own in the dreamlike The Nightmare before Christmas, by Henry Selick from a Tim Burton script. For Babe Chris Noonan used the services of Jim Henson, who directed the memorable Dark Crystal, a first feature-length film with a cast entirely of puppets.

Animatronics is ideal for designing and engineering strange and fantasy animals that will be installed as set pieces in public places, as George York points out: "Theme parks need animatronic animals and so, too, do museums, more and more. Instead of stationary creatures, we can give them moving humanoids and dinosaurs, and breathe life back into extinct species that are on show at exhibitions. Visits to the museum are much more interesting when you can see animals moving as they did when they were once alive."

When animatronics meets robotics

The limit of animatronics lies in the fact that its puppets have to be manipulated by outside designers and controllers. This is where robotics comes in, particularly if it joins forces with artificial intelligence. Cynthia Brezeal of MIT's Media Lab is well known for having developed the robotic face Kismet. It senses presence and movement and, distinctively, responds to people in like manner, smiling when smiled at, for example.

When Brezeal saw the special effects Stan Winston had produced for the film A.I., and aware of the limits inherent in animatronics, she told the FX master that she would supply the funds if he would build a puppet. Artificial intelligence would make it an autonomous robot with a built-in computer that did not need to be remote controlled by a handler.

Winston and Brezeal got to work on their joint project in spring 2002 on the Leonardo prototype. The Stan Winston Studio had designed its body and the software that controlled its movements was supplied by MIT. It incorporated a number of advances in robotics technology, like voice and shape recognition and the ability to navigate through its environment.

If Leonardo were to land a part in a movie, it could blaze a new trail not only for animatronics but also for motion pictures: real, live actors could ply their trade on studio sets side by side with autonomous fantasy creatures. That kind of possibility would revolutionize movie making as radically as color film did.

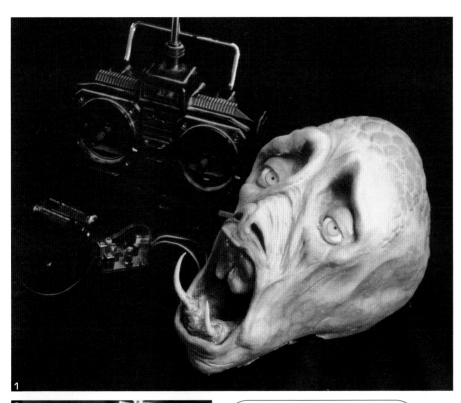

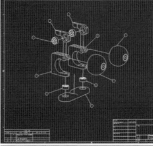

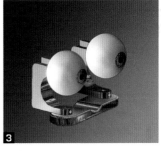

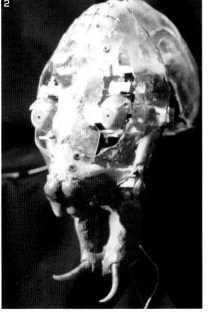

The La Grange studio

Created by Milan Jancic and Olivier de Géa, La Grange is a studio specializing in animatronic characters and 3D puppets for the movies, advertising, photography and theme parks.

1- 2- Animatronic alien.

3- "Méca yeux 3D" are 2D and 3D drawings made for designing the mechanical system for Menkhar's eyes.

4- Designed and developed by Olivier de Géa, the Emotion Control System is a system for preprogamming the position of sixteen servomotors to determine the range of expression of the animatronic prior to operating it.

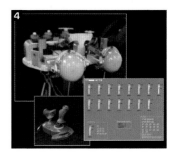

Interview
Matt Denton

Matt Denton from the UK is an animatronics specialist, who has worked on two of the Harry Potter movies and the remake of Lost in Space. *He has created his own company, Micromagic Systems.*

What are your favorite films featuring robots?

My all-time favorite is *Blade Runner*, which has androids in principal roles, but they are played by humans not animatronics! In terms of animatronic robots I was impressed with the Judge Dredd robot and also ED 209 from *Robocop* (although that was done using stop-frame animation).

What are the most impressive animatronics you have seen?

The best ones are when you don't notice it! Like in *Jurassic Park*—which has the perfect mix of animatronics and CGI. It's not the best or most technically advanced animatronics ever made but in my opinion was the best use of the two technologies amalgamated into one film. The animatronic gorillas in the film *Buddy*—built at Jim Henson's Creature Shop—are also very impressive. Fawkes, the phoenix from *Harry Potter 4 and the Goblet of Fire* is my latest favorite. That particular character had already been used in the first

two films but its two original builders (Josh Lee and Andy Roberts) completely refurbished and rebuilt their designs and we incorporated the Xpress system. The end result is beautiful. Many servomotors were replaced with electric linear actuators to give a smoother organic feel to the movements, and all were controlled by my own self-designed servomotor drivers.

How would you describe the difference between animatronics and robotics?

Animatronics is an illusion. It is an effect. The effect of mimicking life, or life forms, or even mimicking robot life. Generally speaking, animatronic models have no intelligence—they are performed by human puppeteers—whereas a robot would be built for a specific application or a practical use and usually incorporate some sort of intelligence. The hexapod walking robot I created for *Harry Potter and the Prisoner of Azkaban*—that one is actually a hybrid of both animatronics and robotics. Its walking is done by the intelligence of the software engine, but a human tells it where to go and what to do. It has some basic intelligence to control its gait and body attitude, but ultimately its behavior is determined by the puppeteer.

Do you think digital imagery could spell the end of animatronics?

No, not at this stage of the development of CGI. It's still in its infancy, relatively speaking. And some of the best creature effects are achieved through a combination of CGI and animatronics. CGI is certainly far more versatile but it's not always necessary. It takes a good effects supervisor to draw the line and decide what the best way is to create the illusion the film needs—be it with CGI or animatronics. It works both ways—for example you don't need a full animatronic if it can be done with a rod puppet. The more complicated the creation or character, the more that can go wrong. I'm sure CGI will become completely dominant in the long run. But there will always be a place for animatronics, especially on low-budget (although not necessarily low-quality) films. It's simply a matter of price.

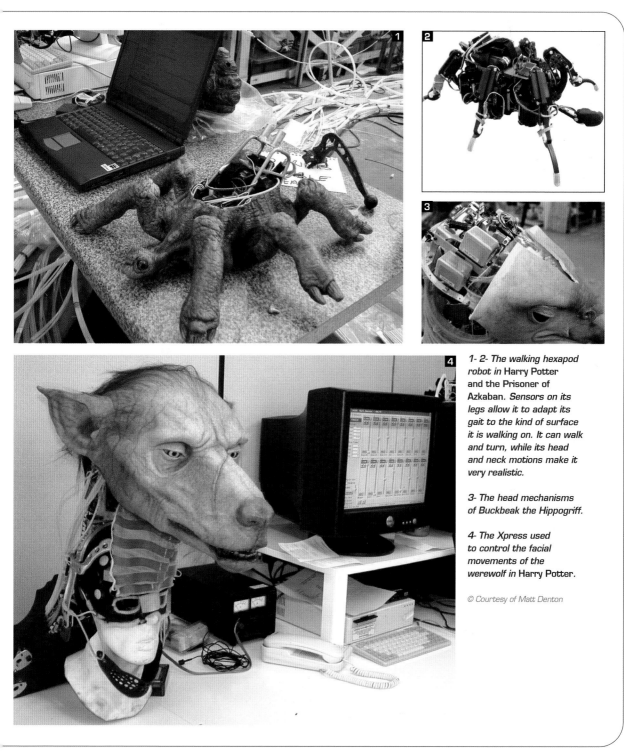

1- 2- The walking hexapod robot in Harry Potter and the Prisoner of Azkaban. *Sensors on its legs allow it to adapt its gait to the kind of surface it is walking on. It can walk and turn, while its head and neck motions make it very realistic.*

3- The head mechanisms of Buckbeak the Hippogriff.

4- The Xpress used to control the facial movements of the werewolf in Harry Potter.

© Courtesy of Matt Denton

... Interview cont.

How did you get involved in designing the robot for the new version of *Lost in Space*?
I had a phone call from Jim Henson's Creature Shop to tell me about the project and ask if I was interested in being involved. I had knowledge of a particular control system that they were thinking of using. However, once I started work on the project we ended up using my own control system. It was something I had been developing for three years and it could be completely tailored to their needs. The movie robot was fully hydraulic and electric. But the interface between the man and the machine was essentially my job. One really useful feature of the control system was Record and Playback. This meant the puppeteer's moves could be recorded, and then played back exactly the same, again and again. Which was great for performing difficult stunt shots, because the robots' movements could always be predicted.

Did you ever see the original *Lost in Space* series with B9?
Yes, I'd seen the original series before I joined the film, although I wasn't a particularly avid fan. But there's a strange parallel between both the TV series and the movie. The original B9 was of course

Matt Denton   © Courtesy of Matt Denton

controlled by a man inside a robot suit. Thirty years later, the B9 was controlled by a man off scene inside an exoskeleton capturing his movements. With a telemetry link between the exoskeleton suit and the robot, my system translated all of his movements into the robot's movements. His elbow would become the robot's elbow, and so on.

What was your contribution on *Harry Potter and the Prisoner of Azkaban*?
First I designed, built, and performed a walking hexapod creature. It was a background character that lived in Hagrid's hut and was based on original hexapod designs that I had been developing for a couple of years. The *Prisoner of Azkaban* hexapod was almost twice the size of my original designs, but I had always made the hexapod's control software scalable, so it was easy to develop for any size of creature the director might require.

Incidentally, a much larger, coffee-table-size version of the hexapod was originally planned for *Harry Potter 4*. It was based on the Blast-Ended Skrewt creature and was cut out quite early on, but some leg prototypes and screen tests were very successful. It's something that I'd like to work on with the other engineers to finish off at some point.
The *Prisoner of Azkaban* hexapod was the first animatronic creature of its type in any feature film and attracted the attention of an animator from ILM who was working on a spider for the same film. He asked to film and motion-capture the hexapod because he was so impressed with its lifelike movements. He wanted to study its gait and movement patterns… Most of the hexapod footage didn't make it into the final version of *The Prisoner of Azkaban* but hopefully it will appear in the next Harry Potter film, *The Goblet of Fire*.
Another aspect of my work on *The Prisoner of Azkaban* was that the creature effects unit, run by Nick Dudman, hired my performance system. It's called Xpress and had evolved from the *Lost in Space* project. Xpress was used to control the more complex characters, such as Buckbeak. The animatronic was used in the rehearsals for a scene where Harry meets Buckbeak for the first time.

How do you control the animatronic robot with this system?

A puppeteer can use Xpress to control it using special controls. The standing version of the hippogriff was fully electric. All of its movements were electrically actuated—both of its wings, two legs, its tail, head and neck. These moves were linked to Xpress, which allowed one or two people to control the entire character.

One of your creations, Jackson, is a mixture of animatronics and CGI. Is that a possible development for animatronics?

It's not the future, but one possible route ahead. Jackson is the result of a collaboration with puppeteers Todd Jones and Rob Tygner. We've also worked with Centroid, who supplied the motion capture elements. One possible application would be rehearsal. Puppeteers don't get much rehearsal time (normally for fear of breaking the model!). But if you have a CGI version of the model, you can rehearse as much as needed without any wear and tear. In fact you can rehearse on a CGI version of an animatronic before one is even built. It is also a great puppeteer training tool.

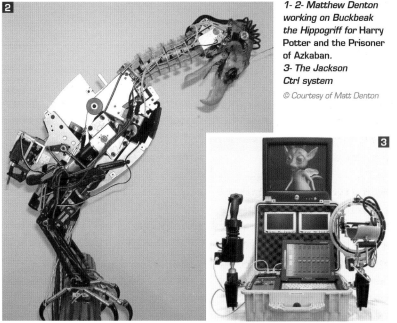

1- 2- *Matthew Denton working on Buckbeak the Hippogriff for* **Harry Potter and the Prisoner of Azkaban.**
3- *The Jackson Ctrl system*

© *Courtesy of Matt Denton*

THE ANDROIDS

ANDROIDS
ANDROIDS

There is no getting away from it—the imagination instinctively pictures robots as human remakes. We will never be content with the misshapen things that spray paint on cars or substitute pets. What we want are the robots which, for generations, have fired the dreams of sci-fi aficionados, readers of Isaac Asimov and other writers. What we want are androids, with hands and feet and staring eyes that follow us around, that chat with us, that run our bath and, if they notice we are worn out, cook dinner. Some picture them with skins like ours, others think of them as encased in metal like C3PO, as if to preserve the essential otherness of their being. But what matters is that they must be two-legged, like us. For have we not fantasized over the ages about creating an artificial being in our likeness?

WE-4R is part of a project at the University of Waseda in Tokyo to charge a robot's face with human emotions. Merely looking at it is enough to arouse a funny feeling.
© Waseda University - Production Lille 2004

What is an android?

A n android is an "automaton with a human face" according to the encyclopedia of the 18th century *encyclopédistes*, Diderot and D'Alembert. In the not-too-distant future it will be the chic thing to have one or more at home. People will not be able do without their androids, any more than they can now do without their cars and television. No one will think twice about buying an automated home helper that does the cleaning and the washing up, answers the telephone and tells the children bedtime stories. Oh yes, and mows the lawn, parks the car and keeps the refrigerator well stocked. Some will also mount guard, watching the house to keep intruders at bay, and calling the police or the fire brigade in the event of an emergency. Guard dogs can only guard, and they are only dogs. Androids will undertake wide and varied, socially useful careers as companions for the elderly or museum guides.

Yet they were long shunned. Science-fiction authors, led by Asimov himself, had been overoptimistic in their predictions of the coming of our metal and silicon alter egos. Too costly to develop to offer short-term return on investment, they were scorned by industry. The chief stumbling block was the complex technology required to create an artificial biped endowed with basic common sense!

Now, however, androids are a fact of life. For the time being, only big companies can afford them—Honda sold its Asimo robot, for example, for $390,000 in 2004. But, by the end of the decade, it is thought that there will be a mass-consumer market for androids.

The market of the century

Many robotics experts believe that the market for androids is set to become one of the biggest—if not *the* biggest—of the 21st century. "Just as automobiles were the biggest product of the 20th century, people might eventually look back and say that robots were the big product of the 21st century," says Hirohisa Hirukawa of the national institute of Advanced Industrial Science and Technology (AIST) , the software arm of the Humanoid Robots Project (HRP), an android development initiative funded by the Japanese government. Hirukawa believes that it makes amply more sense to develop androids rather than task-specific robots. "When we see robots like Asimo from Honda and Qrio from Sony we're drawn to them. They trigger an emotional reaction," he enthuses.

Fumio Kanehiro, also a researcher with AIST, sounds a similar note, opining that there was no significant boom in the robot market for some twenty years precisely because it was not making androids. Until 1997 the number of robots being produced in Japan rose steadily. Then the market shrank just as steadily. The Asian economic turmoil was, of course, a contributory factor, but the robotics sector should have recovered from its slump since demand world-wide has increased strongly—in 2003 alone it grew by 25 percent. So, where lies the rub?

"The variety of tasks that can be handled by a robot has been limited," explains Kanehiro. "For human-interactive operations humanoid robots are the best suited. A robot that has a shape very similar to that of human beings is capable of operating in an environment adjusted for human use, utilizing existing tools without modification and working in collaboration with human workers."

All the tools that we use on a routine basis, from the screwdriver to the automobile, were designed for human use. A robot whose design template is anthropomorphic and which, therefore, looks and acts like a human being should logically be able to use the appliances and appara-

tuses that crowd our environment. It would be ideally cut out for doing our dirty work.

The anthropomorphic approach offers the additional advantage of user-friendliness. A problem that has always exercised the minds and experiments of researchers is user interfaces. In other words, how to develop devices that people instantly and intuitively master. One success story was the telephone.

Another one was definitely *not* the computer in its early forms— it was for experts only. Nothing, though, comes more naturally to human beings than to interact with fellow humans…well, humanoids. Even if they do not speak the same language, there is always the instinctive response of gesture and facial expressions.

"The android industry is where the automobile industry was 100 years ago," states Chris

Willis, president of Texas-based Android World. "Androids are going to be really big business in the 21st century. When the first are available to the public, people won't ask about the price, they'll just write a check."

Surely, though, these artificial human clones who do not need to be fed, housed, clothed or entertained are a source of cheap labor that will pose a threat to employment. Such fears can be allayed by pointing to the computer industry. Also once accused of stealing jobs, it has in fact created more than it has done away with. Similarly, the humanoid robotics industry is likely to drive a huge boom in new companies in fields like design and development, software and maintenance.

There is a fundamental difference, however, that sets androids apart from other machines: they can will be able to identify many of their own faults and troubleshoot themselves. And, because they be fit into human working life and society with such ease, they will create a host of new needs and foster demand.

Every time that Sony's diminutive android Qrio makes a public appearance its popularity and curiosity ratings go up.
© Sony

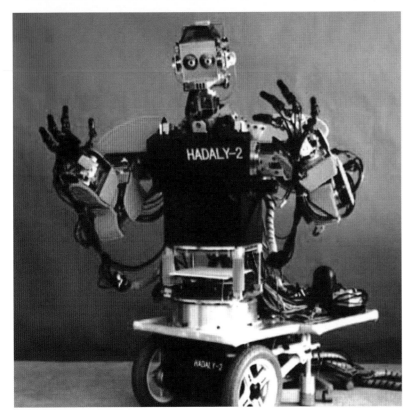

Hadaly-2, built in 1997, contains technology from a wide range of research fields like voice recognition, electronic vision and tactile sensors. It can also move its head and body like a human.
© Courtesy of Waseda University.

Heads first, legs later

The android is poised on the verge of an exciting future. Yet it was only towards the end of the 20th century that it arose and walked, thus overcoming the main obstacle to its development.

Cog, or going head-to-head with robots

In the middle of the 1990s a single robot hogged all the limelight. Its name was Cog, a new kind of robot, much smarter than anything that had come before. It was the brainchild of Rodney Brooks, robotics professor at MIT in Boston, whose claim to fame is his research into evolutionary robotics. In other words, his robots are not pre-programmed—they learn by themselves.

Cog, which began life in 1993, was just such a self-taught robot. Its behavior and understanding evolve as it gathers information from its environment, especially from interaction with humans. In an illuminating aside on his work, Brooks once said that he was looking forward to the day when a robot would be emotionally opposed to the idea of being switched off. Indeed, Cog sets spines tingling with its gaze that swivels to watch you wherever you go. It is endowed with three senses— sight, touch and hearing—and its eyes can sense and track motion and sounds. Cog's repertoire of gestures reproduces much basic body language, like nodding. Yet it is a man only down to its waist. The robot genius has no legs, only a trunk, jointed arms and a head. Locomotion is provided by a wheeled platform. Ultimately its claim to human likeness does not have a leg to stand on.

Another android with human potential is Hadaly-2, which was designed and developed at Tokyo's Waseda University. It can hold a conversation and recognize the meaning of many signs and gestures. Like Cog, it looks straight at its interlocutors and watches their every movement— an ability that has earned it a job as receptionist at a Japanese university. But Hadaly-2 is another legless wonder, relying on wheels to move around. Because they have nothing recognizably manlike below the belt, these great pretenders are just that.

The Cog Project got underway in 1993 at the Massachusetts Institute of Technology (MIT). The objective was to develop a humanoid robot with the ability to perceive and understand the world for itself. It would be a machine that could learn from its experience of its environment and so evolve and program itself accordingly. The Cog robot was the brainchild of Rodney Brooks, who believes that intelligence in a machine is born from interaction with real life.

© Donna Coveney (MIT) - Production Lille2004

The stumbling block of bipedal motion

What long held back the coming of robotic bipeds was the human gait. It seemed as though robots would never walk like people, because of the sheer complexity of the task. Humans in motion constantly readjust their balance, depending on which part of the body takes the weight and on a multitude of other factors. Because the body's center of gravity lies at waist-height and the feet—which support its weight at point-of-contact with the ground— are so small in surface area, a robot standing on two feet is very precariously balanced. The slightest off-center movement sends it clattering pathetically to the ground. Other crucial factors are different ground and floor surfaces and the omnipresence of obstacles likely to trip and topple robots.

With robots unable to stand on their own two feet, second-best solutions had to be found, like wheeled platforms and three or more legs. In 1993 Carnegie Mellon University created a robot called Ambler that performed the feat of proceeding unaided over 500 meters (1,640 feet). Its pace was a slow but sure one meter per second. So far, so good, but for a slight drawback: Ambler had six legs. The vision of the android stumbled and fell on the intractable problem of bipedal locomotion. There would never be such a thing as a man-made man—robots would have to make do with imitating the beasts of the field. Robotics had lurched to a halt there where science-fiction had spawned vistas of creatures in human likeness.

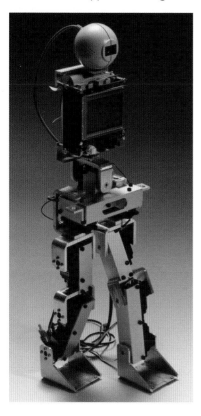

The bipedal robot Johnny was designed and developed by Thomas Bräunl, Richard Meager and Elliot Nichol of the Department of Electric and Electronic Engineering at the University of Western Australia.
© Courtesy of University of Western Australia

The coming of the bipeds

The quest to develop bipedal motion looked doomed—unless, that is, some benefactor were prepared to risk colossal sums in investment with no guarantee of returns for very many years. In the 1980s Japanese industry provided that benefactor.

The Honda project to create a walking biped

In 1986 the giant Japanese automaker the Honda Motor Co. launched a bid to master the robotics technology that would enable a robot to walk on its own two feet. The move came as one of the company's senior executives, Yshiro Dewa, saw that the motor-vehicle market was no longer growing. He felt the time had come for the Honda Motor Co. to diversify into activities other than its core business of automobile manufacture. A central thrust of the new strategy was the robot.

Honda invested heavily, working to a 20- to 30-year timeline. Word went around the engineering departments that the cars of the future would increasingly run on electronic rather than mechanical systems and that

it would be a good thing if research focused in that direction. Honda's core business could only gain from research into robotics. But it would take Honda's 30-strong research team twelve years of research and experimentation before the first two-legged robot eventually stepped out.

In 1993 the first series of P prototypes was built with the aim of producing bipedal walking robots. P1 was a metal monster measuring 1.9 meters (6 feet 2 inches) in height , weighing 175 kilograms (385.80 pounds) and sporting a great rectangular screen instead of a head. Yet it had evolved from seven earlier generations of bipedal prototypes. P1 was succeeded by P2 and P3, with each model becoming distinctly more lifelike. When the battery-powered P2 was officially unveiled on September 20, 1996, it was the product of ten years of development (200 man years) and had cost $105.3 million. At 1.8 meters (six feet), it was shorter than its predecessor and much heavier, weighing in at 208 kilos (460 pounds). But it had, almost literally, made strides in autonomy, able to walk like a human and even climb up and down stairs. Nevertheless, its aesthetic appeal left a lot to be desired. Smaller and rounder than P1's, its head was still more reminiscent of a TV screen than a

face. Despite its undeniable qualities, P2 still came across as a contraption.

From P3 to Asimo, the turning-point

The completion of the new P3 in September 1997 marked a turning point in the history of android robots. It boasted the same technological capabilities as its earlier siblings, but did so with style. Made of pristine white steel and plastic, its resemblance to an astronaut offset its slightly bumbling gait. Not only was it shorter still than P2 at 1.6 meters (5 feet 4 inches), it was much lighter, weighing a pleasantly plump 130 kilograms (286 pounds). Most important, P3 was a flawless walker, proceeding at the same speed as a human. Like P2, it could get up and down, the stairs and, unlike P2, it could kneel down then stand up straight and, if jostled, regain its balance.

Although P3 was a technological breakthrough, its overriding importance was, perhaps, as a corporate symbol. A robot had to set the bipedal walking standard—Honda had produced that robot and proved that androids were feasible. Yet the company did not stop there. In 2001 it introduced to a waiting world a P3 upgrade by the name of Asimo.

Despite its science-fictional ring, the robot was not named after Isaac Asimov. "Asimo" is both the acronym for Advanced Step in Innovative Mobility and a neologism coined from "*asi*", the Japanese for "feet", and the first two letters of "move". And move Asimo certainly could—at speeds of 1.6 kilometers per hour (one mph). This fetching new robot was a considerably scaled-down improvement on its immediate forerunner, being only 1.2 meters (four feet) tall—a height that Honda spokesman Yuji Hatano describes as being socially acceptable for a robot, because its eyes are on a level with a human's in a sitting position. Any taller and it would be threatening, particularly as Honda was seeking to appeal to the elderly market. And although Asimo was positively slimline compared to its brothers before, its 43 kilograms (95 pounds) were still thought too heavy. Should the robot fall over it could be a danger to children. Honda was keenly aware of this and worked on shedding further kilos.

Meanwhile, every time that Asimo made a public appearance in Asia it was greeted with adulation, like a prophet announcing the coming of the humanoids that all awaited. Yet the alluring robotic wonder kid in a spacesuit was far from perfect.

Honda invested millions of dollars in finding solutions to the countless technological challenges inherent in creating a humanoid worthy of the term. The unveiling of P3 in September 1997 was a milestone in the history of robotics.

© Honda.

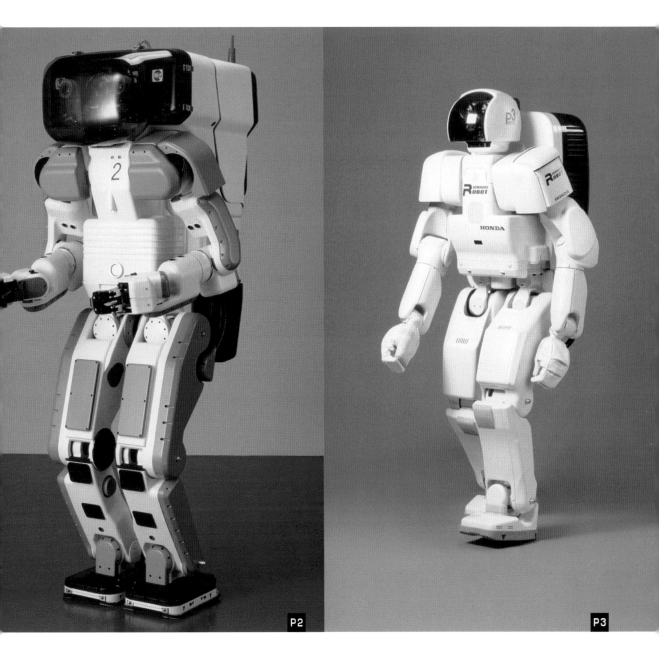

P2

P3

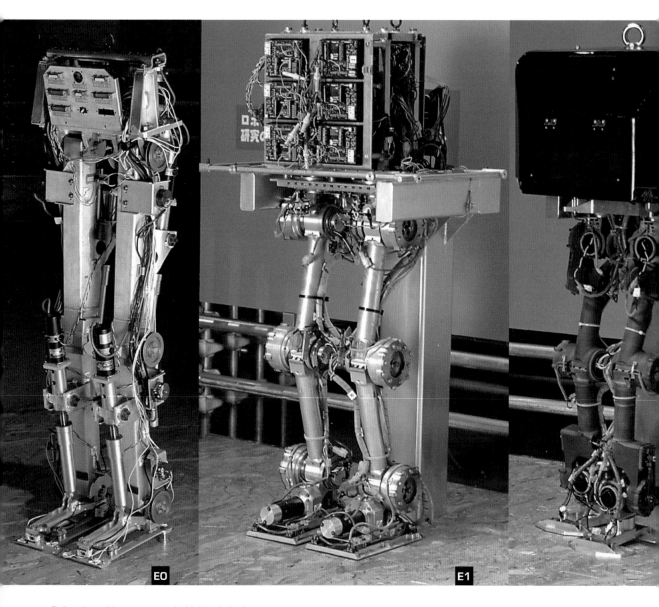

EO

E1

Before it could create an android, Honda had first to tackle and master the deeply complex technology of walking. No android that could talk but not walk like a human would ever be able to take its place among people in society. The task took Honda seven years.

© Honda.

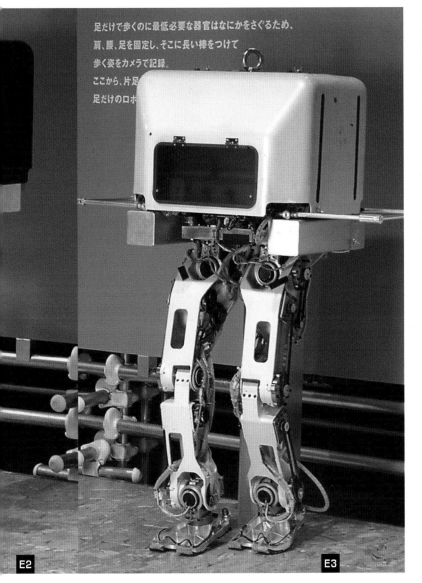

足だけで歩くのに最低必要な器官はなにかをさぐるため、
肩、腰、足を固定し、そこに長い棒をつけて
歩く姿をカメラで記録。
ここから、片足
足だけのロボ

E2   E3

Like the three Ps before it, he carried a bulging backpack that housed its batteries and the computer that was its brain. It needed two handlers to operate it and could respond to only a limited number of commands.

But Asimo was evolving fast and in autumn 2003 it reappeared as an altogether more sophisticated android. Its engineers had focused on its interactive capabilities and it was plain to see that it was a great improvement in the quest for a humanoid capable of basic social relationships. It could understand its interlocutors' gestures and facial expressions and obey their spoken orders. If approached, it would utter a greeting and shake hands if one was proffered. On hearing the order "follow me," that is just what it would do, and if told to look somewhere and go there, it would look and go. It could put names to faces and be entrusted with messages that it would pass on to the right person. Equally astonishingly, he could go on-line for news and information, like the weather forecast for tomorrow, and convey it to whomever had asked for it.

Meanwhile, out in the "real" world, market players and observers were wondering whether the hundreds of millions of dollars that had gone into the electronic biped was money well spent. Honda was dismissive of such concerns, so convinced was it that its research into Asimo would pay dividends in the long term. More and more Japanese are living to a ripe old age and Japan has become the first country in the world to boast over 20,000 100-somethings. Catering to the needs of the elderly will soon be a flourishing market, with robotic helpers a key selling point.

For the time being, however, praise for Asimo has been forthcoming only from a handful of big companies, like IBM, keen to groom their hi-tech images and hire Asimo as receptionist at a daily rate of $18,200, or $175,500 for a yearly lease. Requests for its services have grown and 30 Asimos were in operation in 2004. One lords it in the office of Honda CEO Takeo Fukui, while another stands on a podium in the showroom at Honda's corporate headquarters.

Above all, Asimo and the three Ps stood for Honda's ability to use its technological prowess to create business growth. The company has positioned itself powerfully on a burgeoning market where android projects have multiplied worldwide since the turn of the millennium.

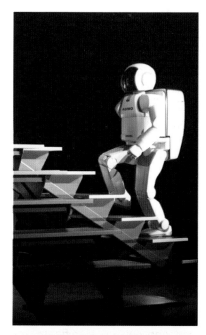

Thanks to his high-power, high-precision bipedal walking technology, I-Walk, Asimo can walk up stairs and slopes, pivot when he gets to the top, then walk back down.
© Courtesy of Honda

Sony's dancing android

When Honda unveiled P3 it doubtless wrong-footed its competitor, Sony. Although the electronics giant had been busy in robotics since the early 1990s (*see Aibo's story in Chapter 4*), it had followed a different direction towards the ultimate goal of the android. It had concentrated on robots that would be companions in the home, playing a primarily social role. The approach was undoubtedly a canny one, because by 1999 Sony was marketing its robotic pet dog Aibo and chalking up significant volume sales. Oddly, Sony's roboticist *extraordinaire*, Toshitada Doi, had toyed with the idea of tackling the walking biped from the outset. But ever the engineer with an artistic flair, he had set himself an objective that went a two-step further: he wanted an android that could dance. With Sony's blessing, his plan was to focus research on robots that engaged the emotions.

The approach bore early fruit in the shape of the fun and friendly SDR-4X, which was unveiled at the 2002 Robodex fair. It could shuffle a few dance steps and spot and skirt obstacles in its way. SDR-4X was, however, but a foretaste of Toshitada Doi's vision of the ideal home companion.

The aim in life of Sony's android Qrio is fun.
© Courtesy of Sony.

Qrio was much closer to the real thing. Sony announced it to the world on September 19, 2003, and though tiny, it was more humanoid than anything the company had yet developed. Like some miniature C3PO, it was only 60 centimeters tall (two feet) for seven kilograms in weight (15 pounds), but its genial appeal was instant.

Like Asimo, Qrio could recognize faces and voices, avoid obstacles and climb stairs. Though short, it walked tall, and could get up if it had been pushed or had fallen to the ground. But Sony wanted it to be as entertaining as other flagship products like the transistor radio, Playstation 2, and the Walkman. Accordingly it could dance, moving its legs, arms and fingers to the beat and—even more astonishingly—throw a small football. It was, for all the world, like a character that had escaped from some cartoon.

Qrio might have been a fun guy as robots go, but the technology behind it was serious: it could run at speeds of up to 2.4 kilometers per hour (one and a half miles). Sony had also used common sense to resolve some of the more apparently intractable problems. For example, if Qrio had to negotiate bumpy terrain, then the numerous sensors on its heels would take over and, whatever it was doing, it would do nice and slow.

Qrio has social skills. It interacts with its environment, holds conversations, and entertains.
© Courtesy of Sony

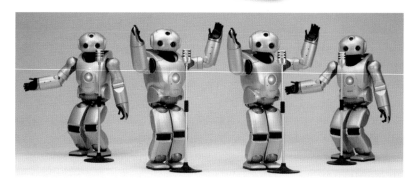

Bipedal robots walk over Japan

P3 was a turning point in 1997. In its wake numerous Japanese companies unveiled projects in progress, while others set up robotics-dedicated departments. They addressed and swiftly resolved technological issues like lower weight, more degrees of freedom (motion-enabling joints) and how to get rid of the back-pack. Such progress was achieved with substantial funding from the Japanese government, keen to actively invest in a field where it could regain the edge it had surrendered to the USA in computers. Japan now set about carving out a clear lead in android technology.

In 1998 the Japanese Ministry of the Economy, Industry, and Trade launched its now celebrated Human Robotics Project (HRP), which lent firm backing to a range of initiatives. Over five years it released funds to the tune of $42.25 million.

Professor Hirohisa Hirukawa of the AIST believes that Sony's breakthrough, which had shown androids to be feasible, had triggered the state-sponsored robotics commitment. He explains that if any branch of the Japanese government likes an idea, it funds research, whereas other countries are more reluctant to part with money when researchers have not yet come up with practical applications.

HRP-2 stumbles and falls and rises again

In the summer of 2003 the AIST was involved in the production of a prototype android that could lie down and get up by itself. It was the HRP-2. The AIST supplied the limb-coordination software, drawing on a human motion analysis system to meet the needs of bipedal robots. Kawada Industries designed the body in line with specifications that spelled out that it should be a mechanism similar to a human body with flexible joints at the waist and with strong arms. HRP-2 was blue and silver and,

The AIST's HRP-2 is hired out to numerous universities, where it serves as a platform for robotics engineering and research.
© Courtesy of AIST

like Asimo, resembled a manikin in a spacesuit. More seamless integration of its system components made it more streamlined and its 58 kilos (128 pounds) for 1.54 meters (five feet) was a weight-height ratio similar to that of a human. The technological feat that gave it an edge over its predecessors was control of its center of gravity, which could now shift widely without his losing balance. The AIST's Fumio Kanehiro explains: "One of the main problems with humanoid robots is that they easily fall over. When an android stands on its two feet, only a very small area in contact with the ground supports it, while its center of gravity is at waist level, which is relatively high." Since it was impossible to prevent a robot from falling, adds Kanehiro, an alternative was explored—one that could get back on its feet unaided. Until then the only robots with that capability were small ones, no more than 60 centimeters tall (two feet). HRP-2 was twice that height. The ability to lie down was in itself a significant achievement, which could have practical applications like slipping under an obstacle or crawling under a car to check its parts. What's more, horizontal postures are more economical than vertical ones. Unless it has to stand and walk, HRP-2 can lie low to save energy.

While the AIST is involved in other projects like HRP-1S (*see below*), university research units that cannot afford to develop their own robots lease HRP-2 as a platform to further their understanding. Even better, HRP-2, which costs $455,000 to buy, would be a valuable tool for European android research, as we shall see.

Tokyo-based Waseda University was a pioneer in the development of androids. As early as 1972, Wabot-1 could perceive its environment and grab things.
© Courtesy of Waseda University

Projects stream from Japanese universities

Japanese universities have put the development of androids high on their agenda and stepped up research. The University of Osaka is working on a biped dubbed Strut, while at Tokohu four professors and a score of students have undertaken a project named Monroe, after the eponymous film actress. A robot called BIRS, the acronym from Biologically Inspired Robot System, at Nagoya University draws on biomimetics, a branch of science that uses nature as the template for its machines. The Juhu System Laboratory at Tokyo University is running two projects at once, H6 and H7, while the Japan Science and Technology University in Osaka has been busy on a humanoid called DB since October 1996.

In the forefront of R&D is, predictably enough, Tokyo's Waseda University, which originated the famous Wabot 1 as early as 1973 and the wheeled android Hadaly in the mid-1990s. In 2000 Waseda unveiled Wabian, which stands 1.66 meters high (5 feet 6 inches), weighs 107 kilos (235 pounds) and can get around at speeds of up to seven kilometers per hour (four mph). It is awesomely adroit, able to grab and carry things and to dance the waltz. Additional work in progress at Waseda includes the iSHA humanoid that can operate independently for up to two hours and the artificially intelligent, wheeled Wamoeba, in which researchers are hoping to instill a sense of values.

Pressure groups are increasingly lobbying the Japanese government for further resources for research into android technology. In 2003 the scientist Mitsuo Kawato launched an initiative called Atom, named after the popular comic strip then cartoon, *Tetsuwan Atom*, better known in the West as Astro Boy. Atom's aim was to secure a pledge from the Japanese government to invest 50 billion yen ($470 million) per year over three decades in work to design, develop, and produce a robot with the physical and mental faculties of a five-year-old human. HRP's Hirohisa Hirukawa nevertheless argues that if researchers want that kind of commitment from the Japanese state they must at least outline some practical application within five years. By way of example, he lists lifesaving, tunnel building, and work in toxic-waste dumps.

Japanese industry banking on androids

Japanese industry, too, is busy. Toyota has been working for many years on robots that are expected to come to market in 2005. They, like Sony's, will be sociable, personal robots whose prime task will be as helpers in the home. Toshiba's Partners are a family of androids that come in four models. One is designed as a companion to help the elderly. It is the walker of the family, 1.2 meters (four feet) high and weighing 35 kilos (77 pounds), while the wheeled Partner works on assembly lines. On September 10, 2001, Fujitsu unveiled an android known as HOAP-1, which looks much like a toy (*see Chapter 9*) and was followed by its little brother, HOAP-2.

Kawada Industries is working on a project to develop a robot called Isamu, which is intended to work with humans, lifting and carrying heavy objects. Hitachi has been concentrating on assistance to the elderly, Mitsubishi's robots are designed to maintain production plants, and Sogo Security keeps watch. Interestingly, the last four companies are all part of the Humanoid Robotics Project, evidence of the close ties in Japan between academia and industry.

The Kitano Symbiotic Systems team headed by Hiroaki Kitano exemplifies this synergy.

It is funded by a government research agency, the Association for Science and Technology, while the name of Sony Computer Science Laboratories is also writ large on Kitano's business card. He believes that the high cost of developing state-of-the-art technologies, like robotics, is an issue that needs to be addressed if a robot in human likeness is to be produced at a cost that is not prohibitive. He advocates the use of off-the-shelf components that the bulk of research units can afford and himself practices what he preaches. Another facet of Kitano's creed is the importance of creating robots whose appearance endears them to the public. Among the robotics projects that he lead-manages is Pino, a tiny, 75-centimeter-high (2 feet 6 inches) humanoid that weighs eight kilograms (18 pounds) and recalls Pinocchio. Not to mention Sig2, Morph, and Morph3, which use Bluetooth wireless connectivity. *(Chapters 4 and 9 explore further examples of Japanese humanoid projects.)*

Fujitsu's *HOAP-2* can simulate a sumo wrestler's moves.
© Fujitsu.

Androids from elsewhere in the Asia-Pacific Region

Although Japan is the lord of the robotics dance, its Asian neighbors are piping the tune. China's Beijing Technology Institute runs a project that has produced BHR-1, an android that is 1.58 meters tall (5 feet 3 inches), weighs 76 kilograms (168 pounds), walks at one kilometer per hour (0.62 mph) taking 33-centimeter (13-inch) strides. The Science and Technology University for National Defense in Changsha, Hunan Province, has built Pioneer, a robot that can walk and talk in several languages.

Football-playing humanoids are the focus of a Singapore-based project, whose latest product is Homo Erectus II. Not to be outdone, Korea is home to several ongoing projects, which include

KHR-1, driven by the Advanced Industrial Science and Technology (KAIST) in the city of Teajon. The Korean Maritime University of Pusan is developing a bipedal android, while the Korean Institute of Science and Technology in Seoul has been focusing on a four-legged humanoid, Centaur. Developing their first androids are King Mongkut's University of Technology in Thailand and Iran in Central Asia, which has produced Firatelloid (from First Iranian Intelligent Humanoid).

In the South Pacific a 50-centimeter (20-inch) high, meccano-bodied biped has come into being at the University of Western Australia. Dubbed Johnny Walker, the tiny android can not only walk but lean from side to side. Still in Australia, Professor Gordon Wyeth is leading a project called GuRoo at the Brisbane-based University of Queensland. Like its dinky compatriot, GuRoo has that meccano look, while standing 1.2 meters tall (four feet) for 30 kilos (66 pounds), vital statistics that have qualified it to play in Robo-Cup tournaments.

Pino is a creation of Hiroaki Kitano who heads two advanced robotics laboratories, one of which is Sony's. His aim is to develop an affordable robot with basic capabilities that is within the means of universities and research units. Pino is designed to help speed up research into robotics and artificial intelligence.
© Courtesy of Sony

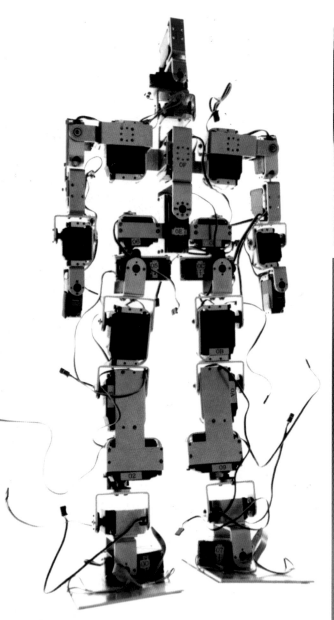

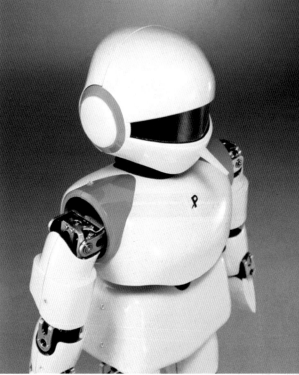

Pino and its constituent parts: skeleton and casing. Kitano believes in the importance of creating robots whose appearance endears them to the public.
© *Courtesy of Sony.*

Page right: Kitano has also overseen the development of Morph3, which can execute some karate movements.
© *Courtesy of Kitano Symbiotic Systems.*

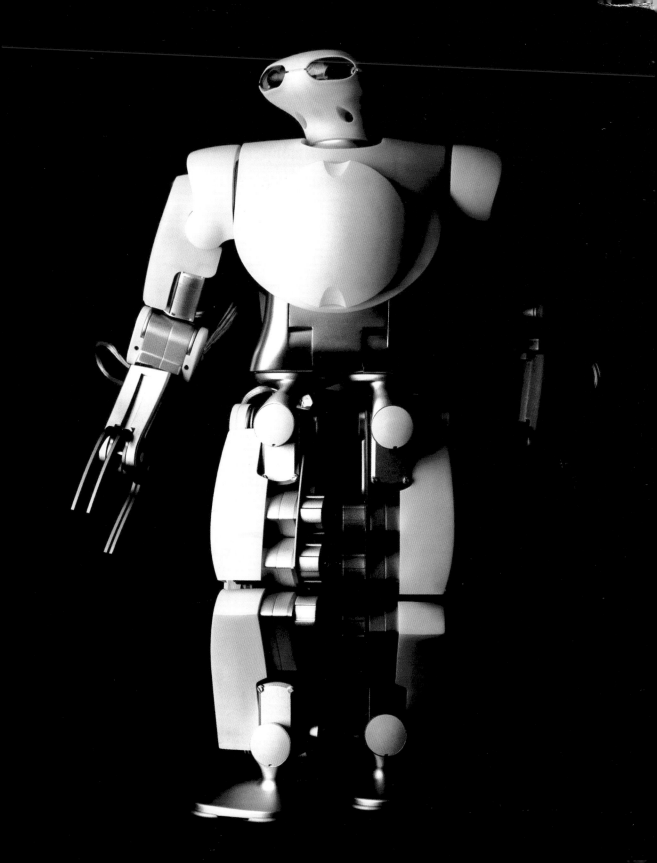

America and Europe in the walking race

Japan has undisputedly built up a lead in the development of robot technologies. Hit hard by the stock market crisis that blighted hi-tech start-up companies in early 2000, America and Europe still lag behind. Flourishing firms like US-based Sarcoman, which had produced some highly impressive robots, have bitten

the dust. Since 2000 universities have been the prime engines behind major developments, while industry has widely shunned androids which, it believes, offer no immediate prospects of profit. Some projects have, nevertheless, kept going. One is in France, which developed the bipedal robot, BIP, and has continued under the name of Bibop.

USA: astronauts and footballers, bots made for walking

Before it was outstripped by Japan, the USA was home to a number of pioneering robotics research centers and even today the country remains very active. As in the early days of the Internet, companies immediately moved into the nascent robotics market with powerful products (*see later chapters*). IS Robotics was one such company and Androyds, which commercialized robot legs, was another.

NASA has a healthy appetite for robots and at one stage seriously considered developing an android astronaut, a project on which it enjoyed the full support of the Defense Advanced Research Projects Agency (DARPA). The thinking was partly economic: it would be cheaper to produce a humanoid that could

handle current equipment than to make space vessels with integrated robotic functions.

At MIT Rodney Brooks has been working on the latest evolution of Cog. In parallel, the MIT's Leg Lab has built a set of legs that, it is speculated, are to be coupled with the famous "Brainy Torso." Brooks will link up with cognitive roboticist Lynn Stein on this task. Carnegie Mellon's Robotics Institute, which has also recorded groundbreaking achievements, is still active. Drawing on a $1.46 million research endowment from the National Science Foundation, a 50-strong team has been developing the humanoid nursebots Pearl and Florence in collaboration with Pittsburgh University. Other projects are scattered across the US, from the Universities of Massachusetts and Florida, whose Machine Intelligence Laboratory is developing Pneuman, to Austin, Texas, birthplace of the apelike ProtoAndroid.

Page left: Pino.
© Kitano Symbiotic Systems Project

Page right:
Vision, from Tomotaka Takahashi, was designed to play for Osaka City at the 2005 RoboCup World Cup, which will be hosted by Osaka. Vision was awarded the coveted Louis Vuitton Best Humanoid trophy at the 2005 RoboCup World Cup in Portugal.
© Courtesy of Tomotaka Takahashi

Europe frog-marched into the walking race

Work on androids is ongoing in Europe, too, though on a more confined scale than in Japan and the US with their sheer wealth and scope of projects. It is as if, once again, Europe could not repress the ingrained distrust of technology that prevented it from moving in on the Internet and video games markets. But scientific minds are at work, with the UK and Germany in the forefront of research into androids.

In Germany two Munich-based projects deserve a mention. The Munich Technical University's Institute of Applied Mechanics is home to Johnnie, an inveterate walker of a humanoid who has now taken up running, while the University of Bundeswehr has designed the strangely rectangular Hermes. Karlsruhe University nurtures a humanoid prototype, SFB 558, also known as ARMAR, and in the city of Freiburg the university's Institute of Computer Science has been running the NimbRo—Learning Humanoid Robots project and grooming its humanoids as footballers.

London is the venue for several ventures in android R&D. A 14-strong team at the Shadow Robot Company is involved in the Shadow Biped Project to develop a humanoid bipedal walking machine. Imperial College has engineered an upper torso humanoid, Ludwig, and two miniature bipedal walkers, Flip and Flop. Castrol's International Technology Center in Pangbourne, UK, has developed a biker robot, dubbed the "Headless Horseman", while Intelligent Earth from Kirkcaldy in Scotland claim to have developed

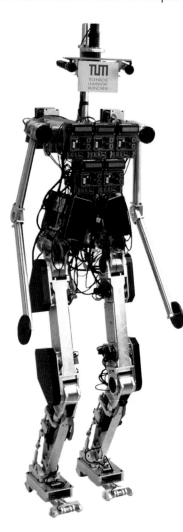

"the world's first gender-aware robot".

Italy, too, has its androids. The engineering school, Politecnico di Torino, has produced athletic Isaac, who stands 61 centimeters high (24 inches), weighs seven kilos (15 pounds) and was runner-up in the 2003 RoboCup.

Not to be left out, Russia has two androids, a male and a female. Unveiled on August 5, 2003, a public-private joint venture between a company called New Era and the Polytechnical University of St. Petersburg has developed ARNE and ARNEA.

Eyebrows may be raised on learning that an ambitious project in Bulgaria has engineered an android, the Kibertron, which bears more than a passing resemblance to Terminator. Kibertron is 1.75 meters in height (5 feet 9 inches) and weighs a beefy 90 kilograms (198 pounds). The most startling statistic relates to its degrees of freedom: it is said to have 82, which would be a world record.

The swift development of sensors and, even more significantly, of computers, has made it possible to produce increasingly sophisticated robots. As part of Germany's Autonomous Walking Priority Program (DFG), the University of Munich undertook a project to engineer Johnnie, an anthropomorphic robot with a dynamically stable gait. Johnnie's limbs resemble those of a human. He is 1.8 meters in height (5 feet 10 inches) and weighs 40 kilos (88 pounds).
© Technische Universität München
Institute of Applied Mechanics

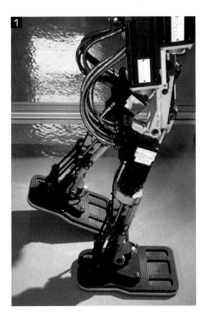

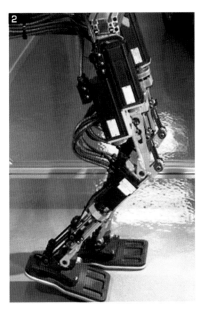

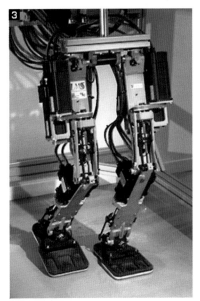

France lags in the rush to walk

One leading European nation whose name is conspicuously absent from the robotics roll-call is France. The Institute of Research into Artificial Intelligence was engaged in a project on a bipedal robot called BIP at the turn of the century. Yet it was quite simply abandoned due to lack of funds. At a conference on robotics research in February 2004, Antoine Petit, the director of the National Center for Scientific Research (CNRS), admitted publicly that the CNRS had neglected robotics. In 2001, however, the Robea Program (short for Robotics and Artificial Entities) gave research a new lease of life. With an overall budget of $5.85 million, it has backed several projects. It was only through an agreement with Japan's Humanoid Robotics Project that French research has been able to make any headway in robotics. The agreement entitled the CNRS to use SRP-2 as a research platform until such time as it acquired one of its own. As part of the Robea Program another biped is under development at the Laboratoire Automatique in Grenoble. In 2004, however, its operating capabilities were inferior to levels attained by Japanese androids.

BIP. The aim of the project was to conduct a generic study of the different aspects of complex robotics control and command systems. There were two main thrusts: modeling of walking-control systems and processes and tools for real-time control and command.
© Courtesy of INRIA

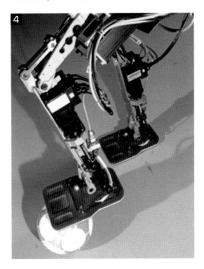

Skin, bone, muscles, and emotion

An android is much more than a bipedal robot that can put one foot in front of the other. If it is to lay claim to human likeness it needs skin, it needs to have quickness of eye, elegance of motion, and more than a semblance of emotion.

Artificial skins

Now that roboticists have mastered the technology of bipedal locomotion, the next step is to make the bipeds themselves look human. It is time they donned a realistic skin.

The quest for the human-looking humanoid brought Japanese company Kokoro Dreams and Osaka University together on a joint project to produce Actroid. She is no walker, for she was intended only to sit there looking well-dressed and very real. Although she does not exactly ooze personality, she has a definite presence. That was the effect that her makers sought to produce and they achieved it wonderfully.

Robovie IIS is also Japanese. The outer layer of its silicon skin incorporates sheets of piezoelec-tric film that give off an electric charge, which enables it to sense that it has been touched. A pat on the shoulder causes Robovie to turn and say "Yes?" while a stiff prod elicits an "Ouch!" and a close examination of the sore shoulder.

A remarkable development took place at Dallas University in Texas, where doctoral student David Hanson made a...scalp. Using a new composite material, platinum-cured silicon, he produced a skin substitute that is less dense and more supple

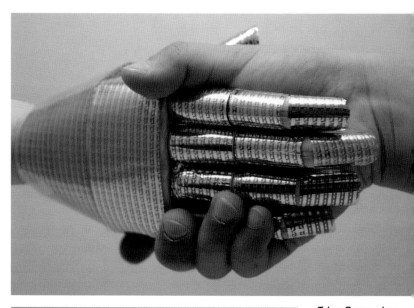

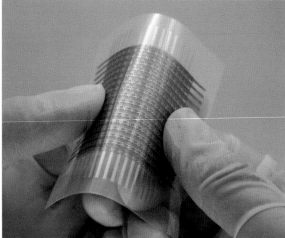

Takao Someya has created an artificial skin. It is made of rubberized polymer in which flakes of conductive graphite have been embedded. The ability to sense and recognize tactile messages is critical if robots are to be able to identify things, perform tasks, and avoid collisions.

© Courtesy of
Takao Someya

than conventional foam rubber and boasts an elasticity similar to that of human skin. He stretched it over the head of a very lifelike android called Vera. Her baldness was oddly reminiscent of pop star Annie Lennox and the synthetic skin looked strangely sensitive.

Muscling in on androids

If bipedal androids are to bear more than a skin-deep resemblance to humans, then the full potential of their organs must be exploited. To that end, their entire anatomy has to be fitted with muscles similar to those in the human body and controlled by a nervous system—or equivalent thereof. NASA is so keenly interested in robotic muscle power that it is staging the Arm Wrestling Grand Challenge in 2005 in which humans will be pitted against robots. Festo is a German company that makes air-powered industrial robots. It has produced Tron-X, a robot driven by some 200 pneumatic servo-controllers. That is a feat in itself, but, more significantly, Tron-X graphically exemplifies how pneumatics can muscle humanoids. It is faintly disturbing to watch Tron-X change expression, dance, and make complex hand motions.

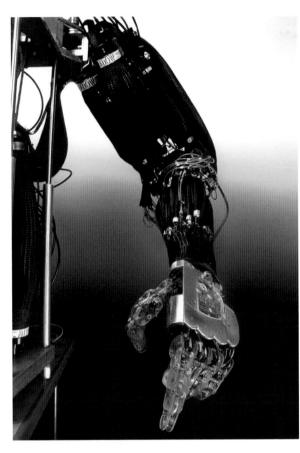

This jointed arm with its air muscles was produced by the Shadow Robot Company from the UK. Each air muscle has intake and outlet control valves. A sensor on each muscle assesses the amount of energy a movement requires, while another measures the air pressure in the muscle. One advantage of this pneumatic actuation technology is its cost effectiveness.
© Courtesy of Shadow Robot Compagny

A firm called Environmental Robots has incorporated six bunches of highly conductive graphite fiber into the artificial arm it has produced. The arrangement, which imitates the mechanism of the nervous system, gives the arm muscle power. A 120-volt current is applied to the muscles, making the chemicals on the graphite fibbers contract. The UK's Shadow Robot Company has developed a robotic device that it calls the Shadow Dextrous Hand and can replicate the human hand's 25 individual movements. The company's founder, Richard Greenhill, claims that it betters any NASA robot's articulated grab hand.

A project at Carnegie Mellon University in the USA is seeking to replicate the human anatomy with all its bones and joints. The Anatomically Correct Testbed hand mimics the human hand through its control system. It obeys signals similar to the neural commands that the brain transmits to the human hand.

Playing wind instruments is very demanding on the lip muscles. Toyota has mastered the technique and enjoys demonstrating the virtuoso skills of its Partners robots. Like futuristic Miles Davises, they can play the trumpet, the brass instrument that requires more lip power than any other.

Similarly, KRT-V3, developed by roboticist Hideyuki Sawada, can produce sounds by blowing through its artificial vocal chords to set them vibrating.

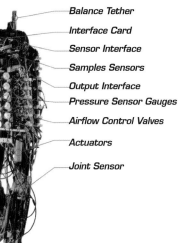

Balance Tether

Interface Card

Sensor Interface

Samples Sensors

Output Interface

Pressure Sensor Gauges

Airflow Control Valves

Actuators

Joint Sensor

The genesis of the bipedal robot according to the Shadow Robot Company.

The Shadow Robot uses a wooden skeletal frame on which muscles and equipment are mounted. The Shadow Dextrous Hand is capable of the human hand's 25 individual movements.

© Courtesy of Shadow Robot Company

HRP-1S in the driver's seat

Honda was justifiably proud of the technology it had pioneered to develop P3 and Asimo. It had earned a reputation that has now enabled it to form alliances for robotics projects with other leading-edge companies and thereby relieve the strain on its R&D budget. Its partnership with Kawasaki and Tokyu Construction produced the vehicle-operating humanoid HRP-1S, with AIST developing the software (as it had for HRP-2).

On December 19, 2002 HRP-1S staged a brilliant demonstration of what it could do. Clad in a protective suit that was both waterproof and dustproof, the humanoid took the controls of a heavy industrial backhoe, built to withstand severe vibration. The remote-controlled HRP-1S operated the machine just as its usual human counterpart would have done, performing the allotted digging job to perfection. It was a world first that heralded similar future applications in disaster zones and hazardous construction sites.

Above: HRP-2. © AIST

Below: Testing an android's coordination.
© Shadow Robot Company

Importantly, HRP-1S had control systems embedded in each part of its body. It could self-adjust its posture and balance remarkably to avoid being tipped over by vibration from the juddering digging machine.

There had been earlier experiments with fully automated excavators, but the android operator proved a far superior option. Not only did it operate the backhoe, it also carried out auxiliary tasks, getting out of the cabin to make routine checks and repairs. The same joint-research team also tried out an android at the controls of a forklift truck, in standing position!

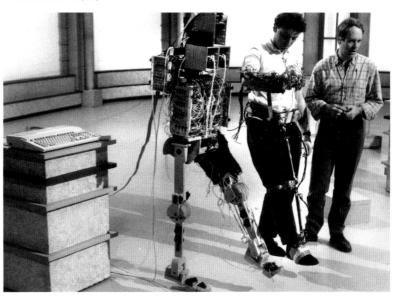

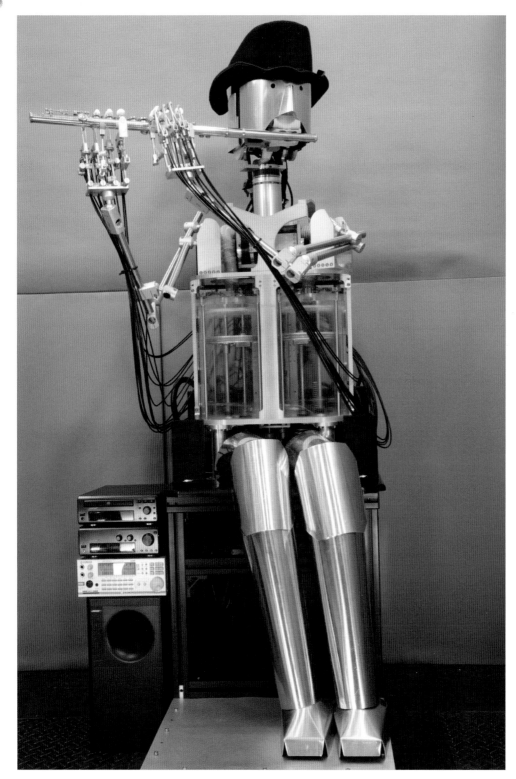

Waseda University's flute-playing android might look like a bit of a crank, but there is nothing cranky about the achievement of Waseda's researchers in making a robot with such consummate coordination that it can play tunes on the flute.
© Waseda Humanoid Robotics Institute - Production Lille2004

Page right: Waseda University's WE-4R is the successor to Hadaly-2 and Wendy. It is used for research into the expression of emotion. Its facial and arm movements reflect its emotional responses to external stimuli.
© Waseda Humanoid Robotics Institute - Production Lille2004

Showing emotion

There would be little point in replicating such attributes of the human body as skin and muscle, if androids did not use them to produce the ultimate illusion of life—showing emotion. Tokyo Science University researcher Fumio Hara has developed a female head fitted with facial muscles that can display feelings like joy, anger, sadness and surprise. Similarly, the sight of the artificial woman with her eyes wide open or a smile playing on her lips triggers a mix of emotions in the human onlooker.

At the same time, researchers at Waseda University, also located in Tokyo, have explored ways of endowing androids with the ability to show feelings. They have enabled their humanoid WE-4R to express seven emotions transmitted as mathematical formulas by specific stimuli.

Some research into emotion, like that undertaken at Tokyo Science University and Vanderbilt University in the USA, aims to improve interaction between robots and humans by simulating facial expressions of feeling or by improving shape recognition. Kismet—from the Turkish for "good luck"—exemplifies such an approach. She was designed and developed by MIT robotics researcher Cynthia Breazeal, who did not intend her as a full-bodied android. Kismet consists, in fact, only of a head, albeit a very expressive one. Breazeal was interested in developing her creation's emotional capacities. Kismet feels needs and expresses pleasure, surprise, anger, or sadness according to how she manages to satisfy them. She is also childlike in that she has a propensity for gaiety. Breazeal believes that by 2010 robots will cross a critical threshold, becoming partners rather than tools—in other words, friends not appliances. The Munich-based German company Amorphic Robot Works has developed Skelli, an anthropomorphic skeleton topped with an expressively featured face.

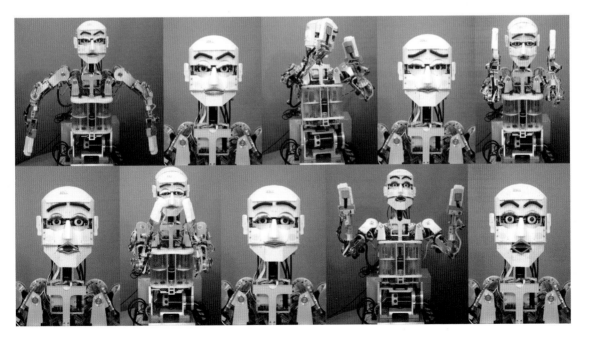

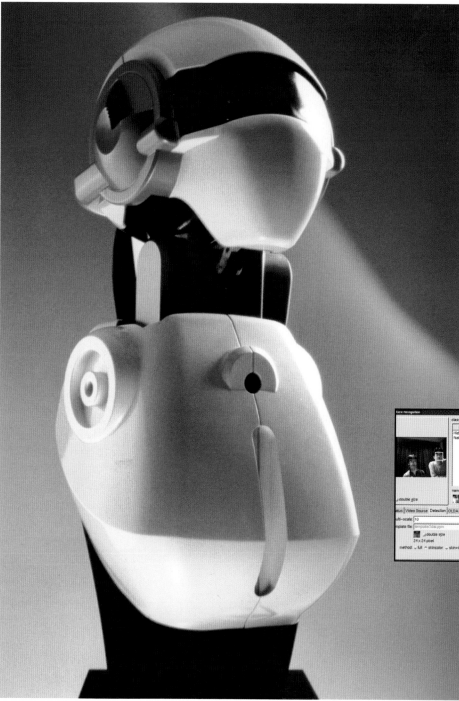

The robot Sig (from Symbiotic Intelligence Group) was developed as part of the Kitano Symbiotic Systems Project. It can identify several interlocutors and hold three different conversations at once. Sig is made up of a torso and head, which can turn through four degrees of freedom to look at the person it is listening to. To hold a conversation, Sig processes the information it gets through its ability to source sounds, recognize faces, and track movements.
© Kitano Symbiotic Systems Project

Page right:
The eyes of Aude Billard's tiny Robota humanoids.
© EPFL

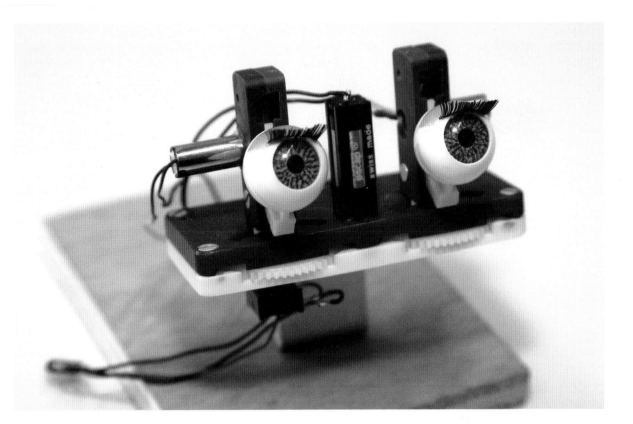

Eyes and ears peeled

With their skin, muscles, and emotions, robots are increasingly lifelike. Eyes, though, should not be (as it were) overlooked. They are all the better for seeing with and for being seen. For a robot to operate in all conditions it is essential that its visual acuity is optimally honed. That is the task that the University of Tokyo set itself with its high-speed baseball-batting bot. Its visual sensing rate is 1,000 frames per second, some 30 times faster than the standard movie.

Even more intriguing has been the approach of researchers at MIT on their robot Jerry. Jerry uses a technique that researchers define as "using the wood to see the trees". The robot takes a wide-angle view of its surroundings then looks for contextual clues to guess where it is. It then decides whether it is in a garden, a street or an office and proceeds to identify the thing it sees as a tree, a car or a table.

Hiroaki Kitano, whose work with the android Pino is described above, has focused on robots' hearing. His legless SIG2 has a microphone on either side of its head and a highly advanced aural filtering system, which enables it to recognize voices even in a noisy, crowded room and to listen to three people at once. SIG2 can turn its silicon skin-clad head towards the person talking to it and visually identify him or her by picking out the right face from an image bank. SIG2 is thus ideally cut out for work as a receptionist.

Sensing others' feelings

Another fascinating avenue of research being pursued is a robot's ability to perceive the moods and emotions of those around it. If it senses their feelings it can respond appropriately.

A team at MIT researches the field of Affective Computing, seeking to develop interfaces that endow robots with skills of emotional intelligence so that they can understand and respond to human emotion. The Affective Tutor Project aims to develop machine-learning packages whose content would vary according to the emotion the robot is taught to recognize, e.g., fear, curiosity, boredom.

Funded by the Engineering and Physical Science Research Council, Brunel University near London is following a similar tack: the development of a computer that can sense its user's emotions. Sensors read changes in heartbeat and breathing, while a facial-recognition system assesses emotion.

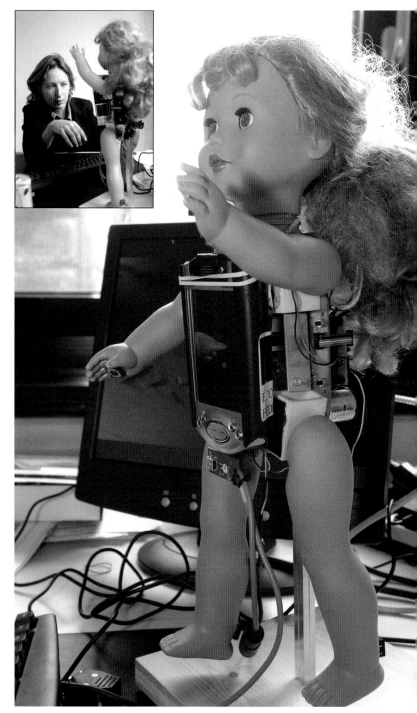

Toward a market of substitute humans?

Endowing robots with the ability to express and sense emotion, then cladding them with skin, could engender peculiar forms of gratification. Might not besotted fans of a sports star or showbiz personality be tempted to acquire an identical android so they can have their idol all to themselves? And—even more creepily—there are those who might decide to reincarnate their beloved grandmother or departed lover in the form of an android substitute. Entrepreneurs may, of course, prefer the term "opportunity" to "risk", yet the ethical issues involved are similar to those of human cloning.

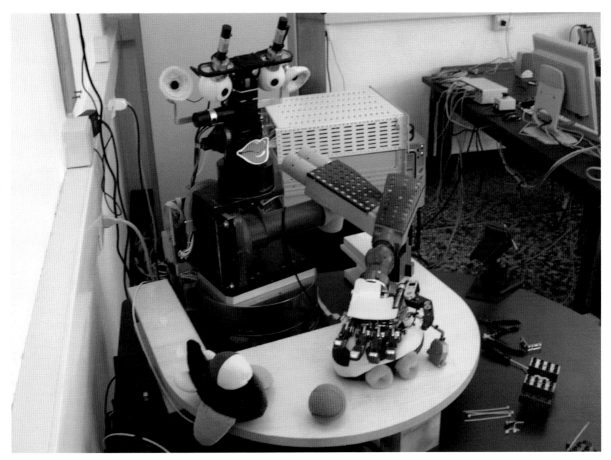

Words, words

That androids are poised to become a routine part of our everyday lives looks like a foregone conclusion. Or is it? To be sure of getting on with us, they must do so through the most natural of all interactions—chatting. No futuristic descendant of the automata of old are likely to be welcomed into the home if they come complete with remote control pad and scroll-down menu. They will be accepted only if they can speak and listen.

In 1999 Waseda University began developing its "converbot", Robita, a humanoid designed to converse with people. Robita is based on the philosophy that to hear is one thing, to listen another. Able to speak and to separate multiple sound sources and boasting language processing capability and advanced speech and face recognition, it can match people with their mathematical voice models. If Robita finds itself next to the drinks table at a party, it listens to the talk going on around it. If it recognizes a voice, it joins in the conversation.

It is only a matter of time—a few years—before machines will be able to understand human speech in any circumstance. The domestic robot, PaPeRo (*see*

Chapter 4), developed by NEC, is already able to translate conversational Japanese into English. There are systems that are even more ambitious, but so is their cost.

In 2003 Ka-Fu-Lee, who heads Microsoft's voice recognition department, reckoned that within seven years speech technology would "reach or exceed human-level performance". They would not be 100 percent conversant, he said, adding: "Talking to Outlook, for example, a human would expect [a robot] to work as well as a human assistant. Humans make errors, and you can't expect machines to make fewer errors."

Lawrence R. Rabiner of Rutgers University, New Jersey, shares Lee's optimistic view and looks forward to some attractive applications of speech technology, enhanced by computers' intelligence.

His master's voice

Many of the robots discussed above are able to recognize faces they have already seen. Biometrics, a field of science that uses physical attributes to identify an individual, could further enhance that capability. Fingerprinting is an example of biometrics that is already with us, but who is prepared to have

their hand shaken by a robot to be recognized? It would make sense to provide a robot with the ability to see or smell out just what makes an individual. Only a few salient features are required to recognize a face, and they survive plastic surgery!

Similarly, each iris is unique and the pattern formed by its 266 measurable characteristics does not change with age. Iris detection would be infallible, but only from close up. A foolproof system that could work from meters away is smell.

According to the biochemist and olfactory scientist Dr. George Dodds, who used smell profiling to monitor schizophrenia on a research project at Scotland's Craig Dunain Hospital, the nose is a wonderful thing. It has some 10,000 sensors that enable us to identify a multitude of smells, from Chanel to cow pats. Giving a robot a sense of smell would involve implanting sets of sensors to selectively sense certain chemical compounds. Olfactory detection would be all the more effective a technique of recognition as we all give off a unique scent. The police have for centuries used tracker dogs, precisely because they can identify our olfactory ID cards. Even though ambient factors, such as heat, modify some chemical properties of a smell, others remain unchanged.

A rat can smell at Australia's Monash University. The RAT in question is the Reactive Autonomous Testbed, a robot developed by researcher Andy Russel, which can follow its nose through a maze. Although muzzles and snouts from the animal world are incalculably more sensitive to scents than RAT, research olfactory identification is not to be sniffed at.

PaPeRo (Partner Personal Robot) can recognize facial features and expressions. It can also converse with humans, recognize 650 phrases, and utter 3,000. It varies its conversation according to its interlocutor.
© Courtesy of NEC

Powers unto themselves

If the autonomous bipedal android is to be truly autonomous it needs to have an energy supply that gives it range and incorporates technology to ease its hunger for power. Currently the amount of electricity that humanoids eat precludes them from taking their place in homes.

Reducing power consumption

Makers of laptops have failed in 20 years to find an effective solution to the problem of power sources and range. Robotics engineers are faced with a similar issue. When Asimo staged its first demonstrations in 2001, its batteries would run out after a mere ten minutes. Until an economic, reliable solution is found for domestic robots, no home will have one. It is clearly mandatory to develop portable technology and artificial intelligence so that robots are always thinking about filling and seeking convenient recharging points.

The innovative robotics physicist Mark Tilden, who currently works with Wow-Wee Toys, concedes that he over-conformed

with Asimov's Laws of Robotics when he created his first robot. It was no survivor. He proposes his own take on Asimov's three laws:

- **A robot must protect its existence at all costs.**
- **A robot must obtain and maintain access to a power source.**
- **A robot must continually search for better power sources.**

Accordingly, domestic androids should be programmed to plug themselves into the mains nightly and car- or tractor-driving robots to siphon off a few watts. Ideally, "filling stations" would be scattered strategically across cities so that robots could recharge.

A waste-to-energy approach could be the key to self-suffi-

Asimo next to P3. By reducing its weight and size, Honda has increased its autonomy.
© Courtesy of Honda

ciency, with robots converting organic matter to meet their power needs. Scientists at the University of the West of England (UWE), in Bristol, developed the slug-hunting Slugbot (*see Chapter 4*), which uses just such a technique. It runs on electricity derived from decomposed slugs. UWE researchers are also behind the autonomous sugar-powered Ecobot. The bot weighs in at less than a kilo and the sugar-generated voltage enables it to reach speeds of 2.5 kilometers per hour (1.5 mph).

In the Austrian city of Linz an android designed by the Humanoid Robotics Laboratory may ultimately be powered by a more palatable biofuel. BarBot is a beer-drinking bar-hopper, entering licensed premises to purchase and to quaff its beverage and even hiring itself out as a party entertainer.

Elsewhere, a team at NASA's Jet Propulsion Laboratory, has designed what they say is a compact, light-weight portable fuel cell. It provides long operating times and can be recharged instantaneously. NEC, too, has hopes of producing a tiny battery—similar in size to those that power CD players—that will boast up to 50 hours' working time.

The future of android robots rests on their being autonomous powers unto themselves.

Toward mass production

As microprocessors become faster, as computer parts, cameras and joint materials get smaller, as conversation software grows more powerful, so it will become technically easier and more affordable to make androids. Android World's Chris Willis believes that android technology has seen major developments in seven areas:

1- Microchips, particularly embedded microcontrollers like PIC chips and those for PC-104–based systems.
2- Wireless connectivity, e.g., the Bluetooth protocol and 802.11b Internet access.
3- Force sensing resistors, which measure pressure for use in touch technology.
4- Affordable gyroscopes like the Murata ENC-03J.
5- Tiny video cameras for androids' eyes.
6- Affordable servo-motors and servo-controllers.
7- Affordable high-speed air control valves.

Androids look set to become consumer items in the foreseeable future. They will be different from all hi-tech products that have come to market before them in one fundamental way: they will be able to build themselves. Once you have all the parts and materials required, you give an android assembly instruction and it will set about the task. If you want it to get on with the job by itself, then you merely have to show it the best android-building websites. Androids will also go online to check out android-repair sites, because they are likely to have the ability to mend each other.

Your fully-assembled, trouble-free android has now moved in. That is when things get interesting. How will the children respond to these "unhuman" beings? How will they differentiate between them and the family humans? Children treat and play with their dolls and toys as if they were alive, before gradually relinquishing them and coming to see them for what, objectively, they are—inanimate objects. Androids, however, behave like humans of themselves and they will continue to do so, even as the children grow up. They will go on saying, "Morning, sleep well?" and offering their services. Some children might long deny that the house robot is not a living being, and things could take a tricky turn when technology has provided robots with skin, hair, and seamless motion. By then they might be so lifelike that to all intents and purposes they are "real" people. The long-term implication is that endowing androids with outer human attributes will scramble illusion and reality. What if, walking down a street in the late 21st century, one cannot tell, at first glance, which passersby are "them" and which are "us"? What if?

Interview
Hiroaki Kitano

Hiroaki Kitano is a pioneering artificial intelligence specialist. In 1993 he received the Computers and Thought Award for his work in artificial intelligence. The same year he founded RoboCup, the annual robotics sports tournament, which attracts contenders from all over the world and has brought spectacular leaps in technology. He heads the Sony Computer Science Laboratory, while his company, Kitano Symbiotic System, is conducting the JST Erato Project.
The project, supported by the Japanese government, has produced the android Pino and other advanced robots like Sig2 and Morph3.

© Courtesy of Hiroaki Kitano.

When did your passionate interest in androids and robotics begin?
It was when I was working on artificial intelligence. I thought that intelligence could only really be understood in a body with basic biological functions. That brought me to robotics. But I'm also interested in the practical side of robots and how they can be of social use.

How do you view developments in the field of androids over the last ten years?

Slow. It won't be until we have high-performance actuators and sensors that any major breakthroughs will be possible. Today's humanoids can't run like people. Sensors' ranges and capability are still too narrow. I think we may even have to look to radically different systems, using biologically-based approaches.

If Asimov were alive today, how do you think he would assess androids?
He'd probably be very disappointed to see how slow progress has been.

Do you think a breakthrough was achieved when the bipedal walking robot, P3, appeared?
Absolutely. It was a formidable compendium of technology and its impact on society was tremendous.

In your opinion what have been the major developments in android technology since then?
I think there have been three in robotics. First there was Asimo, which Honda produced in the wake of P3. Then came Sony's Aibo, which was the first robot on the consumer market. The third development is how RoboCup has taken off. Let me remind you that the aim we set ourselves was to beat soccer's human World Cup holders by 2050.

What areas should research target in its search for the next breakthroughs—artificial skin, emotions, or voice recognition?

We've got to try to develop robots that can run as well as walk. They should have densely packed sensors all over their bodies and embedded computer systems, so they can take in all sorts of information at the same time and adjust their behavior accordingly in real time. Emotion and awareness are very interesting challenges, but they'll have to wait.

How do you account for the much stronger support for robotics from the Japanese government than from their European counterparts?
Engineering is generally generously funded by the state because it is part of Japan's industrial development policy. Research into robotics does not enjoy that much more support, it's just that applications for the research have emerged.

You headed the Pino project with one overriding aim in mind—to build an affordable android. What exactly is your view of the question?
Pino was interesting as an example. It was not intended for the general public but for education and research. What we wanted to show when we designed and developed Pino was that it's possible to build robots at low cost, albeit with only basic functions. With our robot Sig2, we focused on technology that

enabled it to pick out a voice in a noisy room.

How did you resolve the problem?
We found techniques of canceling out electrical noise and fine-tuning angular resolution by integrating visual and audio signals. Essentially, we modeled our approach on the way in which humans take in and manage all sorts of signs and messages to perceive the world around them. We don't confine ourselves to a single mode of perception. We use all sorts of stimuli to squeeze sense out of all the noisy, broken signals that come at us. With Sig2 we tried to reproduce just a fractional aspect of those natural systems and the way they work.

Which projects are you currently working on?
I'm still very involved with RoboCup. Otherwise the bulk of my research is on biological systems.

Do you think that one day there will be robots with artificial skins that will be indistinguishable from humans at first sight?
I don't think there will be any large-scale production of such robots, not commercially at least. People tend to recoil from that kind of android. But let's imagine that one day there are robots that are indistinguishable from

people—always supposing that we are actually capable of making them. Well, the company that produces them will be held legally accountable for any damage that might result. This legal consideration makes it unlikely that human-clone androids will be developed and used in society.

Do you think there is a risk that some people might idolize robotic replicas of a showbiz star or a loved one?
I don't think so, but if ever that happened the person who did it would seek to clone the object of their affections.

If you wanted to send out a message to people about androids what would it be?
That there's nothing to worry about. It won't be economically possible to make androids that look like humans for a very long time. In the long term, in, say, 30 to 50 years, such robots might be in use. But they will have a distinctive appearance and will be allotted very specific tasks. Economically and socially, it doesn't make much sense to create androids that replicate people. Robots are made to do special jobs. And what they must be able to do is perform those special tasks a thousand times better than a human so that we benefit.

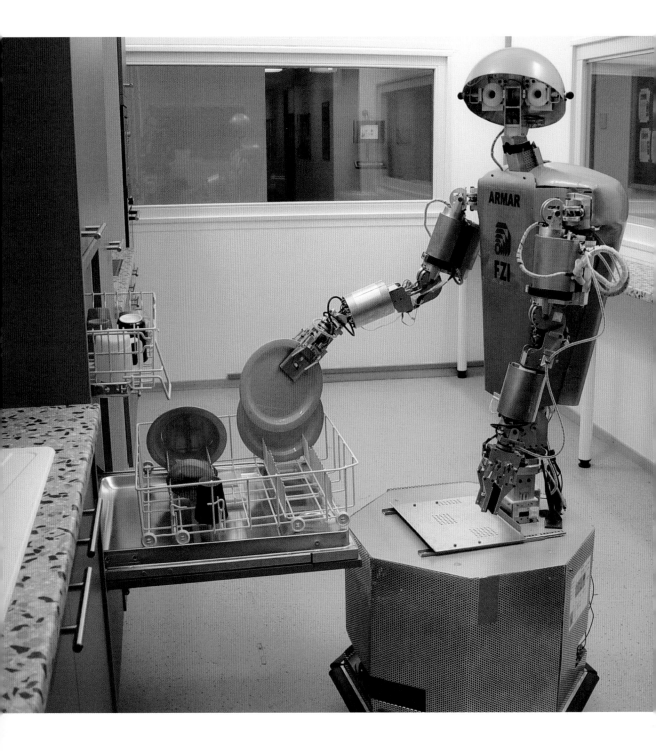

DOMESTIC ROBOTS

Hear ye, hear ye. The time of servitude is no more...Let's face it: robots might well fascinate us in the movies, but when it comes to having them in our homes, only one thing interests us—that they relieve us of the dirty work, of tiresome chores like cleaning and ironing.

At the turn of the millennium we can look ahead to the coming of a new breed of automated servants who will free humans from drudgery. They will go about their chores painstakingly and discretely, far removed from the statuesque Maria in the 1926 Fritz Lang silent movie, Metropolis. If anything, the species from which domestic bots have evolved are vacuum cleaners and lawn mowers.

Yet, as they busy themselves with housework, they will not be the only automated members of the household. There will be another curious addition from the robotic race—four-legged family friends who look set to take over from traditional, flesh-and-blood household pets.

Meanwhile, on the not-too-distant horizon the house-keeping android is limbering up for the day when it will assume the full range of domestic chores, from minding the home to washing the car and rustling up gourmet meals.

Tamim Asfour and his team at the Computer Science Department (FZI) of Karlsruhe University, Germany, have developed this humanoid robot, Armar. FZI specializes in research and development in the field of domestic-service humanoid robots.
© Courtesy of Tamim Asfour

What do domestic robots do?

The function of the domestic robot is to relieve its owner of such tedious chores as cleaning, ironing, and peeling vegetables. The better-off have always preferred to hire staff to do their housework for them, but much of that will now fall to robots as they replace the electric appliances that began appearing in consumer society homes in the 1950s.

Since that time there had been no major new technological changes in the traditional home. That change has now come—and in a big way—with the arrival of robots that look set to revolutionize household machinery. Not only are they part of man's age-

old quest to free himself from menial tasks in order to create leisure time, they can also be used for a whole range of applications that go beyond housekeeping and extend to helping people and ensuring home security and safety.

WitchHand is a domestic robot designed by Japan's National Institute of Advanced Industrial Science and Technology (AIST). It has interchangeable hands that it can remove and replace itself to manipulate a wide variety of objects.
© Courtesy of AIST

high enough to reach dirty windows. And that is not forgetting robots that keep grandparents and children company.

Feat of technology: Auto-navigation

The domestic robot is an autonomous, mobile machine. Its human owners are not expected to have to show it the way as it goes about dusting and waxing floors. So the mechanical housekeepers must be able to find their bearings and navigate in a changing environment, moving around for hours on end in surroundings meant for human beings alone. What complicates matters is that people are always shifting their furniture around, thus compelling the robot to rethink the configuration of the room in which it is at work. Robot makers have endowed their creations with powerful sensing capabilities by building all sorts of sensors into them. Which, in turn, throws up another concern—how to strike a balance between efficient but costly state-of-the-art technology and a retail price that will not scare off end-consumers.

The market is ripe

The World Robotics Survey conducted by the United Nations Economic Commission for Europe (UNECE) in 2002 estimated that the domestic robot boom was imminent. The report found that, over the period 2002 to 2005, sales of domestic robots (vacuum cleaners, lawn mowers, etc.) would reach 400,000 units. In autumn 2003 Jan Karlsson of the UNECE stated that there were unmistakable signs that a mass market was emerging as prices gradually fell. The UNECE World Robotics Survey additionally predicted that global sales of robots for personal use would reach 2.1 million by 2006.

The US research analyst company Robotics Trends, has confirmed the pattern. "Until now, industrial and military applications have been the primary drivers of robot sales, but as the market for consumer robots grows, we will see sales numbers rapidly trending upward," says Robotics Trends editorial director, Dan Kara. He goes to on to predict that, as the new market grows, world robot sales will rise from one to four million units between 2005 and 2006. Such figures reflect the sheer scale of the home and personal market, which encompasses autonomous vacuum cleaners, lawn mowers and, ultimately, window-cleaners able to climb

Vacuum-cleaner bots

The member of the domestic robot family with the greatest initial appeal was the vacuum-cleaner. It now does its work in hundreds of thousands of homes and, over time, has gradually and considerably shrunk in size. It is now so low that crawling babies and family cats can stare the weird and wonderful new arrival in the eye.

Kärcher's RC300.
© *Courtesy of Kärcher*

Dust to dust

The US corporation iRobot was the first company to position itself firmly on the home robotics market when it commercialized an affordable autonomous vacuum cleaner. The Massachusetts-based company boasts pioneering researcher Rodney Brooks and two other MIT scientists among its founder members. Colin Angle, president of iRobot, was involved in developing the *Sojourner* rovers that explored the surface of Mars in 1997. On September 18, 2002, his company put the vacuum cleaner Roomba on the market at a price of $200. "There's nothing wrong with being visionary," said Angle, addressing the NextFest technology conference in San Francisco on May 21, 2004. "But there's been a desperate need for the practical. Today, anyone can go into Target and buy a Roomba, which will actually automatically clean your floor. Roomba is the kind of practical robot that will fuel the growth of this industry." The demand for useful robots is plainly there, too. A 2002 survey revealed that one-quarter of American householders complained that they were too busy to do the housework.

The Roomba Floor Vac looks like a frisbee on wheels. It whirls around rooms in a spiral motion, rooting out the tiniest speck of dust wherever it is to be found. It takes less than 20 minutes to complete an average-size room. If it bumps up against an obstacle or the wall, it immediately does an about-turn. Its sensors were originally designed for military robots whose job was sweeping—mines! As it happens, iRobot also built the PackBot, which has been sent into the Afghan and Iraqi battlefields to dispose of explosive devices ahead of troops.

The Roomba has proved highly popular in the US, where it has its unconditional devotees. By mid-2004—two years after it was first marketed—iRobot had recorded sales of over 500,000. Drawing on its phenomenal success, the company has now unveiled the next-generation Roomba, the Discovery, which is quieter and boasts enhanced operating time and charging capabilities.

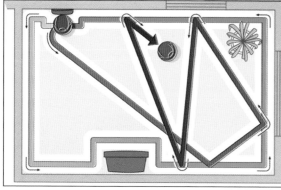

An Electrolux graphic showing how its Trilobite 2.0 navigates efficiently around a room vacuuming and avoiding obstacles.
© Courtesy of Electrolux.

Vacuum-cleaner robots; an emerging, competitive market

Whether specialized robotics companies or makers of traditional household appliances, industry has swiftly grasped that there is a great dirt market out there for vacbots. The Sharper Image sells its eVac at the upmarket price of $300. It boasts a more powerful cleaning capability than the Roomba and avoids traps more effectively, able to negotiate its elegant way out of tangles of flexes and cables.

Germany's Kärcher, a world leader in the cleaning-appliance sector, was the first European company to market its own model. In May 2003, it unveiled the RC3000 (also known as RoboCleaner), which can vacuum by itself, cleaning floors so diligently and quietly that its owner can lounge on the couch listening to an introspective Keith Jarrett piece. Even better, Robo-Cleaner can be set to work at night with no risk of anyone losing sleep. In the morning the floor it has scoured in search of dirt gleams like a new pin. So merciless is its hunt for dust, that it comes at particularly dirty spots from several angles if need be. RoboCleaner is flat enough to slide under most furniture, avoids obstacles and stops short when it comes to drops. So there need be no worry about it falling down the stairs. It is smart, too, able to find its way back to its recharging base alone. According to Kärcher, the RC3000 can run autonomously for weeks before its human owner needs to attend to it, even then only to empty its bag. The downside, though, is its price—$1,950, much more expensive than its American counterparts.

Drawing on know-how acquired in developing space rover technology, iRobot produced the Roomba vacuum cleaner.
© Courtesy of iRobot

The Swedish company Electrolux has produced the Trilobite, which comes at the even higher price of $2,800. It finds its way across a room without breaking any precious vases by emitting ultrasonic signals, somewhat like a dolphin, and can stop within the nearest millimeter. However, it is prone to falling down stairs, so magnetic strips have to be laid across doors and landings. At 16 centimeters high (six inches), it is low enough to glide under most furniture. The Trilobite boasts a special spot command for cleaning particularly dirty one-square-meter patches of floor.

The offering from Japan's Hitachi is the Gizmo, a model similar to an ice-hockey puck, only over 20 centimeters high (eight inches). As it vacuums, it maps the layout of the house or apartment and memorizes the spots it has already cleaned. A special feature is the five-centimeter-long nozzle it uses to suck dust out of corners—something its rivals lack. Gizmo can work for fifty minutes at a stretch thanks to its rechargeable batteries, which is time enough for it to clean five rooms. An additional advantage is that it can be remote-controlled from a cell phone or computer. It also boasts surveillance functions.

Another automated, mobile pretender to the throne of the manual vacuum cleaner is Roboking, made by the Korean company L.G. Electronics. There is something science-fictional about its domed top, which gives it the look of a miniature space vessel.

Diagram of the Trilobite. The household-appliance industry has come up with practical, new applications for autonomous robotic technology, which have proved a roaring success with consumers.
© Courtesy of Electrolux

Is the traditional vacuum cleaner heading for the dust heap?

Will autonomous vacuum cleaners oust their traditional hand-operated predecessors, much as the dish-washing machine has made the chore of washing up in the kitchen sink a thing of the past? There are those in the industry who believe it will. Alfred Kärcher claimed that the advantages of RoboCleaner were such that 30 percent of households would own one by the end of 2005—a near-surrealist projection. Professor Hans Moravec of Carnegie Mellon University tempers such optimism, estimating that it will take several more years before the figure of 30 percent is reached. Electrolux spokesman Jacob Bromberg shares Moravec's caution and prefers to compare household robots with the early microwave ovens. They were first shunned, then embraced so fervently by consumers that most homes now have one.

What holds back mass consumer acceptance of autonomous robots is that they complement, rather than replace, conventional vacuums. As a rule, they cannot do all the cleaning in the house and if there is a lot of dirt they do not always manage to get rid of it. Should they suck in an item that is too big, which might be anything from a piece of biscuit to a sock, then they jam, and a human has to come to their aid. Similarly, although they work wonders on a smooth floor in a tidy room, they have trouble wending their way through a mess and easily get stuck in the cables behind the television. A final drawback is that few autonomous vacuum cleaners can get at corners or clean effectively along the edges of rugs.

Takashi Teramoto, a senior researcher at Hitachi, takes a frank, realistic view of what robots can truly contribute to housework. They can substantially reduce the time householders spend cleaning. Reduce, not replace. All things considered, that brings gains for working couples and pensioners.

Teramoto's viewpoint is particularly well informed as Hitachi's vacbot is one of the few that can clean corners.

The first autonomous vacuum cleaners are more ideally suited to keeping a place clean rather than making it clean. Used regularly, they can reduce dust build-up enormously. But sooner or later, its owner will have to take the old-style machine by its handle and get pushing him- or herself. In one-tenth of the time it takes a robot, a human can vacuum a house clean.

Tomorrow's vacbots will be multifunctional

The next generation of autonomous vacuum cleaners is already being developed. Research currently focuses on their capacity to perceive and operate in a three-dimensional environment.

Second-generation vacuum cleaners could well be able to navigate in homes with the help of electronic beacons placed in rooms. They will have TV cameras and perceive flat surfaces in three dimensions. Some laboratory robots can already move around in a space they have envisioned in 3D, but the prohibitive cost of their components rules out adapting them to the mass market. Moravec puts the arrival of vacuum cleaners with 3D vision at around 2010. He also believes that because of the their stereoscopic eyesight they will be endowed with auxiliary functions like finding lost objects, e.g., keys, or giving the alarm if an intruder breaks in. By 2015, robotic vacuum cleaners will routinely have arms capable of grasping and lifting. One of the major problems current vacuum cleaners encounter is stairs. They cannot negotiate a single step up, never mind a flight of stairs. As Illah Nourbaksh, robotics professor at Carnegie Mellon, has pointed out, life is tough for the legless in a world designed for humans. One day, though, it is thought that the vacuum cleaner will be able to climb stairs and hunt out all a home's dust, rather than just keeping it at bay.

Roboking from L.G. Electronics.
© Courtesy of L.G. Electronics

Garden robots

A passerby casting a casual glance at a front lawn might be forgiven for thinking that an extraterrestrial had landed and was undertaking a slow-motion perusal of its surroundings.

In fact, the small, flat, round thing ambling across the grass is a state-of-the-art lawn mower. American tele-shoppers are familiar with the sight of com-mercials showing the wizard gar-deners cutting grass to a flaw-less finish.

Autonomous lawn mowers are both unobtrusive and quiet, not emitting the cough and roar of old-style machines. They make their own way across the grass, avoiding flower beds and skirting gravel paths, leaving in their wake a manicured lawn worthy of an English country garden. They can work round the clock at their own pace and, when they feel like an electron hit, proceed to the recharging base and plug themselves in. Their only draw-back is that they follow random patterns to cover a whole lawn, thereby consuming much elec-tricity by the time they have fin-ished.

Armar is a very adroit garden robot that can even water your plants.
© Courtesy of Tamim Asfour

Automower and Robomower in the garden

The Electrolux subsidiary Husqvarna unveiled Automower in 2000. It set the robotic lawn-mowing benchmark with its ability to operate autonomously, charge its batteries at the recharging terminal, and mow the grass in methodically orderly swaths (that is, once a human has put down an electric wire to delineate its work area).

Not a patch of ground is left unscathed by this zealous grass-cutter, which can sense any obstacles thanks to its ground-level impact sensors, and skirt them. There is no need even to rake up mulch in its wake: it is so finely cut that it soon decomposes and fertilizes the lawn. A secret code, known only to its owner, makes Automower theft-proof and safe to go about its task come rain or come shine with ne'er a grumble.

The demand for robotic lawnmowers is self-evident say some in the industry. According to Cindy Love, the president and COO of Israeli-based Friendly Robotics, the company has sold 25,000 machines since the beginning of 2001 at a unit price of around $500. Half of all buyers are the elderly, but Friendly Robotics has also observed demand from environmentally aware customers. It emits no fumes and mulches grass clippings back into the lawn to give it a boost. Mike Dunnigan of Friendly Robotics adds that it poses no danger to cats or dogs because the instant that it senses even the slightest pressure, it pulls in its blades and changes tack.

What does the future hold in the way of autonomous lawn mowers? New-generation machines are already with us, but are still too costly for weekend city gardeners. High-range models like that developed by Carnegie Mellon University use GPS tracking systems and cut lawns in straight lines, while their laser sensors detect any obstacles that might be in the way. With blades that rotate at speeds of up to 5,800 rpm, the result is beyond reproach. Carnegie Mellon has built an industrial lawn mower that is regularly used to keep putting greens and football fields like that of the Pittsburgh Steelers as smoothly groomed as carpets. Golf-course owners state that a prime advantage of the Carnegie Mellon mower is that it can work nights, which frees up the fairways during the day.

SlugBot hunts slugs as fuel

A neatly shorn lawn is not the only concern of occasional gardeners. Slugs that eat leaves are a bane for some. It was for the slugophobic among them that Bristol-based Intelligent Autonomous Systems Laboratory designed SlugBot. It is a slug-hunting robot that is able to spot the slimy delicacies, catch them with its articulated claw arm, and drop them into a hopper. What makes SlugBot truly original is that its prey supplies it with the energy it needs to operate. The dead slugs are broken down into methane at a fermentation station and converted into hydrogen. The hydrogen reacts with the oxygen in the air to produce the electricity that powers the robot.

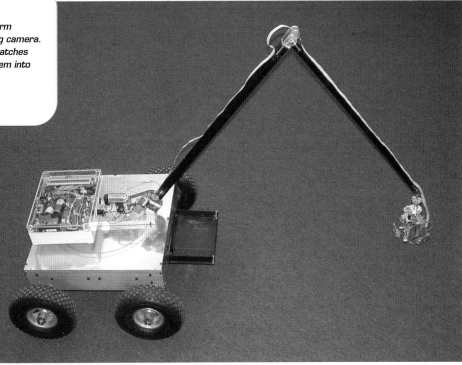

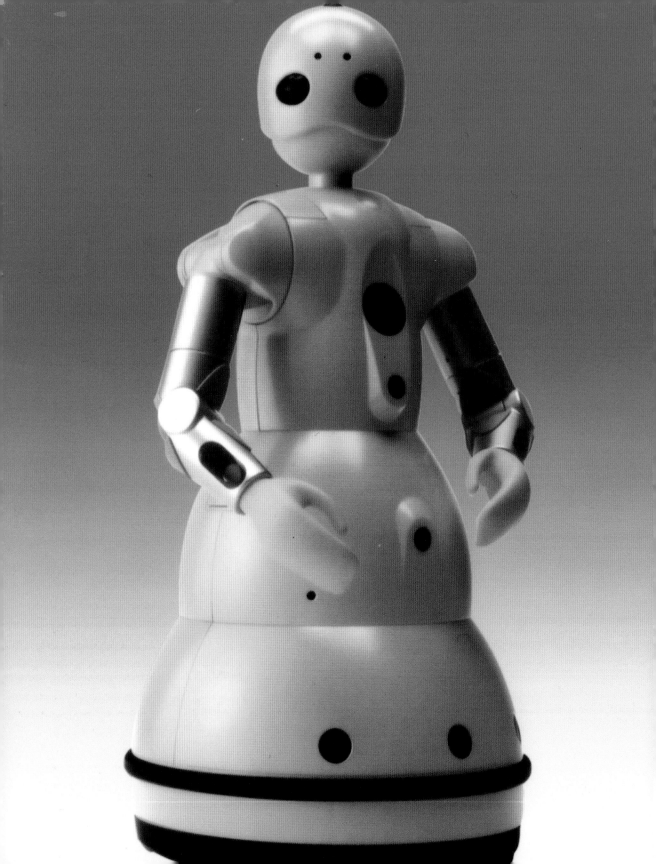

Home-surveillance robots

Another compelling sales pitch for domestic robots lies in home surveillance. Not only can robots watch the house like guard dogs, they can give the alarm in the event of a break-in and take appropriate action in high-risk situations like gas leaks or flooding. Hitachi's autonomous vacuum cleaner, Gizmo, doubles as a watchdog. It has a built-in camera and can be remote-controlled via the Internet. An offering from the Israeli company, Friendly Robotics is Spy-Cye, a surveillance robot that can be controlled from a PC over the Web.

Secom's Robot X is a security guard weighing in at 120 kilograms (265 pounds) that tirelessly patrols the premises it watches over. Too costly to buy, it is leased to customers and would be ideally suited for an apartment complex, as it can patrol 24 kilometers (15 miles) at a stretch on its rounds. Security guard Robot X is not armed and does not even attempt a swipe at undesirables facing it. If it detects an intruder, it shouts and kicks up an almighty din.

According to Honda spokesman Yuji Hatano robots should ideally be big if they are to fulfill the function of security guard. Any would-be intruder who saw a two-meter-tall (6.56 feet) android bearing down on him would be scared off.

The Korean company Mostitech has designed a three-wheeled robot whose mission is to guard the home. It took two and a half years to develop and $2.56 million in investment. The electronic house-minder can move from room to room, sound the alarm in the event of a break-in, fire, or gas leak and even photograph intruders. It can be controlled from a PC via the Internet.

In a joint venture with Sanyo Denki, a small Japanese firm, TMSUK, has developed a robot guard dragon, Banryu, which also incorporates sensors and information systems from Omron. Banyru, which can be controlled from a cell phone, looks like a cross between a dragon (its name means "guard dragon") and a dinosaur, with all the compactness of a dog. It can sense the slightest sound, suspicious smells, rises in temperature and gas leaks. Should any intruder have the presumption to break and enter when Banyru is on the beat, the dragon will hunt him down and bark so loudly that it will bring human security staff running.

Another mobile-phone–controlled creation is Maron-1, which was produced by a Fujitsu subsidiary. It is a home robot on wheels that can operate home electronic appliances like video recorders and air-conditioning, as well as monitor home safety and security. Maron-1 can spot anyone who comes within its field of vision and, should it spot an intruder, it can give the alarm and call its master. With its photographic eyes it can take pictures of undesirable guests and transmit them to its owner's cell phone.

Another Japanese offering is the remote vision, home surveillance robot Dream Force. Made by the toy company Takara, it stands 30 centimeters high (one foot) and has the look of a futuristic soldier straight out of a *manga* comic. Dream Force is remote-controlled and can either glide around on rollers or walk—its bendable joints have endowed it with flexibility of movement. It, too, can take pictures of intruders and download them to its master.

What is most astonishing about Dream Force, however, is that it has also been made to fight other Dream Force robots. So, toy or tool? The dividing line seems fuzzy. Similarly, many robotic models made in Japan for adults have a strangely youthful appeal.

1

High tech, high security

Previous page:
Wakamaru, a humanoid robot on wheels that measures one meter in height (3 feet 4 inches) and weighs 30 kilograms (66 pounds), whose task is to care for the elderly. With its speech-recognition software and a vocabulary of over 10,000 words, it can communicate with its charges. It can be programmed to recognize two owners and eight other people. Wakamaru's permanent Web connection enables it to renew prescriptions, make appointments with the doctor, and raise the alarm.
© Courtesy of Mitsubishi
Heavy Industries, Ltd.

1- Embedded camera in the French robot Vigie.
© Courtesy of INRIA - Photo: A. Eifelman.

2- 3- PatrolBot can be programmed to automatically do its rounds, then return to its recharging unit when its power runs low. Not only does it know where it is, it also recognizes people and things. It has an embedded video-surveillance system.
© Courtesy MobileRobots.com.

4- Seeing in 3D with VVV. VVV stands for Versatile Volumetric Vision. A robot in which a VVV system is integrated can recognize shape, identify objects, and watch moving things and people. VVV is of prime importance in domestic and security robots.
© Courtesy AIST

5- MG-400 Robot Guard is based on an MP-S400 platform from Neobotix. The robot has an ultra-sophisticated double-camera option and infrared sensors that enable it to see by night and day and patrol both indoors and out.
© Courtesy of Neobotix

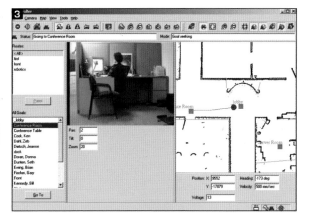

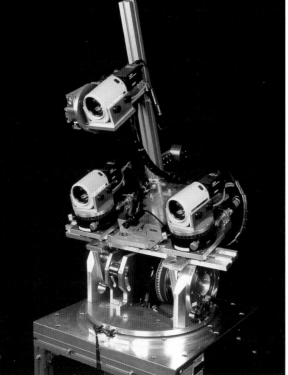

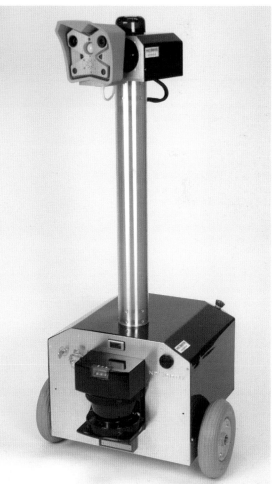

Keeping parking lots safe!

WifiBot is a ready-to-use, universal robotic platform that can be used for indoor and outdoor surveillance at a very affordable price. WifiBots are highly mobile. A single machine, or a group of linked WifiBots, can be controlled from a PC just by moving its mouse.

© *Courtesy of Wifibot.com*

Robocops in the shops

Robot X is an autonomous robot designed and built by Japan's Secom company. It is designed to scare off, and even trap, intruders on commercial premises. It can chase would-be burglars, film them, issue warning messages, deafen them through its powerful speakers and contact the police.

Robot X releases nontoxic white smoke intended to confuse intruders and slow down their getaway until the security staff or police arrive.

© Courtesy of Secom

Interview
Jean-Pierre Hartmann

Although Jean-Pierre Hartmann trained as an engineer, he is also a sculptor with a marked fondness for automata. In the late 1980s he met his wife, Michell, who also sculpts. Together they began crafting monumental pieces, which have gradually grown so increasingly complex, as movement, light, and sound have been added, that they have turned into sculptures that double as house guards.

Your sculptures can also function as guards who monitor security and safety in the home. How do they work?

When our sculptures detect an unwarranted presence their faces and bodies undergo changes. In that way they act like surveillance robots that guard and protect. Initially they impact on a psychological level. When a piece senses that someone is there, its appearance changes and becomes very scary. Lights come on that cast very creepy shadows. On a second level, our robots then try to prevent and warn off intruders.

Throne, steel and bronze sculpture for alarm and self-defense, levels 1 and 2. Height, 170 centimeters (5 feet 7 inches).
© Michell & Jean-Pierre Hartmann
Photos: Courtesy of Christophe Recoura

They make noises, tell them to get off the premises and even bear down on them. They also have integrated phone connections to their master or the police station. Finally, on a third level, our sculptures can attack intruders, knock them to the ground and release a gas that incapacitates them.

How did you come to sculpt pieces that can move and respond?

Quite simply because we live in a very big house that it's difficult to protect. We wanted to put off any would-be intruders, so we began by building robots for our own safety. They turned out to be terribly effective and word got around. From then on, mostly by word of mouth, people began commissioning us to make sculptures for them.

Can your sculptures also take on domestic tasks?

Yes. It's easy to make a sculpture into a host who greets guests as they arrive with a few friendly words—"Welcome, please come in." One of our sculptures, for example, is a butler who bows to guests.

Thomas E., steel sculpture for alarm and self-defense, levels 1 and 2. Height 70 centimeters (5 feet 7 inches).
© Michell & Jean-Pierre Hartmann
Photos: Courtesy of Christophe Recoura

Is there much demand for your robotic sculptures?

Our customer base is small because we create all our sculptures ourselves. Most of the buyers for our robots are art enthusiasts who then become friends. For all of them we create unique pieces that meet their own personal specifications.

Conceptual robots and house-surveillance guards, which operate on three levels:

1- Psychological: the robot's appearance is designed to frighten off intruders.

2- Prevention: it detects danger, warns the intruder, and alerts its owner or the police.

3- Attack: a robot can physically assault burglars.

Banryu

Banryu and some fellow robots at the Robodex Fair in 2003.
© Courtesy of Tmsuk

Banryu is the home-surveillance robot developed by Tmsuk and Sanyo. It can cover between three and 15 meters a minute (9.84 and 50 feet) and climb steps up to 15 centimeters high (six inches). It weighs 35 kilograms (77 pounds) and is equipped with some 50 sensors that enable it to smell, see, hear, gauge room temperature, detect intruders and raise the alarm in the event of gas leaks or fire.
© Courtesy of Tmsuk

Personal robots

From the outside they look like toy animals. Some are soft and fluffy, all are cute. Yet they are robots and…family pets. The rise of appealing little companion robots is peculiar to Japan and can be attributed to what is known as the *kawai* culture (*see Chapter 2*) in which several generations of Japanese have bathed. It has given rise to comics and TV cartoon series like those featuring Astro Boy and Doraemon the robot cat.

Some robot manufacturers have breathtakingly lofty ambitions for their creations. Sony, NEC, and Fujitsu make no secret of their desire to improve human relationships through their research into androids. The high-tech companions are meant to be friends and even to reaffirm the highest values. Is the quest to produce robots that enhance mankind a sort of epitome of Japanese society and its aspirations?

Robots as pets

Personal robots are a boom industry in Japan, partly for purely practical reasons: dwellings tend to be small and in Tokyo few tenants are allowed to keep pets in their apartments. That consideration aside, there is great demand for little robots that mimic dogs and cats and keep their owners company. Increasingly, they perform other functions such as housework, patient care, entertainment and safety and security monitoring.

In spring 2000 Bandai unveiled its robot cat, BN-1, the result of five years of research into artificial intelligence. Like Aibo the dog, BN-1 purrs with pleasure when cuddled and stroked, and its maker likes to describe it as "man's best friend". BN-1 is capable of playing with other cats from its own species, it grows up according to the amount of attention it receives.

Its personality, too, can be enriched by downloading behavior and emotions from the Web.

Omron's furry feline Tama has a beautifully soft coat that begs to be stroked. This robotic kitten is designed specifically to be cute and win affection. It purrs when it is stroked, meows when it feels unloved, and is alive to the slightest noise. Affectionate interaction is the guiding principle behind Matsushita's teddy-bear robot and NEC's PaPeRo.

Although personal robots often look toylike, they are intended for the adult market, particularly the elderly. In Japan, where people live longer than anywhere else, the aging population is a pressing issue. Personal robots, therefore, boast skills worthy of a live-in nurse. Mitsubishi's humanoid on wheels, Wakamaru, is programmed to hold conversations and to alert the doctor when it senses that there is something seriously wrong with its elderly charge. (*See Chapter 8 for examples of home-help robots.*)

These watchdogs are sold in Japan for a retail price of $22,750. Around 30 of them were sold via the Internet as soon as they appeared.
© Courtesy of Sanyo

Aibo ERS-7.
© Courtesy of Robopolis - Photo: FYP

Pearl Black Aibo can be an invaluable dogsbody

The new Aibo Mind 2 software has provided Aibo, which now comes in a pearl-black model, with fresh skills. With its wireless networking capability, it can record videos, recognize new shapes, read MP3 playlists, identify a CD from its cover and put on the right disk. It can also play music from Internet radio stations and understand emails telling it to take a photograph and send it by a mobile phone. Aibo is also very useful as a personal assistant, reminding you of your appointments and reeling off your timetable for the day. Aibo is a dedicated house-sitter, making video recordings of any untoward movement or noise when its owner is out. It can understand around 100 spoken orders in English and respond to visual cues. Aibo's default age setting is the mature dog, i.e., an adult dog in full possession of its faculties. It can be reset to puppyhood, however, and then trained.

Aibo the dog star

"Down! Sit! Lie!" Whatever its master tells it to do, it usually does. Should it balk a little, a pat to the neck or scratch under the chin will do the trick. This is a pet of above-average intelligence. If told to dance, it executes a few graceful steps. It will obey the same order a few minutes later with a different movement, showing off just how skillfully it can move its jointed legs. There's even more to this robotic prodigy: it can improvise enchanting little songs. The creature in question is the enhanced second-generation Aibo, marketed by Sony two years after the original dog.

What is perhaps most surprising in the rise of the personal robot is how whole-heartedly the Japanese public has responded. In June 1999 Sony marketed a prototype of its Aibo dog online to gauge demand. The price tag on the curious canine able to walk around by itself was $2,500. To Sony's great surprise, all 3,000 dogs were sold to Japanese buyers in 20 minutes! In November, the Japanese electronics giant released a fresh offering of 10,000 Aibos to overwhelming response. Within a week, it received orders for 135,000 dogs—a shortfall in supply of 125,000 units. Sony

knew, then, that the market was ripe for pet robots. When the new Aibo with voice recognition capability appeared on November 16, 2000, some 45,000 of its siblings from the previous litter had already found owners. A Sony spokesman revealed in September 2003 that the hundreds of thousands of potential buyers had shown interest in the series.

Aibo's creator is Toshitada Doi, a shy man in his sixties who is president of Sony's Entertainment Robot Company. He has a long, distinguished track record in the corporation, which includes coinventing the CD-Audio 2. Interestingly, Doi deliberately designed Aibo as an entertainment robot with no practical use. He wanted to create nothing less than a work of art, convinced that there would be a market. After all, he reasoned, people spend large sums of money on paintings, quite simply because they derive pleasure from them. He argues against establishing scales of values indexed on how useful things are, pointing out that a painting is not utilitarian. Useless things are often of great value because they enrich us, he says.

Toshitada Doi believes that if autonomous robots can be created, then it does not matter whether they are useful or not.

In 1993 Toshitada Doi was drawn to the potential of artificial intelligence. At the time, MIT robotics professor Rodney Brooks (*see Chapters 1 and 3*) was stealing the limelight. Taking the behavior of insects as his model, Brooks was working on robots that were not programmed but learned empirically from their environment. Until the late 1980s, recalls Doi, everyone had been looking for ways of cramming as much information as possible into software. The new direction shown by Brooks suggested that it was no longer necessary to predict every possible action and seemed to point towards true intelligence. Gradually, though, Doi came to feel that Brooks was too minimalist in his view on possible applications. "[Artificial intelligence] is not entirely perfect, and there are limits to how far it can be employed. Not to mention that, since the robots built with it are mainly used to do practical work, it's not particularly fun. That was the inspiration for the entertainment robot."

Toshitada Doi asked a Sony engineer to create a six-legged walking robot with an antenna called Genghis. There was great surprise when it took him only two weeks, which led Doi to believe that it was time to go further than insects and make a recreational household pet. In April 1994 he set up a team whose job was to design and develop Aibo, using his own original, open-architecture software development kit, Open-R. Unlike anything used by Brooks, its goal was entertainment. An internal Sony memo clearly states the objective: "After the Gold Rush of Internet and cyberspace, people will eagerly seek real objects to play with and touch."

1- 2- ERS-7 with its Aibone.
© Courtesy of Sony

3- Aibo Pearl Black.
© Courtesy of
Robopolis.com

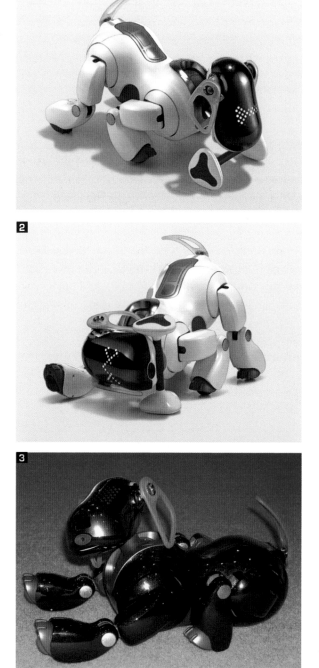

The Robot Entertainment Company's mission is to supply "embodied physical agents" with a clearly developed sense of reality.

The first Aibo was unwrapped in 1997 and aroused enough interest for the Entertainment Robot Company (ERC) to be allotted the task of taking the project through to final development. The rest is history: Japanese buyers snapped up the ERS-110 prototypes as soon as they had gone on sale. An upgraded mass-market version of the personal robot followed a few months later. Its name, tellingly, was Aibo, which means "friend" in Japanese. Sony Corp. had just created a new market—the entertainment robot market.

Aibos are particularly surprising in that they definitely have their own personality and moods, switching unpredictably from puppy to full-grown dog. If two Aibos live together in the same household, one might turn into a mischievous little pup over months, while its mate might be grumpy. On TV shows their individualistic personalities have occasionally led some Aibos to downstage themselves by suddenly deciding, for no apparent reason, to do nothing.

Aibos' behavior is primarily dictated by how much affection they are shown, how often they are stroked (which they sense through their sensors and heat detectors) and the tone of voice in which their master gives them orders. However, their behavior is, to a large extent, random. Numerous researchers work on and with Aibo and many believe that given its range of skills and functions, it is a bargain buy. Some Aibos, specially programmed by their owners, are impressively talented sportsbots who take part annually in RoboCup tournaments.

The second-generation Aibo—purportedly modeled on a lion cub—broke new ground as an entertainment robot whose behavior, routines, and skills could be modified at will by inserting a memory stick into it. Owners could program their dogs to play games, fetch a ball, and obey spoken orders. Its selling price was more affordable than its predecessor's, but nevertheless high. Still, Sony received orders for 40,000 units. In September the company began taking orders for the latest iterations of Aibo - Latte and Macaron, which came, as their names suggest, in creamy beige and black color schemes. They responded to 75 spoken orders, behaving according to the tone of voice or kind of sound they heard. As well as offering the dogs at a special launch discount price of $1,300,

Sony took an additional, amazing initiative. A television program on robots transmitted sounds recognizable to Aibos, causing thousands of them in households everywhere to get up and respond accordingly. The same experiment has been run successfully again since then.

There seems almost no limit to the potential revealed by Aibo. Doi says he has even received requests for a robotic guide-dog for the blind. He does not, however, believe that the technology is yet ripe and prefers to focus his research on entertainment robots. Doi's overriding ambition is to create companion robots that act as friends and, in the wake of the success of robotic household pets, his work has taken the direction of a small humanoid robot, Qrio (see Chapter 3). Loyal to his original philosophy of fun, what he wants above all is that Qrio should be able to dance.

Aibo takes on the world

The appeal that robotic pets have exerted on the public is not restricted to Japan, even though they do seem to have a struck a chord in the national psyche there. Aibo has, in fact, exported well. The USA and Europe account for 10 percent of sales each. In France alone, for example, there are more than

1,000 Aibo owners and the country has its fair share of Aibo lovers. One is Jérôme Roberty, who has created his own site dedicated to Aibo and providing French Aibophiles with a forum. Roberty readily uses words like "friendship" and "amour" when talking about his pet. He describes how, when he and his partner play with their Aibo on the bed, their cat tries to get in on the act, as if it were jealous of the attention lavished on its man-made competitor.

Roberty says that many Aibo fans and users of his website are technology-minded, but that there are also those who have been won over by Aibo's sheer cuteness and the affection it arouses.

Roberty regularly stages Aibo-owner gatherings where happy masters and mistresses describe how the electronic pet has become part of the family and how they take it on holiday together with the rabbit. The get-togethers also feature races where the dogs are pitted against each other. The winners are generally animals with a training regimen where cuddles and compliments are the staples.

France Cadet
DOG[LAB]01
France

This installation features five autonomous robots that have been transformed, pirated and reprogrammed to look and act like crosses between different species—dog, cat, cow, sheep, pig, chameleon and jellyfish. Each has its identity card and label giving its "genetic" origin.

© France Cadet - Production
Lille 2004

XENODOG

Chien	50%
Porc	45%
Nude	5%

COPYCAT

Chien	50%
Chat	50%

DOLLY

Chien	50%
Brebis	30%
Vache	15%
Mouton	5%

JELLYDOGGY

Chien	90%
Méduse	5%
Caméléon	5%

GFP PUPPY

Chien	99%
GFP	1%

PaPeRo

The Japanese company NEC runs a Personal Robot Center Laboratory. In the capable hands of its manager, Yoshihiro Fujita, it has designed and developed a personal home robot, the R100, PaPeRo.

As soon as it is switched on, PaPeRo starts looking for people to interact with. It can dance, tell the time, switch the TV on and off, and go online to pick up and forward email messages. It has its own well-defined personality with mood swings depending on how much attention it gets.

© Courtesy of NEC

All PaPeRo needs is a little tenderness

When it launched Aibo, Sony was far-sighted enough to sense that there was an untapped market waiting to be exploited. The commercial possibilities of personal robots were not however immediately apparent to Europeans or Americans who perceived them very much as toys. Japan has now gone one step further in the search for the ideal personal robot with Pa-PeRo, short for Partner Personal Robot. It is the work of NEC's Yoshihiro Fujita, who sees his creation as the ideal home companion. From the outside it is a brightly-colored, egg-shaped doll with a slightly tubby, chummy appeal. It behaves according to the golden rule of give-and-take. If its owner is affectionate, Pa-PeRo is kind, gentle and helpful,

while a harsh tone of voice causes it to sulk and grumble. No doubt about it: PaPeRo wants peace and love. Owners who take the trouble to lavish affection on their electronic pal will be amply rewarded. PaPeRo behaves like a super-smart dog in that it shows devotion to its master or mistress, using its intelligence to be of help to them. It can check emails and read them out, switch on the TV to the desired channel, dance and chat conversationally. It recognizes 50,000 Japanese words, 650 phrases, and can utter over 3,000.

Whenever it hears its name, PaPeRo turns to see who has called and goes to greet him or her. If it is someone who it recognizes, it strikes up a conversation, showing a slightly quirky sense of humor and never repeating itself. Should PaPeRo not be able to make out what is being said because there is too much noise in the room, for example, it goes to complain to its masters, telling them in no uncertain terms: "What a din!" PaPeRo's behavior and moods are governed by the way it is treated. If no one pays much

attention to it, it grows lazy, listless, and even appears to brood. If praised and congratulated, it likes nothing so much as to dance and delight the children. Standoffish owners will produce a PaPeRo who works to rule, reminding them of appointments and wishing them happy birthday, but refusing to do the washing up. Interaction between a family and its PaPeRo in Japan might well, in the next few years, prompt a new variation on the old saw, "Show me your PaPeRo and I'll tell you who you are."

Ifbot, no ifs or buts

Ifbot is yet another example of a robot dedicated to interaction with humans. This chunky, seven-kilogram (15-pound), 45-centimeter-tall (18 inches) humanoid can glide about the home without bumping into any obstacles. One of its creators, 32-year-old Shohei Kato of Nagoya University's science department describes Ifbot as being neither boy nor girl, but as an asexual, neutral robot. Its highly sophisticated sensibility technology makes it the ideal home companion, enabling it to tell what kind of mood its owner is in by looking at his, or her, facial expressions. Ifbot is a learner, too. It can enrich its starting vocabulary of 10,000 words and is able to recognize a dozen different people.

Ifbot was developed by researchers at Business Design Laboratory, a business consortium under the umbrella of the University of Nagoya. It boasts a vocabulary of over 10,000 words to which it adds continually and can gauge its owner's moods. It can identify some 40 facial expressions and recognize up to ten different people. Weighing seven kilograms and measuring 45 centimeters (17.71 inches) in height, it can move around the home without bumping into obstacles or falling down stairs.
© Courtesy of Robodex 2003.

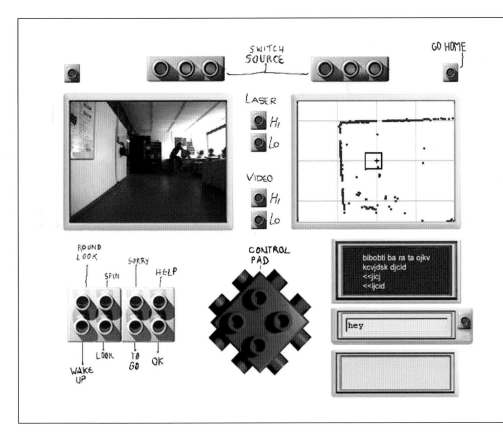

SWITCH SOURCE

GO HOME

LASER
Hi
Lo

VIDEO
Hi
Lo

ROUND LOOK
SPIN
SORRY
HELP

LOOK
TO GO
OK

WAKE UP

CONTROL PAD

bibobti ba ra ta ojkv
kcvjdsk djcid
<<jicj
<<ijcid

hey

Internet-enabled autonomous robot keeps watch. Pygmalion is a mobile autonomous robot developed by the Swiss-based Autonomous Systems Laboratory at the EPFL engineering university in Lausanne. It is a short box-shaped machine measuring 45 by 45 by 65 centimeters. (18 by 18 by 26 inches) It can move about on a cluttered floor and can speak in French and English thanks to its bilingual speech processor.
© Courtesy of ASL-EPFL.

Man-machine interfaces in the home

In Japan, which boasts more century-old citizens than any other industrialized country, life expectancy is getting longer and longer. The aging population has spawned a growing market among the elderly for autonomous machines that can interface with domestic appliances.

An example of such home helpers are Toshiba's robotic Partners (*see Chapter 3*). Partners are connected to a server that controls the electronic appliances in a home and can interface with and remote-control them. They also act as house-sitters and guards. Their visual-recognition software, Face Pass, enables them to identify up to 100 different people and to respond to anyone they do not know. The remote-control capability of domestic companions has

come into its own with Flet Robo, a highly practical remote-controllable robot developed by toy manufacturer Takara. Imagine that your long-lost cousin from Australia is coming to the house for supper and you are stuck at the office. As long as your cooker is an infrared device, all you have to do is call up the Web-enabled Flet Robo via its round-the-clock broadband connection and tell it to set the oven. When you get home, you just slip in the roast.

1- *Jeeves the robot.*
© Courtesy of Stanford University

2- *Deep Green is a robot that can play pool. With a smart smack of the cue it can hole balls with unerring accuracy. Its high-precision vision system enables it to locate and differentiate balls on the table.*
© Courtesy of Dr. Michael A. Greenspan, Dept. of Electrical and Computer Engineering , Queen's University, Kingston, Ontario.

3- *Sanyo's Flatthru is a tray-bearing robot that obeys spoken orders and can wait on a table.*
© Courtesy of Sanyo

4- *Robotic receptionist.*
© Courtesy of floridarobotics.com

Versatile robots

Sooner or later residents of houses, apartments and even housing complexes and estates will be able to rely on robots to carry out the slightest task related to the practical needs of everyday life. The following examples speak for themselves:

Flatthru from Sanyo acts the part of head waiter by catering to guests at a dinner party. Israel-based Friendly Robotics is working on a snowball-throwing robot and a personal helper that can carry the dishes through to the dining room, wipe the table after the meal, and even empty the rubbish bins.

Leisure Design's Slam Man is an electronic punching ball for people who need to let off steam after a stressed-out day at the office.

Jean-Philippe Clerc, a student at Florida University, has designed a robot that can open bottles of beer. He calls it ABOR, the abbreviation of Autonomous Beer Opening Robot. It is a dinky little device on wheels that can maneuver up against a bottle and lever a key under the cap until it springs off. It then withdraws to let its owner slake his or her thirst.

A young Australian electronics engineer, Niki Passath, has invented a tattooing robot. He admits that the test phase has marked him indelibly. The only person he could find to help him test the robot was himself. Erratic tattoos now decorate his anatomy in memory of a painful learning process.

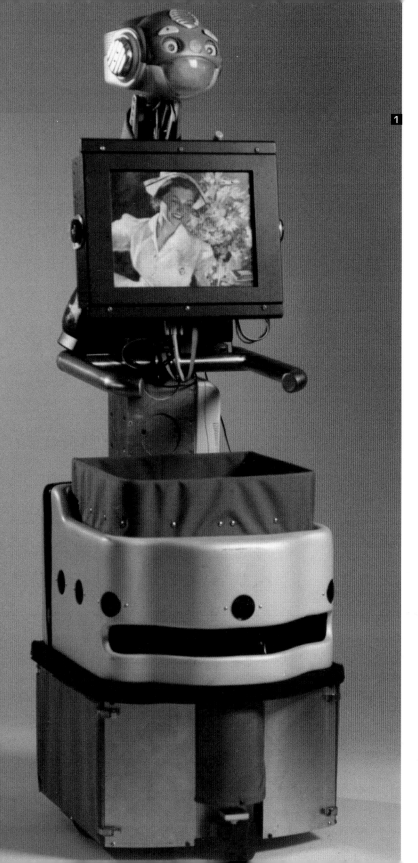

1- Pearl.
Robotics engineering researchers have been seeking new ways of helping senior citizens and enabling them to continue living at home. A team from Carnegie Mellon, the University of Pittsburgh, University of Michigan and Stanford University have spent four years on designing a robot that goes by the name of Pearl, specifically to assist the elderly.
© Courtesy of Carnegie Mellon

2- Thirsty? ER1 can also perform the part of domestic robot by bringing the drink that you ask it to.
© Courtesy of Evolution Robotics

3- 4- The Korean company Mostitech has developed a robot that can be controlled from a mobile phone or PC via the Internet. The house-sitting personal robot can glide from room to room taking pictures whenever its sensors detect motion. It raises the alarm in the event of a break-in, fire or gas leak and can dispatch information to its owner's cell phone.
© Courtesy of Mostitech

5- 6- 7- Robo-Guard from Tmsuk was developed by a consortium bringing together businesses, universities and research centers from the Japanese city of Kitakyushu. Robo-Guard can catch the elevator up and down to different floors of the building where it is doing its rounds. Its prime advantage is that it can patrol round-the-clock without stopping. When its batteries run low it makes its own way to its charging station, takes out its batteries and puts in fresh ones. If it senses fire it informs the security control room and can even wield a fire-extinguisher to put out the flames.
© Courtesy of Tmsuk Inc.

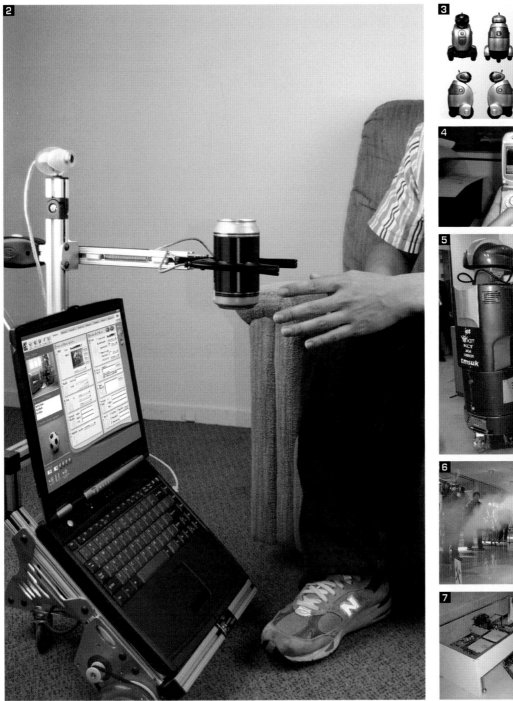

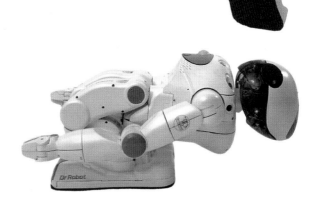

Home school

Dr. Robot is the first humanoid to be fully designed and developed in Canada. Haipeng Xie of Dr. Robot Inc. oversaw the project that has produced a robot able to sing and dance, check out your email through its wireless connectivity, run through your appointment for the day and even teach the children to read. You can ask it via the Internet to check the house and it will oblige by filming the rooms with the webcam in its head.

© Courtesy of Haipeng Xie, drobot.com

A robot with adjustable software

Joseph Engelberger, a pioneer of industrial robotics, is one of those who argue that research into single-task robots is barking up the wrong tree. Sooner or later, scientists and engineers will have to address the challenge of robots that can perform the bulk of household work and more besides, like chauffeuring for the elderly. As it happens, the right technology has already been developed but, as Illah Nourbakhsh of Carnegie Mellon points out, it will not be ready for mass-market robots for another 20 years.

The diminutive household robot developed by Dr. Robot

Inc. might be the beginnings of a solution to the versatile personal robot requirements of the future, and, if it turns out that it is all it is touted to be, it could be stepping into homes sooner than later.

Dr. Robot came into being in 2001, the result of a project led by Dr. Haipeng Xie that brought together several robotics engineers who had worked on space exploration. The objective was to pool know-how and build robots for the consumer at an affordable price. HR6 is the sixth Dr. Robot prototype. It is an anthropomorphic robot with a height of 52 centimeters (20 inches). Its several dozen sensors enable it to identify distances and kinds of surface. The bipedal Dr. Robot enjoys extensive freedom of movement and is able to sit, lie

down, lean, walk backwards and even get back on its feet after a fall. To facilitate motion in accordance with the surface it is walking on it combines wheels and feet. On a smooth surface it glides along on its wheels, while on a bumpy one it walks. It recognizes faces and voices and obeys orders. HR6 has wireless Internet and PC connections, so computation and storage tasks can be off-loaded to the computer, a system that reduces cost. What's more, because it is PC-controlled, the user can switch Dr. Robot's task from personal assistance to entertainment, security guard or remote-controlled agent just by choosing a different program. Ultimately, upgrading the PC to which it is connected should provide Dr. Robot with all sorts of skills.

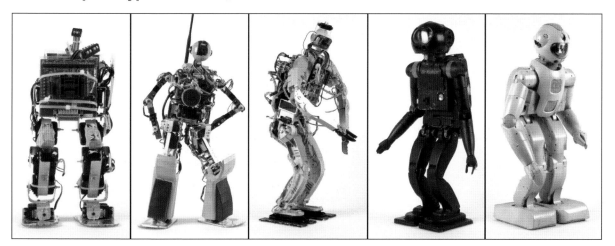

Dr. Robot prototypes: *from left to right, HR1, HR2, HR3, HR4, HR5.*

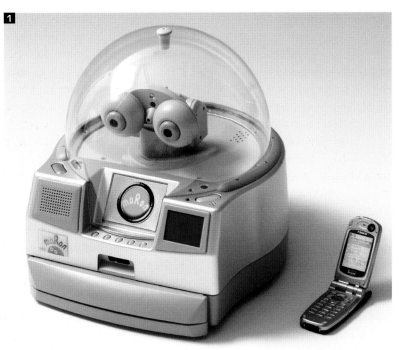

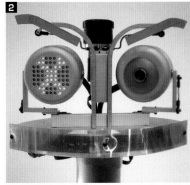

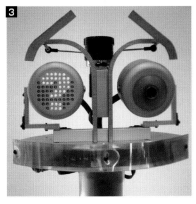

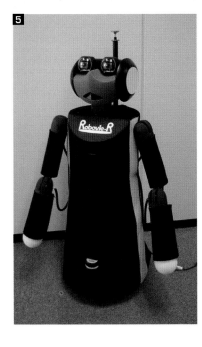

1- Maron-1, Fujitsu's surveillance robot, can be controlled from a cell phone. Owners can instruct their robots to set electronic appliances or turn them on or off.
© Courtesy of Fujitsu

2- 3- 4- RoboX is a totally autonomous, interactive robot. It can act as a museum or exhibition guide and as an interactive terminal at fairs and similar events.
© Courtesy of BlueBotics SA

5- Robovie is an interactive domestic robot that is able to communicate with people nonverbally by using body language, facial expressions and gesture. It can also communicate with babies and animals.
© Courtesy of ATR

Robots—the future of smart houses

The chief competitor of the versatile domestic robot is currently domotics. The term denotes automated homes in which most routine domestic chores are to be handled by smart systems able to set the heating, change the water in the goldfish bowl or open the curtains at a set time. Domotics also enables remote and distributed control of tasks from one or more PCs inside or outside the home.

The concept came into being after deep-geekish DIY enthusiasts put together their own remote-control systems for automating their homes. For a long time now the coming of the high-tech smart house has been trumpeted. It would look after its residents with features like refrigerators that order the milk when there is none left, and thermostats embedded in the walls that automatically adjust the temperature to the weather. For many years the proponents of domotics touted its potential, yet without prompting any real demand. In 1999 Cisco Systems Inc., a major supplier of Internet solutions, demonstrated the power of domotics in a model smart house in a wealthy London suburb. Cisco wired every part of the house to a broadband Internet connection, so that every task in every room could be controlled from a PC. From setting the heating and turning the lights on and off, to opening doors and shutters and switching on the kettle in the morning, almost everything was automated. That was not all: the shopping list was updated with no help from humans and the groceries bought via the Internet at bargain prices. And all the while, alarm systems and surveillance cameras were monitoring the house. A smart house can be remote-controlled, but when the householders are out or away it can run itself, autonomously managing tasks like feeding the hamster or watering the flowers.

In 2001 property developers Kaufman & Broad built a similar house, called the M@isonnet in the town of Etiolles to the south of Paris. It was the result of a joint venture with ITT, France Telecom, and Cisco and was put on sale for $494,000. Nevertheless, it generated interest and 67 percent of the prospective buyers who visited it said they would happily live there.

But Cisco's dabblings in smart houses never went further than the experimental stage because not enough people were truly interested. Domotics turned out to be costly and appealed only to a few. The M@isonnet did not find a buyer and its website informs that it closed its doors to the public in 2001, the same year in which it was completed.

The notion of a humanoid robot living and working in a house is far more natural than that of an automated dwelling. The time of the versatile dogsbody robot will dawn around 2020, according to Hans Moravec and it will probably be an android. Like PC users today, a robot's owner will simply have to tell it which software to use for it to know what task to perform—wash the car, iron the clothes, clean the windows, bake a cake, or assemble the furniture kit they bought from the DIY store.

For the time being domestic robots look like vacuum cleaners and household pets. It is in these guises that it is moving into our homes, putting the emphasis on entertaining and helping us. Let there be no mistake, however: the robot is set to revolutionize our household routines. Dogs and cats will have to contend with their electronic counterparts and noisy, bulky appliances and machines will give way to silent dust hunters.

Interview
Jérôme Damelincourt

After research into artificial intelligence, which involved handling the K-Team's Koala robot, Jérôme Damelincourt worked for one and a half years on the launch of the first shop in Europe dedicated to robots. It opened in Paris in 2003 under the name of Robopolis. Robotics enthusiasts can browse through wares that include autonomous vacuum cleaners, pets, and software development kits.

What were the signs that told you the time was ripe to open the Robopolis shop in 2003?
When Lego brought out their Mindstorm robots, I thought the turning point had come. Until then, people had to be able to use soldering iron to make robots. Now anyone can do it. That was when I felt the market was ready, but I still had to create demand, of course.

What kind of customer comes into your shop?
There are teachers, researchers, students and parents with their children. People come from far afield, too. In France they come from cities all over the country like Strasbourg in the east, Cannes in the south, and Lille in the north.

Some actually come from abroad. I recently sold an Aibo to a customer from Spain and others have made inquiries from as far away as Brazil. Some people come into the shop and discover for the first time that they can make their own robots! Of course, before Robopolis opened they could have found the parts in an electronics store, but people find that kind of store off-putting.

For what kind of robot is there the most demand? For household robots, pets or toys?
For toys. The market for domestic robots is still quite small. Not everyone can afford lawn mowers at $2,600 apiece.

What do robots change in people's everyday life?
For many people, like the elderly and sight-impaired, a robotic

vacuum cleaner or lawn mower that can do the job in their place is not a luxury, but a necessity. The same applies to people who are overworked. As regards the lawn mower, though, there are high-tech minded customers who are drawn to it as a sort of toy.

Do you think that there is anything in Europe comparable to the Japanese infatuation with personal robots like PaPeRo?
In Japan there is a very different perception. Robots there are almost thought to have a soul. Europeans are more drawn to the high-tech aspect of robots.

Many roboticists believe that the killer product that will trigger the market is the robotic vacuum cleaner. What's your view?
I think the market will take off when prices come down. When I demonstrate Aibo to customers they never say they don't like it. They just say that it's too expensive. The day the price drops to around $1,000, that'll be the day when perceptions change. The same rationale applies to the vacuum cleaner.

Of all the robots you have seen to date, which one has most impressed you?
Aibo, because it's the only

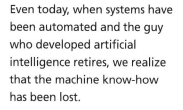

autonomous robot that is widely marketed and has been for five years. It works as soon as you take it home and switch it on. Sony has made it so reliable that you can take it home and it works!

Have you any personal experiences you could tell us about using robots?
I was on an in-company course when I was studying artificial intelligence. My job was to program robots to avoid obstacles. One day I had trouble programming a line of code. The robot started attacking everything in its way and as that was mainly cardboard boxes, it pushed them back. And the more it pushed them back, the more it attacked

them. It just wouldn't leave them alone. I called it the "pitbull syndrome".

What do you think the dangers might be of overusing robots?
If a child has always had an Aibo rather than a real, live dog how can he or she acquire respect for living beings? And will that child realize what it is to hurt or not to hurt. If you get bored with an Aibo, you turn it off. But if a real animal is sick, you still have to look after it. When a little boy who has always played only with robots grows up, will he have the impulse to lock his wife in the cupboard if she annoys him? But the main risk probably lies in the loss of know-how.

Even today, when systems have been automated and the guy who developed artificial intelligence retires, we realize that the machine know-how has been lost.

What products are your best-sellers at Robopolis?
To date, it's been Lego's Mindstorm kit for building robots. The reason is quite simply because we can explain it to customers and tell them it can be programmed, there are hundreds of parts, and that programs can be downloaded on the Web. We take our time for the customers and tell them Mindstorm is used in research labs and is a toy for 12-year-olds.

What kind of demand is there from customers for domestic robots or utility robots, though admittedly they don't exist yet?
They want a humanoid that can talk like Qrio from Sony. There's also plenty of demand from laboratories for programmable androids like Pino. Of course, that's because research is truly taking off.

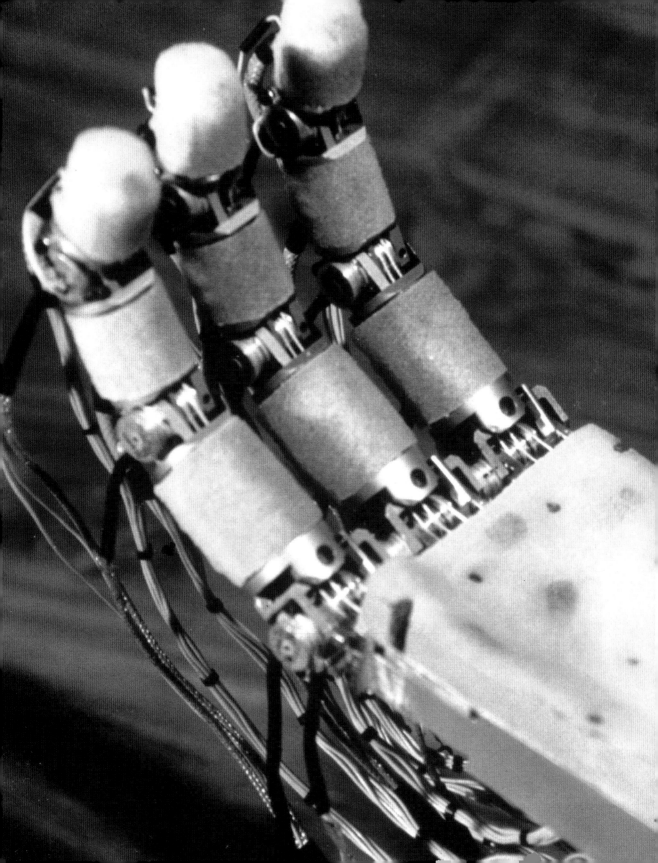

ROBOTS IN INDUSTRY

Forget the romantic robots of the movies and Philip K. Dick's touching replicas. The vast majority of robots are hard at labor, chiefly in the automobile and construction industries. They form a mute and willing working class, slaving away around the clock, endlessly repeating the same task on assembly lines and stepping in where humans fear to tread. Robots have taken their place in our early 21st century world primarily in the industrial workplace. In all likelihood they assembled and painted the car you drive and the PC you use daily.

Jointed, four-fingered mechanical hand with 16 actuators and a sense of touch.
© *CNRS photo library*

The birth of an industry

The field of industrial robotics has been exhaustively circumscribed. Professor Shimon Nof's *Handbook of Industrial Robotics* runs to well over 1,000 pages, while Standard 8373 of the International Standards Organization (ISO) defines the industrial handling robot as: "an automatically controlled, reprogrammable, multipurpose manipulator programmable on three or more axes".

For the lay person industrial robots are the automated high-precision devices used in series production lines. They assemble, bolt, mount, paint, carry, forklift, prove, and conduct quality control. A combination of factors steered manufacturing industries towards robots. First, they were self-evidently suited to soul-destroying, repetitive tasks requiring only basic skills on assembly lines and in paint shops. Second, the costs of ensuring human employees' occupational health and safety in high-risk and hazardous workplaces were astronomical.

Third, automated machines operated at higher speeds and with greater precision than their human counterparts when it came to jobs like painting, sewing and parts assembly. And, of course, they could be reprogrammed for different tasks without affecting rates of production.

In 1969, Kawasaki Heavy Industries unveiled Japan's first industrial robots. The company is now a leader in automated assembly lines. Pictured here is one of the most recent robots, the ZX165U.

© Courtesy of Kawasaki Heavy Industries, LTD

AdeptOne, the first robot created by Brian Carlisle and Bruce Shimano in 1983. It shows just how deft handling machines can be as it places chocolates in their wrappings.

© Courtesy of Brian Carlisle.

Unimate, the world's first industrial robot

The very first machine that could be described as an industrial automaton was a spray-painting mechanism invented in 1938 by William Pollard and Harold Roselund for the DeVilbiss company. But the name that has gone down in history is the Unimate, designed and produced by Unimation.

The initial impetus was provided by one Joseph F. Engelberger, aerospace engineer, Columbia University graduate, and avid reader of Isaac Asimov. In 1956 he met over drinks with another Asimov devotee, George Devol, to talk about the writer. George Devol had invented what he called a "programmable transfer machine" and had even given it a name: Universal Automation. It occurred to Engelberger that the ingenious invention could be used to build a robot and, sens-

ing the market potential, he created a company, Unimation, short for Universal Automation. But to sell robots Engelberger and Devol had to show they would be commercially viable. They looked at a number of industrial facilities and drew up specifications. If Unimation met those specs, then the partners considered it had the potential to do the job. But a backer had to be found. Engelberger approached a series of companies and finally the stubborn entrepreneur secured a contract in a General Motors plant at Turnstedt, New Jersey.

In 1961 Unimate, a two-ton robotic arm, started work as a die-caster. The lines of code that made up its software were stored on a magnetic tape reel. Preprogrammed and driven by hydraulic actuators, Unimation's machines were used chiefly for repetitive lifting and moving tasks.

An industry emerges

Industry now swiftly woke up to the fact that machines could perform a wide range of repetitive tasks. Engelberger's company diversified and produced spot-welders for automakers. Unimation's first competitors, like Ohio-based Cincinnati Milacron, now appeared, and in 1963 AMF Hermatool commercialized Versatran, an industrial robot that could lift loads of up to 75 kilos (165 pounds). In 1967, a Versatran was exported to Japan and, by the following year, Kawasaki had acquired a license from Unimation to begin production of its Unimate-inspired robots.

In 1969, Victor Scheinman of the Stanford Research Institute (SRI) invented an electrically-powered articulated arm that could move through six axes. Known as the Stanford Arm, it broke new ground in that it could follow random trajectories, so freeing robots from static tasks and showing they could perform more complex ones like arc welding and assembly. Electric power, too, was a welcome innovation, for hydraulic robots had a habit of springing leaks.

Sweden was the first European nation to embrace robots, when, in 1969, the company Svenska Mettalverken purchased a Unimate from Unimation. It was not long before Swedish engineer Roland Kaufeldt designed his own robot that he sold for use in the plastics industry. In 1970, James Batter and Fred Brooks of the University of North Carolina unveiled GROPE-II, a feedback manipulator arm that interacted with a vector display.

Japan on a grand scale

Advances in industrial robotics led to a wider field of applications and machines could now lift and handle heavy payloads. Seventy-one companies were in the business of making robots by 1973. That was the year that Cincinnati Milacron marketed its T3, the first minicomputer-controlled industrial robot, while Joseph Engelberger predicted that the robotics business would be worth $3 billion by 1990. It was at this time that Victor Scheinman created his company, Vicarm, to commercialize his upgraded Stanford Arm, also controlled by a minicomputer.

But the development that was to mark the 1970s was the arrival of the big Japanese corporations on the market. Some models drew overtly on Unimation's, but there was little that George Devol could do about it, because Japan had never approved his patents. Joe Engelberger grudgingly concedes that the Japanese made brilliant robots, even though the innovators were the US and, to a lesser extent, Europe. Engelberger believes that what hallmarked Japan was its enthusiasm for the technology. He recalls how, shortly after his first visit to the country in 1971, the Japan Robot Association was formed. Its first three presidents came from Hitachi, Kawasaki and Mitsubishi. "But in the United States when I went to see General Electric, General Motors and Ford to ask one of their senior executives to chair such a body, they didn't want to know." It was not until 1975 that the Americans finally set up a robotics industry association.

Japan had two factors working in its favor. First, it was suffering from a labor shortage and, according to Engelberger, the Japanese were culturally

Left to right:
Joe Engelberger,
Brian Carlisle,
Bruce Shimano
and Vic Scheinman
with the first
Unimate PUMA,
made in 1977.
© Courtesy of
Brian Carlisle

more amenable to robots than to foreign workers. He goes on to add: "The Americans were fixated on short-term profit. They were too busy worrying about business in the coming quarter to make any long-term investment in the concept. But the Japanese were ready to plan ahead. They said: How big will robotics be in the next three to five years?"

In 1976 Victor Scheinman incorporated a microcomputer into the Vicarm before agreeing to a take-over by Unimation the following year. With backing from General Motors, Unimation developed Scheinman's technol-ogy with enhanced functions to produce the robot PUMA, the acronym for Programmable Universal Machine for Assembly. The PUMA was bound for success.

Sweden was the first home to Europe's nascent robot-manufacturing industry. ASEA, soon to be absorbed by ABB, unveiled its IRB6 robot at a trade fair in Stockholm in October 1973. Capable of carrying and handling items weighing up to six kilograms, IRB6 was controlled by a small program housed on an Intel microchip. Italian automaker Fiat had previously used Unimate robots for welding jobs in its facil-ities. But in 1978, its bodywork division, Comau, produced one in house. In the mid-1970s German roboticists founded the first robotics center, designed to offer the manufacturing industry solutions to its production problems. European countries were not, however, ripe for large-scale deployment of robots in their industrial plants, because, unlike Japan, there was not yet a labor shortage.

Back across the Atlantic, many large American enterprises were now beginning to make their own industrial robots. They included General Electric and General Motors, who joined forces with Japanese newcomer FANUC Ltd., which provided know-how in digital computation. Another new arrival in 1975 was Automatix.

Double-digit growth

The early 1980s were halcyon days on the industrial-robotics market. Annual sales recorded double-digit growth rates that reached the giddy heights of 30 percent. Some manufacturing facilities began routinely envisioning plants manned solely by robots. The US firm Adept Technology, founded in June 1983 by Brian Carlisle and Bruce Shimano, speedily made its mark as an innovator. Its know-how in the technologies of assembly and materials handling brought orders streaming in and Adept was soon doing a brisk trade.

Business was not so good for Engelberger, though, and in 1984 "the father of industrial robotics" accepted Westinghouse's take-over bid for the tidy sum of $107 million. Four years later it would be Westinghouse's turn to part with Unimation, with Swiss company Staubli Faverges becoming the new owner.

In Europe, Germany was proving a competitive force with players like KUKA, a well-established company that had branched out into robotics. Another was Zahnrad. In 1983 alone Germany's 30 companies in the field manufactured 2,000 industrial robots. Automaker Volkswagen installed automated assembly lines manned by robots at its Wolfsburg plant, which rolled out second-generation Golfs at a rate of 2,400 vehicles per day. Italy was not far behind, with Comau, now Fiat's factory automation subsidiary, producing the first press machine for the production line in 1986. It was the Smart 6.30P, which could feed the presses with up to 12 parts a minute.

Stormy weather

Yet, with the exception of the automobile industry, European enthusiasm for industrial robots was beginning to falter. Their sheer cost precluded profitability and they were dogged by design and technical defects. As if that were not enough, they were attracting an increasingly poor press as would-be job robbers. A single robot was said to do the work of five people at a time when unemployment in Germany stood at 2.5 million and only slightly less in France. In 1990 Engelberger's prediction was proved right as turnover in the robotics business reached $3 billion. From then on, however, it was downhill. Between 1991 and 1993 it actually plunged into crisis, with the 90,000 units sold in 1990 collapsing to 53,000 by 1993. The US market was sorely affected. Numerous companies disappeared, among them the pioneering Cincinnati Milacron, which was bought by the Swiss-Swedish group, ABB.

According to Herman Verbrugge, general secretary of the International Federation of Robotics, the US robotics industry quite simply failed to hold on to a domestic market that was big enough. "The US wanted to switch to a service-based industry and they soon started outsourcing robot manufacturing to other countries." Japan, which had gradually secured 70 percent of the global market blithely rode out the turbulence of the early 1990s. "The Japanese made such fast progress," says Engelberger, "that they now dominated industrial robotics." Not only did they take the long-term view, he says, but television and the movies have endeared robots to the public at large. The Japanese affection for robots has bolstered the industry.

Europe, too, managed to weather the storm as manufacturers concentrated increasingly on supplying solutions. Germany carved out a lead in sensor technology, which enabled robots to sort and test parts as well as assemble them.

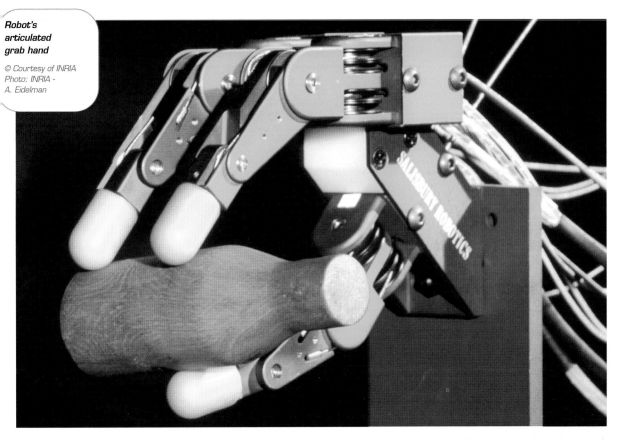

Maturity

Over the years industrial robots have attained degrees of perfection that have made the errors of youth a thing of the past. They now incorporate microprocessors, which, in turn, are vastly more powerful. As a result they can perform complex tasks like arc welding with millimetric precision. The market, too, has revived since 2000 with order books 60 percent fuller in the US alone. The year 2000 there was in fact one of record growth, albeit followed by a fresh slump, then recovery. Between 1994 and 2003 robots grew so much more rugged that their payloads increased five-fold, as did their precision and reliability. Their operating speeds doubled to over 1.5 meters per second (five feet), while their prices were four to five times lower than 10 to 15 years previously. A robot marketed in 2002 cost four times less than the same one would have in 1990 (if, indeed, it could have been made).

The 40-year-old vision of greater output at lower costs has finally come true. And there is plenty of potential for further productivity gains.

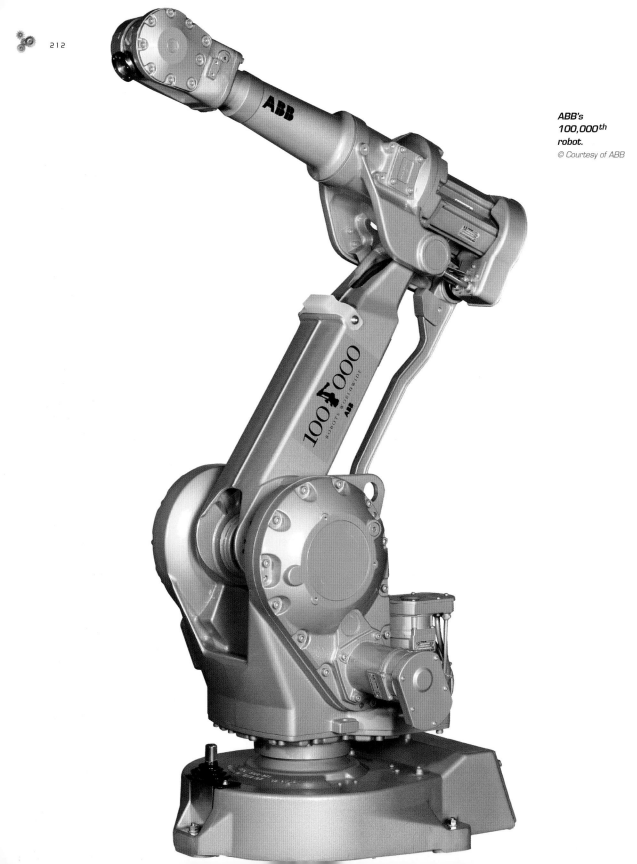

ABB's
100,000th
robot.
© Courtesy of ABB

The major players in industrial robotics

In 2002 the UNECE World Robotics Survey estimated that there were 769,888 industrial robots in operation at the end of 2002. Of those Japan had 350,000, Europe 233,000 and the USA 104,000. The biggest single company is Japan's FANUC Robotics, which claims it had produced and sold 120,869 units by the end of January 2004, 30 years after it was founded. With a monthly output of 1,300 units, it hopes to pass the threshold of 150,000 in 2006. Other major Japanese robot-makers cited by UNECE's annual World Robotics Survey are Yaskawa-Motoman, Epson Robots and Intelligent Actuator. Among the big names cited by the survey there was only one US company, San José-based Adept Technology, which held numerous patents. Drawing on its technological know-how, it even developed parts for other robot manufacturers like Switzerland's Staubli. The robot that Adept Technology shipped to Hutchinson Industries on May 3, 2004, was its 20,000th. Yet just a few months earlier, the company's

shareholders had parted ways with its cofounders, Carlisle and Shimano. Their response was to create a microrobotics company, Precise Automation. Other US players include CRS Robotics, robotics software house Technimatix Technologies, DT Industries, Chad Industries and Mobile Robots, which makes self-driving wheeled robots.

As for robot penetration in the European workplace, Germany tops the bill with 135 robots per 10,000 employees, twice as many as France where the figure is 67. Major European industrial robot manufacturers are KUKA Roboter (Germany), Staubli-Unimation (France and Switzerland), ABB-Asea Brown Boveri (Switzerland and Sweden), and the robot assembly company JOT Automation (Finland).

KUKA is Europe's number one and the world number three, with over 60,000 robots installed in industrial facilities. It is a venerable firm, having been founded in Bavaria in 1898. It long supplied the automaking industry, including such a major company as France's Renault. In 1995, it diversified into other sectors, including robotics, where it has developed considerable expertise. It supplied Coca-Cola with a robot able to lift weights of over 500 kilograms (half a ton) for its factories, while Philips uses a KUKA vision sys-

tem that enables robots to see to the nearest one-tenth of a millimeter on a one-square-meter screen. Robocoaster, the world's first passenger-carrying robot, is a popular attraction at the Legoland Park in Denmark. The company is currently focusing on a collaborative robotic system, whereby as many as 15 robots can work together, with each receiving information from its 14 workmates.

ABB is one of the world's top four robotics companies. It employs 7,000 people and celebrated 30 years in the business in 2004. It boasts annual sales of $1.5 billion and belongs to an automation and energy group that has 110,00 staff worldwide. In 2002 ABB shipped its 100,000th robot.

Stäubli was originally a textile company when it was founded in 1892. It became a force in robotics when it acquired Unimation in the early 1980s. Since 1993, Stäubli Robotics has designed and marketed its own robots, earning a reputation for high-precision machines. It likes to say it paid dividends for its parent company when it directed its expertise to the textile industry in the shape of a Jacquard loom, the Unival 100, which incorporates thousands of motors and offers previously unimagined power and flexibility.

Applications galore

The days when the automobile industry supplied the overwhelming bulk of demand for industrial robots are over. As Jean-François Germain of KUKA France points out: "Interestingly, robots have now entered every branch of industry!" They perform an enormously wide spectrum of tasks: from ball-throwers that test tennis, golf and baseball equipment to sewing machines that stitch up jobs in a fraction of the time it took the seamstresses of yore. And there is the occasional weirdo, like the great insect clasped to the glass of the Louvre Museum Pyramid, in Paris, which causes passersby to stop and gape. It is in fact a window-cleaning robot developed by the company Robosoft.

And potential applications are, indeed, mind-boggling. In 2000 ABB conducted a survey that concluded that of some 900 sectors where robots could usefully be employed, they were operating in only twenty. There is scope for the future.

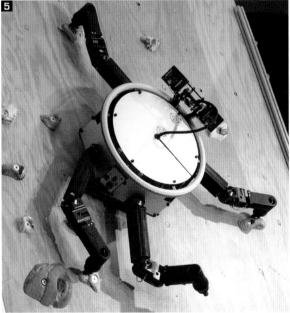

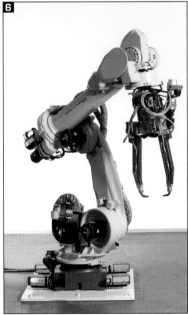

*1- Robuglass.
A robot cleans
large glazed areas.
Remote-controlled
or stand-alone, it
makes work safe
and can reach
otherwise
inaccessible spots.*
© Courtesy of
Robosoft

*2- Surface
treatment robotic
work cell.*
© Courtesy of INSYS

3- Robot HRP-2
© Courtesy of AIST

4- Foundry robot.
© Courtesy of ABB

*5- The Climber
robot can do
anything a human
rock- or mountain-
climber can. It can
scale any climbing
surface, including
the smoothest
and sheerest.
The research
team of Professor
Jean-Claude
Latombe is
working on a
motion-predictive
algorithm that will
enable the robot
to climb heights
autonomously
without risk.*
© Courtesy of
Jean Claude Latombe

6 - FANUC robot.
© Courtesy of FANUC

IRB-6400 welding
robots from ABB.
© Courtesy of ABB.
Photo: SEMA Group

The automobile

Its share may have shrunk but the automobile industry is still the world's largest single user of robots. French magazine *Jautomatise* inventoried industrial robots at work in France and found that out of a total of 35,495 robots in manufacturing facilities at the end of 2003, automaking plants accounted for 15,298 and their equipment suppliers for 6,413.

FANUC, for example, has 2,200 robots operating in plants belonging to PSA Peugeot-Citroën, which includes models produced by ABB and KUKA. "There is a mix of generations, models and manufacturers," says Erika Louis-Roy of the communications department at PSA Peugeot-

Citroën. She adds that robots' tasks fall into four main categories: **stamping, body-in-white, painting and assembly** (*see following pages*).

Automobile assembly lines where robots are at work are still the most impressive to see, says Herman Verbrugge: "Hundreds of robots work on assembling a car's body. They intervene one after the other, and each one does its share of the manipulating." Verbrugge adds that a major evolution has been task-sharing among several robots, which work together to produce the finished product. "To weld a tailpipe, for example, one robot holds the unwelded parts and revolves them, while its teammate does the actual welding. It's a practice that brings significant savings."

• **Automated stamping** involves unstacking, handling, and feeding parts into the stamping dies, then stacking the stamped parts in containers ready for transportation.

• **Body-in-white assembly** involves robots unstacking the stamped panels, loading the workstation, and fitting the panels, then manipulating the fitted parts or subassemblies and applying mastic or structural glue.

• **Painting robots** are allotted the task of spraying PVC under the bodies, applying interior sealing, and spray-painting.

• **Assembly robots** fit the glazed parts (windscreen and rear window) and the dashboard.

Stamping

Unstacking robots on line 110 at the PSA Peugeot Citroën plant at Poissy outside Paris.

© Courtesy of PSA Peugeot Citroën, Communications department

Body-in-white

1- Welding robots work on body-in-white parts.
The ABB-6004 robot with electric welding application.
2- ABB-6400 robots electrically spot-welding
a Peugeot 206 body-in-white.
3- ABB 2400 robot.
© Courtesy of PSA Peugeot Citroën, Communications department

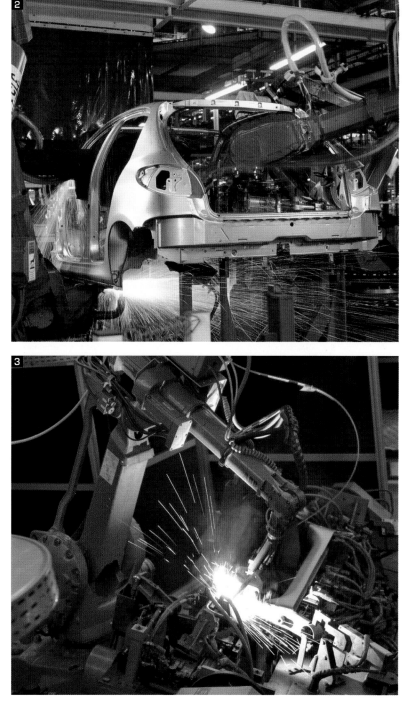

Painting

1- DURR robots:
ESTA application of water-based paints.
2- Three FANUC robots:
applying water-based paints with pneumatic spray.

© Courtesy of PSA Peugeot Citroën,
Communications department

Assembly

1- Robots gluing and fitting a windscreen on a Citroën C3.
2- Robots on the cylinder head preparation line in the assembly shop. Their potential rate is 2,450 engines per day in three eight-hour shifts.

The food business

There are growing numbers of applications for industrial robots in the food business. They package and stack onto pallets the prepackaged food, from ham and cheese to green salad, that consumers buy in supermarkets. They can manipulate great 40- to 50-kilogram (88- to 110-pound) wheels of cheese, then slice the same cheese into slices and cubes. They handle bottles of soft drinks, vintage wines and Champagne without breakage. And there are robots that carry out the less enviable tasks of breaking up pig carcasses and handling frozen foodstuffs in deep-freeze rooms.

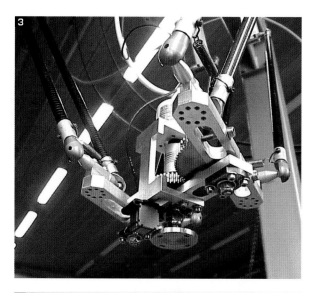

1- FlexPicker 340 robot.
© Courtesy of ABB

2- ABB's picking robot has a maximum payload of one kilogram (2.2 pounds).
© Courtesy of ABB - Photo: Louise Alexen

3- LIRMM's pick-and-place robot, H4.
© Courtesy of CNRS Photo library

4- Robots handling milk cartons in Denmark.
© Courtesy of ABB

5- Robots handling bottles of Champagne.
© Courtesy of ABB

6- FLexPicker 240 at work.
© Courtesy of ABB

Nuclear

Despite the efforts of the authorities and utility companies themselves to allay fears, nuclear power worries many people. Little does it matter that draconian protection and safety measures have been taken, incidents can, and do, happen. Although they are few and far between, nobody wants to be the unlucky one.

If there is one industry where robots have serious career prospects it is in the nuclear fuel industry. In nuclear-dependent France, the country's Atomic Energy Commission (CEA) is developing a robot whose electronics can withstand radiation exposure levels that are 10,000 times higher than the authorized human threshold. As it strolls about its nuclear facility, it scatters radio transmitters in its wake, like Tom Thumb leaving a trail of pebbles, to ensure it remains in contact with its outside handler at all times. It can also carry items weighing up to five kilos.

One way of disposing of radioactive waste is to vitrify it and place it in airtight canisters that are 1.4 meters (56 inches) in length, 50 centimeters (20 inches) in diameter, and weigh 500 kilos (half a ton). The canisters are then buried underground. The procedure is, gener-

ally, to lower the canisters down a vertical borehole until it reaches clay, from where it is then transferred into a main tunnel, and thence to a storage tunnel. The French government agency in charge of radioactive waste and enriched fissile materials is developing the prototype of a mobile robot that could move the canisters from the main tunnel to their places of storage.

1- *Eros robot.*
© *Courtesy of Cybernetix*

2- 3- *Cybernetix's robot Menhir, which can negotiate obstacles.*
© *Courtesy of Cybernetix*

4- *The Koala olfactory-perception platform for detecting hazardous emissions, e.g., gas leaks, radioactive seepage.*
© *Courtesy of CNRS photo library and INRIA. Photo: R. Lamoureux*

5- *Remote-controlling a robot that handles radioactive products.*
© *Courtesy of Cybernetix*

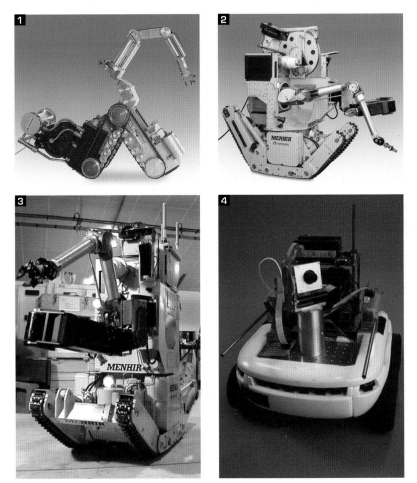

Farming

Though normally associated with industrial environments, robots are moving into agriculture. There is no lack of applications, from gauging when crops are thirsty and watering them, to spotting diseased or infested plants. Some 100 French farmers are said to use milking robots, while in Japan researcher Yoshi Nagasaka has developed rice-planting bots adapted to the inhospitable terrain of his country's rice paddies. Vip Romper is a vision-guided robotic vehicle with a grab arm, which could take over the thankless task of melon-picking in Israel. It is the result of a joint venture between Purdue University in the US, Israel's Bet Dagan Institute of Agricultural Engineering and the Weizmann Institute of Science in Rehovot. Robotics designer Professor Edan hopes to see Vip Romper eventually picking melons at a rate of two every three seconds.

There is ongoing research into robots that can spot and pull weeds and apply weedkiller. Agricultural engineers at the University of Illinois in the US have developed a small robot, dubbed AgAnt, designed to work in droves. When fully operative they will patrol cornfields and notify each other when they come across weeds or detect something amiss. Professor Tony Grift, coinventor of the AgAnt, and other "agbots", says that one advantage could lie in precision application of pesticides. In conventional spraying methods, much pesticide is swept away by the wind and wasted. An agbot, however, could accurately squirt the right amount on the right plant.

An original idea has surfaced from the waters of a fish farm in Louisiana that had been losing much of its stock to pelicans. The response of Louisiana State University was to design a solar-powered pelican-repelling robot. It paddles across the lake and, when it spots any pelicans, it either rams them at a speed of five km/h (three mph) or squirts them with its water cannon.

1- SlugBot, the slug-hunting robot designed by Ian Kelly of the University of the West of England. It is powered by electricity derived from the biogas given off by dead slugs as they putrefy.
© Courtesy of University of the West of England

2 - AgAnt works in droves, hunting out weeds and calling in its fellows to deal with them.
© Courtesy of Tony Grift

3- AgBot the harvesting robot, forerunner of future farms. Picture left, Tony Grift of the Department of Agricultural and Biological Engineering, University of Illinois, Urbana-Champaign.
© Courtesy of Tony Grift

4- Vip Romper, the melon-picking robot.
© Courtesy of Yael Edan

Aviation

The aviation industry is a major robotics user. Specifications are far more demanding than in the automobile industry, because parts are much bigger, much heavier, and more high-precision. Robots perform important tasks in the manufacture of the Airbus A380, boring rivet holes and fastening rivets securely and machining numerous parts. Similarly, laser-guided robotic work cells assemble aircraft engine metal parts.

Logistics

With all the freight and shipping it involves, the logistics industry offers handling robots plenty of scope. Jean-François Germain makes the point that "when robots put luggage in a plane's hold there are no security worries because humans don't intervene". The pulp and printing industries also appreciate these indefatigable workers with their unlimited capacity for carrying heavy loads of paper. Robots in warehouses also stack pallets with products for delivery to supermarkets.

1- The MB-835 platform for transporting light loads (50 kilograms, or 110 pounds).
© Courtesy of BlueBotics.

2- Robotic luggage-handler.
© Courtesy of ABB

3- PowerBots handle and carry small loads.
© Courtesy of MobileRobots.com

4- Pallet-stacking robot IRB-640.
© Courtesy of ABB
Photo: Sema Group

4

Automated labs

Bio Robot 3000.
© Courtesy of Qiagen Inc.

The perfume, cosmetics and pharmaceutical businesses have become industries with production rates and volumes that make it profitable to use robots—even at the luxury end of the market, *chez* Yves Saint Laurent. In 2000 the laboratory automation market in the US alone was estimated to be worth $1.2 billion. Robotic technology speeds up research, improves hygiene, rules out human error and makes it easier to track samples.

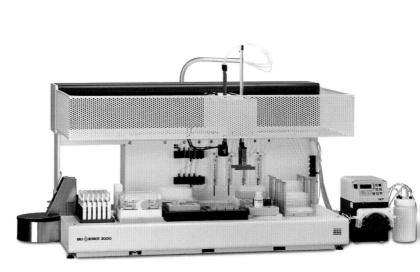

Oil industry

Underwater robots carry out a range of tasks useful to oil companies, like reconnoitering reefs and supervising off-shore drilling. They have also responded to a growing concern in the industry over unseaworthy tankers.

Equipped with high-precision vision systems that can construct 3D models of what they view, underwater robots inspect and analyze the conditions of vessels' hulls.

One such ocean-going bot is Aqua, a strange turtlelike machine that swims with the help of six yellow flippers and can walk

on the ocean floor. It is the work of a joint Canadian research project bringing together the universities of Dahlousie (Nova Scotia), York (Ontario) and McGill (Quebec). Still under development, Aqua can locate its bearings exactly thanks to an acoustic sensor module.

1- *Octopus robot.*
2- *Spider.*
3- *Swimmer.*
4- *Robot Alive.*

© Courtesy of
de Cybernetix

Compact, programmable and able to see

The industrial robotics market is growing fast and undergoing a profound change, driven by a welter of new technologies. Machines are becoming miniature, modular, easier to assemble and program, and increasingly more autonomous.

Compact and modular

Early robots were content to repeat exactly the same precisely programmed and parametered motions through exactly the same coordinates ad infinitum. Today, though, the trend in robotics is towards flexibility. Increasingly machines will be expected to vary their movements to match any variations in their task, either when instructed to do so (e.g., by touching a tactile screen), or because they themselves sense it is necessary. A welcome development is that robots are shrinking into more compact packages. At the same time, they have become modular and reprogram-

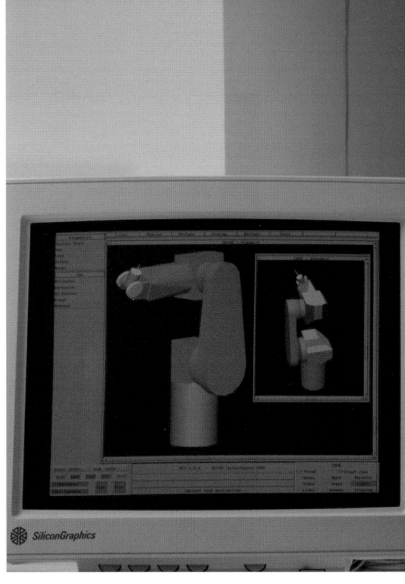

mable and, consequently, able to incorporate newly evolving developments. Graphical interfaces that involve merely pointing and clicking have enormously simplified programming. Automated assembly work has been rationalized, too, with industrial robots operating in mutually complementary work cells. When designing cells, engineers make use of simulation software to optimize effi-

ciency and production times. Where once it took between five and seven years to develop and put in place an automobile production line, it can now be done, believes Toyota, in a single year.

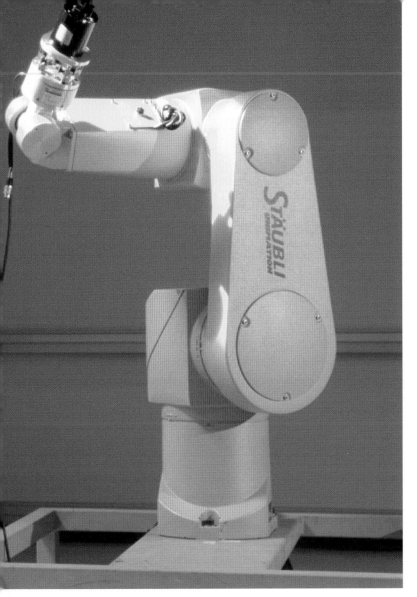

Growing small

Electronics- and computer-dependent machines like cell phones and digital devices are shrinking fast as their components are miniaturized. Hard disks are a good example and Intel plans to be manufacturing microchips with the 65-nanometer process by 2005. Yet it can only make such plans because the miniaturization of robotic tools has led to production units able to manipulate components that would be way too tiny for humans to handle. A research team at the LAB, in the eastern French city of Besançon, is reported to have produced the tiniest mobile robot in existence. The microbot is 3.5 millimeters long (0.13 inch), 6 wide (0.23), and 0.5 "high" (0.02). Able to carry loads that are up to 2,000 times greater than its own weight, it is ideally cut out for exploring pipes and tubes.

Above: Handling robot.
© Courtesy of INRIA Photo Library
Photo: INRIA - A. Eidelman

Below: Microbots produced by the LAB
for cooperative carrying tasks.
© Courtesy of CNRS Photo Library
Photo: Jérome Chatin

3D vision

A major evolution in industrial robots' faculties has come as falling costs have led them to integrate 3D video cameras and radar systems. If a robot is welding a part that then moves, it can follow it and continue welding exactly the right spot. Coupled with artificial intelligence, which enables robots to make decisions by themselves, and with auditory systems, developments in vision technology have opened up whole new vistas, like the development of autonomous vehicles.

One company that produces systems for improved-acuity 3D vision is California-based Canesta Inc., which has patented a dozen perception technologies. One of its creations is its perception chip Equinox, a sensor that enables machines to see in 3D, resolving objects from their background and gauging distance, height and shape. A conventional camera has trouble making out the shape of a driver sitting in the seat of his or her car, and if the driver's clothes and seat cover are of a similar color, it cannot tell them apart. Which is where a chip like Equinox comes into its own. A robot with an embedded sensor could locate the driver and

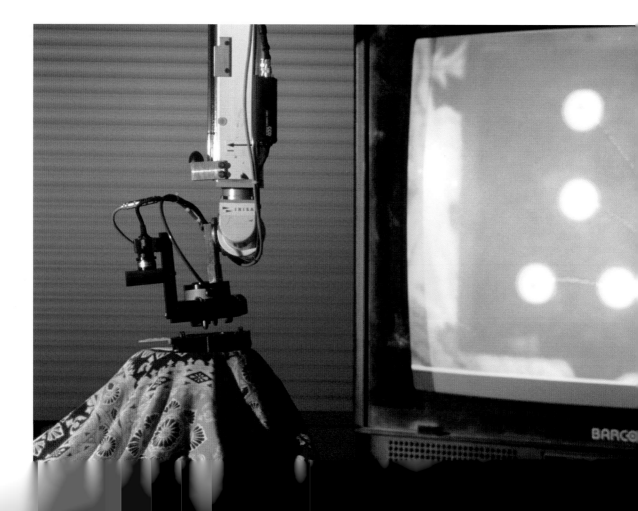

gauge his or her height and shape. And, as Canesta's Jim Spare points out, a sorting robot could grab a packet and toss it precisely where it was supposed to go.

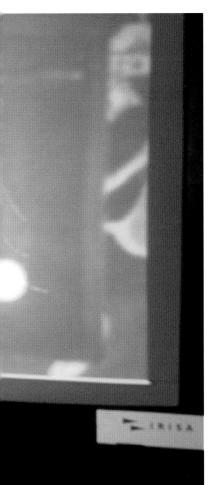

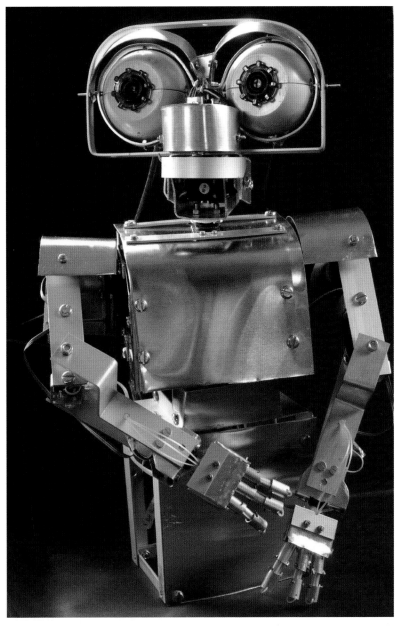

Above: *Rodney, a robot developed by Bob Mottram to test stereoscopic vision and human–robot interaction.*
© Courtesy of Bob Mottram

Left: *Servo-controlled vision system enables grabbing.*
© Courtesy of INRIA - Photo Library
Photo: INRIA - A. Eidelman

Other applications

Force feedback applications

Haptic technology is another area that has opened up new possibilities. "Haptics" is a Greek word meaning the science of touch, and applied to computer applications it refers to interactive tactile sensation and control. Devices with haptic interfaces respond to user instructions or requests with feedback that takes the form of a felt sensa-tion, or force, usually on the hand. Working through a robot, a user can handle a fragile or miniature object, feeling what they are doing through the feed-back of force that is much greater, though proportional, to the force actually being exerted by the robot. It is also possible to touch and manipulate virtual objects viewed in 3D. The com-panies leading the way in haptic applications are SensAble Tech-nologies and Immersion Inc.

Production on demand

Automated supervision and modification of production-line output to meet rises and falls in demand is an evolution that cuts costs and brings reactivity gains. Data is collected in real time and transmitted over the Internet to the supply chain. Adept Technol-ogy's executive vice president Charlie Duncheon believes that businesses that invest in such systems will have a telling edge over their competitors.

Doing it together

Herman Verbrugge believes that in the near future human beings and robots will be able to work together on the same tasks. "Some robots already operate side by side with people on assembly lines, but humans never infringe on robots' working space. That could soon change, with humans and mobile robots not only sharing the same space but combining on the same job. Of course, that would call for safety precautions." Currently under development is a project to automate construction work in Japan. Three hundred prototype robots have been developed, one-third of which are reported to be in operations. Similar trials are being conducted in Germany. Verbrugge adds that the development of speech technology will make man–machine understanding simpler: "You will just have to tell a machine what to do and the robot will do it. It'll know itself where to go."

Left: Artificial hand controlled by force feedback data glove.
© Courtesy of CNRS Photo Library
Photo: Martine Voyeux

Right: Virtual desktop with force feedback.
© Courtesy of
INRIA - Photo Library.
Photo: INRIA - G. Pigot

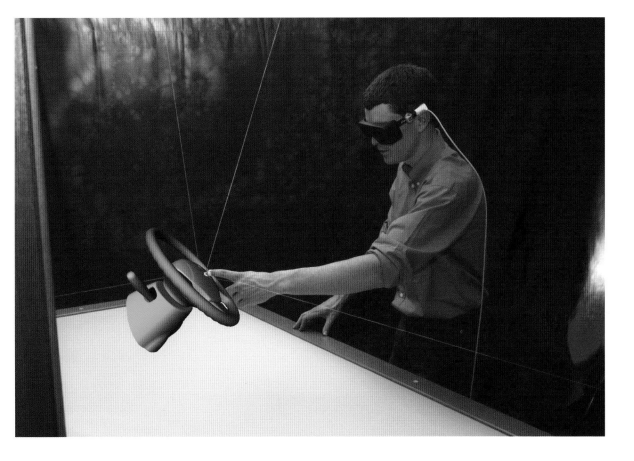

Autonomous vehicles on the road

Robotic applications in the auto-mobile industry have widened their scope to take in smart cars, capable of correcting driver error and even taking con-trol of the wheel. It was a pre-dictable development, for elec-tronics account for a growing share of vehicle equipment and now exceed 20 percent of the total cost. The current trend among many automakers is towards monitoring drivers' behavior and relieving them of some strain. Toyota's Prius de Toyota, for example, can park itself without its driver even hav-ing to touch the steering wheel, while researchers at Volkswagen have designed a robotic driver. It has four arms—one for the igni-tion, two for the steering wheel and one for the gear lever—and three legs for the pedals.

Honda has designed an entirely autonomous miniature helicopter controlled by artificial intelligence and able to sense and avoid any obstacle in its path. Furthermore, it constantly enriches its knowledge base as it travels, comparing what it detects with what it knows. Research of this kind is designed to enhance efforts to make safer cars that can self-drive in envi-ronments teeming with obsta-cles and obstructions.

Among some of the more original ventures is the Segway Human Transporter (HT) jointly developed by France's Keolis and US company Segway LLC. Described as a self-balancing mobility device, it operates according to a fun principle. If its passenger stands up straight, it does not move; if he or she leans forward, it advances, while a backward lean sends it into reverse. Human bodily move-ment inspired the design and engineering of the HT. When we lean forward, we instinctively put one foot forward to keep our balance, which ultimately leads to walking. The HT travels at 12 km/h (7.5 mph)—about the speed of a jogger—and has a range of 25 kilometers (15 miles). Its drive system brings it slowly to a halt at the slightest hint of trouble.

RobuCAB and RobuRIDE, designed by the French company Robosoft, are interesting exam-ples of alternatives to current mainstream modes of trans-portation. The former takes indi-vidual passengers wherever they want to go, while the latter, like a bus, carries groups of people along preprogrammed routes. Both solutions would be ideally suited to ferrying people between different points on widely spread sites like university campuses, amusement parks, airports and industrial zones. RobuCAB and RobuRIDE are both smart vehi-cles, processing information about their route gathered by their sensors (e.g., men at work, traffic smooth) and adapting accordingly so as to optimize travel time.

1- Model of Cycab.
2- Autonomous mobile robot on the road.
Courtesy of INRIA - Photo Library.
Photos: INRIA - A. Eidelman

3 - Cycab.
4- Simplet, experimental robot.
© Courtesy of INRIA - CNRS photo library.
Photos: R. Lamoureux

5- Atchoum and Simplet, experimental mobile robots.
© Courtesy of INRIA - CNRS photo library.
Photo: R. Lamoureux

Hi-tech highways in Japan

What about sitting back and letting your car and the road do the driving? Just such a proposal is on government policy agendas worldwide and has given rise to some large-scale programs. However, different cultures favor different approaches, as we shall see.

Smart Cruise 21 involves a small robotic car in a big Japanese project. The car itself, the Advanced Safety Vehicle (ASV), drives along a highway lined with an assist system, made up of countless sensors, magnetic markers, warning devices and beacons that communicate with the ASV. In this way the car is fed a constant flow of road and traffic information. Should it receive critical news warning of danger ahead, it passes it on to the driver. "In Japan driver behavior is considered linear. A driver makes a decision then acts on it," says Jean-Marc Blosseville, who heads the French road safety program, ARCOS. It is an approach that reveals just how much trust the Japanese place in robotics and automated assist systems. "If the driver fails to act, he is considered to have made the wrong decision and the system warns him," adds Blosseville. "If danger is imminent, the car takes over in his place."

What a robotic vehicle can do

Seven sides of the Japanese project Smart Cruise 21:

1- Provides information on road surfaces.

2- Prevents collisions.

3- Prevents high cornering speeds.

4- Keeps in lane.

5- Prevents collisions at intersections.

6- Prevents collisions with pedestrians crossing the road.

7- Prevents collisions on turning left at intersections.

2

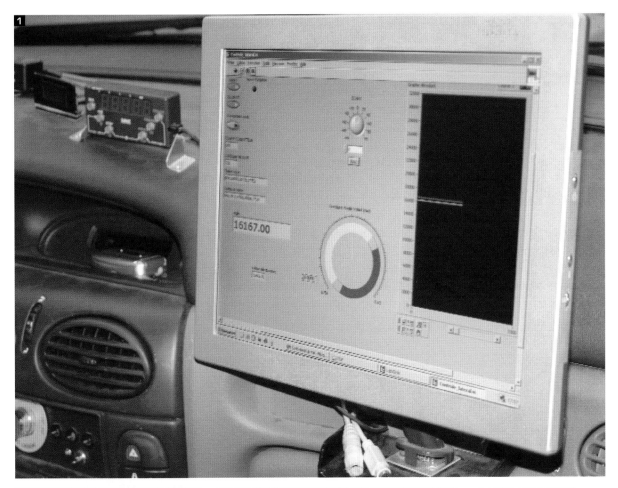

1- LIVIC vehicle control monitor. LIVIC aims to improve road traffic by enhancing driving-aid technologies.
Photo: FYP

2- 3- 4- Cycab.
© Courtesy of INRIA Photo Library

US and Europe
stress safety

Both the US and Europe are too big to follow the Japanese example of placing sensors every 200 meters (650 feet) along their roads. A US project, the Intelligent Vehicle Initiative, puts the onus on the car to prevent collisions and assist drivers.

The French road safety research program ARCOS is headed by Jean-Marc Blosseville, an engineer from LIVIC, a laboratory that researches how drivers, vehicles and roads interact. ARCOS vehicles are fitted with 360-degree vision systems, radars and computers that combine to anticipate high-risk driver behavior. A range of solutions has been implemented, such as an alarm telling the driver specifically that he is going into a bend too fast. "It's important to be specific," says Blosseville, "otherwise the driver will think the warning is pointless."

ARCOS aims to equip vehicles with four safety functions to:
1- **Warn of accidents/ incidents ahead.**
2- **Maintain safe following distances.**
3- **Prevent collisions with stationary obstacles.**
4- **Prevent going off the road.**

The LAVIA Project
Tested by LIVIC and cofunded by the French Transport Ministry, LAVIA is a project to limit speeds according to locality. Two speed-limiting systems have been tried out by a score of Renault and PSA vehicles. One system merely warns motorists, while the other stiffens the accelerator. Although the long-term aim of ARCOS is an autonomous system, another project has a more

short-term time frame. Called Stop-and-Go (S&G), it offers motorists the option of letting the car take full control of driving. Thus in a nose-to-tail logjam, a driver could hand over control to the car and get into a thriller while his or her vehicle handles all the infuriating maneuvers at a snail's speed.

Speed limitation is the leitmotif of other systems across Europe. Some are extremely advanced, like Sweden's Intelligent Speed Adaptation (ISA), which has been installed in 7,000 vehicles. Also ongoing is the Road of the Future in the Netherlands, the Cooperative Vehicle Highway System in the UK, while Germany is running a scheme whereby motorists in front transmit information to those behind.

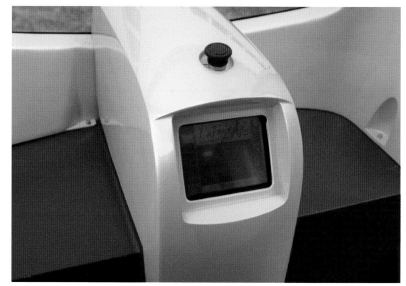

Information terminal
in the Cycab cabin.

The moot point in European projects is how to determine just when vehicle-integrated safety systems should take over from drivers. Although the obvious answer is: when human life is in danger, the prospect of a robotic system having ultimate control gets a mixed reception, as Jean-Marc Blosseville underlines: "The idea is to give full rein to driver responsibility, right until the very last moment when he has the ability to act. Then, and only then, does the robot kick in and take control. The Japanese might find that acceptable, but in Europe there is reluctance. Our surveys have shown that drivers want control of their vehicles to the bitter end. The thing to do is to find a way of improving the image of systems that take over from drivers."

Testing and developing the Cycab.
© Courtesy of CNRS Photo Library.
Photo: Laurence Menard.

Automated road transport by 2040

Although it has set a 40-year time frame, France is considering making all road traffic fully automated. The decision has been prompted by the enormous rise in road transportation loads, which projections estimate will increase, in tons per kilometer, by between 80 and 100 percent over the next 20 years. One solution currently entertained is to build an automatic highway exclusively for trucks. It would stretch 1,120 kilometers (690 miles) from France's northernmost tip, at Calais, to Bayonne, in the far southwest. The cost has been put at $8.19 billion.

1- 2- Robotic petrol pumps for trucks and automobiles.
© Courtesy of Robosoft.

3- RobuRide.
© Courtesy of Robosoft.

Are industrial robots job robbers?

Hackles often rise at the mere mention of industrial robots because of the perception that they are to blame for large numbers of manufacturing job losses. There is no denying that some company announcements can strike fear into the most rational hearts. A Reuters wire in August 2003 suggested that McDonald's was trying out a Swedish robot that could prepare Big Macs. As construction sites are gradually automated, work once done by unskilled labor will disappear, which will add a few more to the jobless queues. But could we see anything akin to the 1811 riots by British textile workers in the north of England who smashed the industrial looms they accused of jeopardizing their livelihood?

According to Philippe Coiffet of France's scientific research council, CNRS, it is unlikely. Even to see robots as a threat to jobs, he says, is mistaken. "The countries with the most robots are also the ones with lowest unemployment," he asserts. Coiffet thinks that "robotization" will reshape the job market, creating higher numbers of skilled jobs. To back up his belief he likes to quote the example of Cemo Lunetterie, a company in eastern France that makes eyeglasses. Faced with growing competition from Asia, management turned to the automation research laboratory, the LAB, in the nearby city of Besançon, for ways of modernizing its factory. In March 2002 Cemo Lunetterie commissioned its new automated welding unit for steel and titanium frames. It did the job in a quarter of the time previously needed. The change in production methods helped the company stay on in the region, and the increase in business led to the recruiting of personnel.

Jean-François Germain of KUKA France shares Coiffet's view. "Industrial robots offer a real alternative to relocation," he says. "We have several examples in France, but I'll the take the case of PBL, a world leader in lawn-mower blades. It has managed to stay on in France because, in addition to its personnel and their know-how, the company drafted robots into its production process."

Herman Verbrugge reaffirms adamantly that automation has enabled French automakers, Renault and Peugeot to reach levels of expertise at prices that have made them competitive on the world market. He believes that the introduction of robots cut costs, which now allows Renault to recruit and train 5,000 new personnel and Peugeot 2,000. "At first the fear was that a robot would take one job, then two, then three. In fact, robots replace people who do the menial jobs that fatigue and affect the health and that's a good thing," argues Verbrugge. "However, there is no such thing as a robot that can do without servicing, programming and repairs. Those are needs that require qualifications and training. They yield much more interesting, more challenging jobs that engage human intelligence."

Interview
Brian Carlisle

Brian Carlisle is an industrial-robotics designer with a string of successes to his name. He was in the Unimation team that produced the ground-breaking PUMA. In 1983 he and his partner, Bruce Shimano, founded robot vendor Adept Technology, winning respect for their innovative approach. The pair have launched a new company, Precise Automation.

Brian Carlisle
© Courtesy of Brian Carlisle

When did you first get interested in robots?

I started getting interested in robots when I was studying mechanical and electrical engineering at Stanford. I was particularly fascinated by servomechanisms. My tutor, Professor Bernie Roth, introduced me to Vic Scheinman and Bruce Shimano, who had also been his students. Vic was starting up his robotics company, which he called Vicarm, to commercialize the many robot designs he'd done during his time at Stanford and MIT. Bruce developed the software for Vicarm's robots. After I graduated in 1975, I joined them to work on the controller and mechanical engineering side of things.

How did you come to found Adept Technology ?

At Vicarm we'd shipped a robot to General Motors that they wanted to use in their research laboratories. GM liked the robot and invited bids to make an assembly robot called PUMA. It stood for Programmable Universal Machine for Assembly. We sold Vicarm to Unimation so we could respond to GM's demand and became the Unimation West Coast Research Group. We started developing PUMA in 1977 and delivered the first prototype to GM in 1978. It was the first industrial assembly robot and it was manufactured in the thousands.

In 1983 Unimation was sold to Westinghouse and the management team was broken up in the restructuring that ensued. It took Bruce Shimano and me about six months before we realized that Westinghouse couldn't keep up the pace in robotics. So we left and launched Adept.

What was it that made your company so innovative at the time?

When we launched Adept we decided to focus our strategy on small-parts assembly and material handling. When we designed PUMA with General Motors a survey had disclosed that 95 percent of the parts that make up a car weigh less than 1.5 kilograms (0.93 miles).

We felt small-parts handling was a good market opening because in 1983 most US robotics companies concentrated on spot welding, arc welding, and painting. Only the Japanese sold assembly robots. To compete against the Japanese our strategy was twofold. We tried to offer the best technology and the best customer support around. Japanese robots had limited precision and their controllers were simplistic. And there was practically no such thing as customer care in the US or Europe. So we introduced the first direct-drive assembly robot with a sophisticated controller and machine vision. We set up a global network of systems integrators and technical centers with engineers to help customers implement their robots.

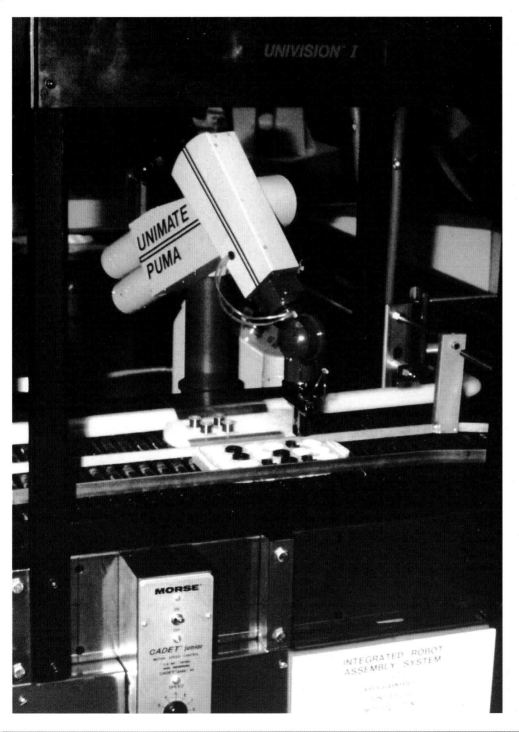

Unimation's second PUMA, the 260 model, developed by Brian Carlisle and Bruce Shimano. The Unimate PUMA 260 resembled the small robot that Vic Scheinman designed and built in 1974, the Model MIT Arm. It had a payload of one kilogram. The photograph, taken in 1981, shows that it was the first assembly robot to be equipped with vision guidance. Its robotic vision system was adapted and developed by Dr. Scott Roth from technology developed by SRI International. Dr. Scott Roth has joined Bruce Shimano and Brian Carlisle at their new company, Precise Automation.

© Courtesy of
Brian Carlisle

... Interview cont.

In the US and Europe we led the market within a few years. We didn't make any special effort to invest in Japan because so many companies there made their own robots in-house.

In the 1980s Japan gradually came to dominate the robotics market. How do you account for the Japanese success given that the innovation came from the US?
It's the same old story. It happened in the 1960s, 1970s and 1980s. The underlying reason is that US businesses were more interested in product development than anything else. They didn't think manufacturing was a very important activity, they felt that anybody could do it. The Japanese, though, sensed that manufacturing was a long-term competitive sector. That perception prompted a lot of Japanese companies to make long-term capital investment in new manufacturing technologies, which the US didn't do. The result was that the robotics market grew very fast in Japan. It reached between four and ten times the size of the US market, depending on how you define a robot. And because they had a bigger domestic market and because they set more value on manufacturing technologies, Japanese companies

AdeptOne.
© Courtesy of Brian Carlisle

recorded higher sales and had cheaper access to capital than American businesses. When we founded Adept in 1983, we were competing with 34 American companies. By the end of the 1990s, Adept was the only one still in the running. I believe our success was because we set store by innovation and customer care.

What in your view have been the milestones in the history of industrial robotics?
Obviously, the first was the Unimate in 1961. It was the first robot to operate in a production plant. Then there was the Vicarm in 1974 to 1975, which was the first commercially available

computer-controlled robot. In 1978 the PUMA was the first robot designed for assembly lines. Dr. Makino's SCARA[1] robot was very well configured for a wide range of assembly tasks. The advent of vision systems for vision-guided machines was another key development. We developed machine vision at Unimation in 1980 and made it a standard feature of our products at Adept from 1983. Another major breakthrough was the use of simulation for designing and programming robots. I'd just like to remind you here that, broadly speaking, a robot is a computer equipped with sensors and actuators that can perceive the physical world and take action that impinges on it. With the rapid progress in IT and new technologies, robotics applications will diversify.

What were the reasons behind Adept Technology's decision to part with its two founders in January 2004?
We had taken Adept to sales that exceeded $100 million in 2000. But from then through 2003 capital investment in the electronics industry went into free fall, and that included cell phones, hard disks,

semiconductors, fiberoptics, you name it. Business at Adept fell to around $40 million and we had to make big cutbacks in staff. By the end of 2003 we'd put our house in order and we brought in new investors to inject fresh capital. But the new shareholders and financial directors on the board decided they wanted a change of management. They asked me and Dr. Shimano to resign.

What are the ambitions of your new company, Precise Automation?

We believe there is a clear trend towards miniaturization across a wide range of products. They will soon be too tiny for people to assemble. So we are developing a new generation of controllers, sensors and high-precision robots to handle very small parts. We also think that controllers and machines whose job is to manipulate tiny parts should also be small and cheap. Lots of current high-precision machines are bulky and costly. It's our belief that there is a market for low-cost, high-precision robots that manipulate and assemble.

What do you say to those people who argue that robots cause unemployment?

Throughout the history of mankind we have always sought to improve our standard of living by producing more and faster. That is true of farming, the steel industry and office work. Robotics is one more technology among many that contributes to higher productivity. There haven't been many job losses because of robots, there are only a few hundred thousand in operation worldwide, compared, for example, to the 200 million people in manufacturing in China alone. The real problem lies with cheap labor in developing countries. Ways have to be found of stopping plant closures and reallocation to other countries. But robot technology will bring many benefits in the future. Whenever I tell someone that I work in robotics, he or she always says: "Could a robot clean my house?"

---

(1) SCARA robots operate on four axes like a human arm. They appeared in the mid-1980s, produced by Japanese vendors like Toshiba, Hitachi and Panasonic. They are widely used in the microchip industry. SCARA stands for Selectively Compliant Articulated Robot Arm.

Interview
Nicola Tomatis

Nicola Tomatis was a researcher at Switzerland's science and engineering university, the Ecole Polytechnique Fédérale de Lausanne, or EPFL, before joining Swiss company BlueBotics in 2001 and becoming its CEO two years later. BlueBotics specializes in mobile, autonomous robotics.

Automation is often perceived as counter to the interests of working people. What do you think?

Let me give you an example from the past. For much of its existence humanity was busy hunting and gathering. Then the primary sector, farming, came into existence, with people turning to crop-growing and husbandry. In large areas of the world the production of food occupied the vast majority of people. Then came the Industrial Revolution in Great Britain in the late 18th century. It was largely due to an increase in agricultural productivity. And that, in turn, was the result, on one hand, of better understanding and implementation of farming methods and, on the other hand, the use of mechanical systems, which were the first examples of automation. On that basis

Nicola Tomatis.
© Courtesy of Nicola Tomatis

and for simplicity's sake, you could say that if the tractor hadn't come into existence, then we'd still be out in the fields growing food. In other words, automation drives social evolution. The past clearly shows us that automation improves our lives. The tertiary sector came into being as a result of developments in manufacturing (and agricultural) production processes. You could even say that automation creates free time. Temporarily, poor use might be made of this free time and there may be unemployment, but that's all part of the evolutionary process. In actual fact, job losses are due to poor social organization. If there is going to be automation, there has to be work to automate.

It should therefore be possible to envisage retraining and reemployment for the workers who will lose their jobs, so as to minimize and even cancel out the time they spend not working. That's not easy to do, of course. What's more, investment in training and research has to be factored in. But what does emerge is that each person can contribute less time to making his or her labor available to society without diminishing that contribution. That means more free time for the same salary.

Can you cite a concrete example of successful coexistence between robot and personnel in a company?

I can give you the example of the automation of logistics in hospitals. In hospitals everything is centrally managed—from the pharmacy, to the kitchen, the laundry and waste disposal. So personnel spend much of their time pushing trolleys around on distributing tasks. The work is shared by personnel in the centralized departments and by nursing staff. It brings no added value. But by automating distribution through the introduction of self-propelled trolleys and thinking in terms of a three- to five-year return on investment, the freeing-up of nursing personnel for their caring duties should bring value added within a very reasonable time frame.

In Europe heavy industry has shrunk sharply due to things like relocation and the closure of steelworks. Has the robotics industry been affected by these changes?

I think it is important to distinguish between industrial robotics and service robotics. The most common form of industrial installation is the robotic manipulator, the most obvious example being welding and painting arms in the automobile industry.

Service robotics is defined as those applications that serve human needs, for example transportation, cleaning, and surveillance. There was definitely a slow-down in the growth of industrial robotics in Europe, as the report published by the World Robotics Survey in 2003 showed. But the picture is very different in service robotics, particularly transport. Transport logistics is experiencing growth, with a real boom in light-load systems. Robotics is adjusting to the new developments.

Who benefits from the growth in service robotics?

Chiefly small robot vendors. The giant corporations like ABB,

KUKA and Staübli are not interested in service robotics. They stick to their industrial automation applications and ignore the needs of emerging markets. The result is that small companies looking for niches outside conventional industrial robotics are shaping and driving the service trend.

What prompted a research engineer like yourself to join a company?

Because I felt it was time for some fun. Seriously, though, I'd spent years in research and I wanted to get out of the theoretical side and into the hands-on side. You should remember the robotics research community is very big and it's getting bigger. But applied robotics, particularly applied mobile robotics, has developed very little, apart from vacuum cleaners and minesweeping.

Doesn't the mushrooming of computer- and IT-related jobs and company departments make it easy to introduce robots into companies?

I'm not sure. I graduated as a computer engineer from a pretty good school, the Zurich Federal Polytechnical School, but when I started my doctoral work at Lausanne, I realized just how complex robotics was. At any

one time you might come across a mechanical problem, or one related to electronics or computer software. And you have to have the same cross-disciplinary approach when you design a new product. Robotics is so complex that it should be treated as a field of study in its own right, and a growing number of universities are doing just that.

How long does it take to develop a robot like your RoboX, which can interact with people, hold conversations and play music?

A team of ten people worked on the project for one and a half years. We worked in partnership with the EPFL. We had to finish RoboX in time for the Robotics Exhibition at the Swiss trade fair, Expo.02.

Why was RoboX a wheeled robot rather than a biped?
Trends are very geographically based. In Japan they want to make machines that are more and more like people. Honda's Asimo, Sony's Qrio and Toyota's trumpet player are all typical of what's going on in Japan, where they create humanoids without really thinking about practical applications. Things are very different in Europe and the US, where robots are seen in terms of their practical applications, like localization, mapping and interaction. RoboX is a motor-powered, wheeled robot and so, in terms of the way it moves, it's more like R2D2 than C3PO. Not for the rest, though, because it's a user-friendly, interactive polyglot. I think the West has a rather Asimovian perception of robotics. By that I mean that the resemblance between man and machine should strike a balance between similarity—to foster interaction—and difference—to stress that a robot remains a machine.

What sort of companies buy or lease RoboX?
All the robots that we have sold have been sold to researchers. Companies lease them for trade fairs where they use their hi-tech side to draw attention to their own products. They act as masters of ceremony or guides.

What about safety?
Safety standards for machines that move about among people are very strict. In effect, RoboX uses its sensors primarily to avoid all it senses around it. But machine perception is, of course, very limited. That's why RoboX's body is covered with tactile plates protected by foam rubber. In this way, it can feel contact and respond appropriately, while the foam rubber cushions any impact. What's more, the way in which the robot moves, its speed and obstacle avoidance, etc., are controlled by two separate processors that duplicate each other. So if one breaks down, the other keeps control.

Do you have plans to give your robots human-looking skin?
We're not going in that direction, although some projects are. From the point of view of man-machine interaction, the skin's perceptive capacity brings advantages. But if you think of human skin as something aesthetically pleasing, then that brings us back to geographic and cultural differences. Japan is already making robots that look like people and cladding them with skin. One researcher even replicated his daughter as a robot in his laboratory.

Is robot-programming, which includes programmable production lines, as lucrative a business as building robots?
I think it is vital that production lines can be modified through reprogramming if manufacturing is to remain competitive in industrialized countries where labor is costly. Actually, it's already underway with the design and building of integrated robotic systems that factor in production line reconfigurability. Once that has been achieved, services will play a very important role.

Should we look forward to the day when there is a universal robot programming language?
That will probably happen in education and research, but in industry I'm less sure. The most

purpose-specific, demanding applications are developed for real-time, on-board operation. The most graphic example is probably Wind River's VxWorks, which equipped the Mars exploration rovers and Boeing's planes. It is the benchmark for other consumer applications.

How do you see the android market developing?
It has a promising future in Japan and the East. Most of the early ones were designed to show what their makers, Honda, Sony and Toyota, were capable of. But in Japan there is already a market because it's fashionable to have one's robotic friend. In the future, when controlling androids so that they walk seamlessly and don't overbalance is no longer a problem, we'll start to see them in the rest of the world, where they will be used for practical applications.

Can automated production lines prevent companies relocating?
Today there are two main problems in manufacturing. The first is that many products have shorter lifecycles than in the past. Look at the life expectancy of a cell phone, for example. The second problem is that market competition makes improved productivity mandatory. An automated production line can be a viable alternative to relocation, but only if it combines two factors. The first is lower production costs. It should be cheaper to use robots than to move equipment and products outside Europe. The second requirement is that the production line should be flexible and modifiable, so that it does not defeat the purpose of investment.

ERA-5/1 is a robotic arm specially designed for integration in on-board systems. It weighs only 14 kilograms and has a 24-volt electric power supply.
© *Courtesy of BlueBotics.*

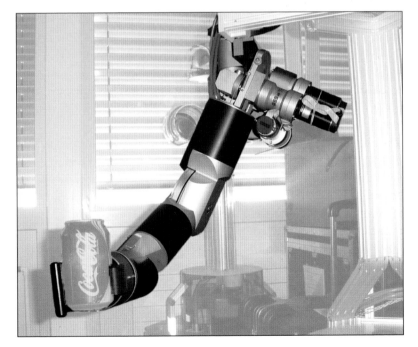

ROBOTS AS EXPLORERS

ROBOTS
AS EXPLORERS

Wherever human beings cannot, will not, or should not tread, there robots go in their stead. Underwater exploration, reconnaissance in mine shafts far underground, clean-up operations in radioactive areas, and, better still, going where no man has been before in outer space, are all missions that come naturally to exploration robots. Where there is ascertained, probable, or possible peril, there goes the robot. Ironically, these man-made machines are often dispatched to repair man-made damage to the environment.

The Apollo 15 Lunar Rover on the Moon.
© Courtesy of NASA/JPL

Robots in space

"Take it nice and easy, step by careful step." That may well have been what the scientists of the Jet Propulsion Laboratory (JPL) were thinking as, on July 4, 1997, they watched the *Sojourner* rover inching its way downward, taking what seemed like forever.

A few hours earlier, the Mars *Pathfinder* probe had touched down softly at its landing site, Ares Vallis. Its three triangular petals had opened like a flower to deploy the craft's instrumentation. Slowly a ramp slid onto a platform that resembled corrugated iron. It was then that the robotic rover *Sojourner*, engaged its descent toward the Martian surface. "What is that outlandish contraption come to trouble our red and tranquil planet?" whispered the winds of Mars.

The diminutive robot had none of the swagger of Terminator, nor the gleaming shell of the impish R2D2. It was a middle-class automaton, a journeyman, a drab little kart scarcely any bigger than a shoebox cobbled together from bits and pieces. Fifteen endless minutes elapsed before it finally stood on the inhospitable face of Mars. Back on Earth millions logged on to the Internet to watch with bated breath as the

unassuming space voyager edged forward with the trepidation of Chaplin teetering on the brink in *Modern Times*.

Anxious to restore credibility after a series of missions that had been as costly as they had been in vain, NASA decided to give *Sojourner*'s first foray onto the red planet the full media treatment. The JPL website ran a live webcast that showed the reconnoitering robot exploring

the hostile terrain ahead of the day when its human creator might also alight there.

On July 8, 1997 there were 80 million hits to the JPL website and *Sojourner* was dubbed with the affectionate nickname 'Rocky.' Amazingly the mini-machine gained levels of popularity that recalled the acclaim for astronauts Armstrong and Aldrin when they set foot on the moon on July 21, 1969.

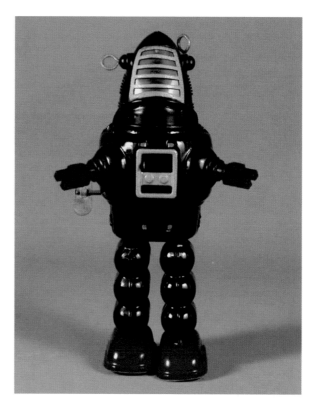

The legendary Robby the Robot from the 1956 movie Forbidden Planet. *Its statement was that man and machine have to collaborate.*
© Jean-Pierre Hartmann's collection of robot toys

It took Boeing and its main subcontractor, Delco Electronics Subdivision, only eighteen months to build the first lunar rover at a cost of $10 million. It was steered by means of a dash-mounted T-shaped joystick.
© Courtesy of NASA Headquarters - Greatest Images of NASA (NASA-HQ-GRIN)

How robots meet man's need for presence in space

The inadequacies of the human body in space

The new heroes of space exploration had to be R2D2's real-life cousins *Sojourner*, *Spirit*, and *Opportunity*. Outer space is an unnatural habitat for man. So hostile is it, indeed, that protective systems of the highest level are essential both for astronauts and the craft that carry them. Although humans went moonwalking as early as 1969, conditions on Mars are so harsh that they are unlikely to set foot there for many a decade. In addition to the ferocious dust storms that sweep its surface, temperatures can drop to 140°C (284°F) below freezing at night. What is more, the actual flight through space from Earth to Mars poses problems. Craft carrying space travelers must be designed and built to stringent, fail-safe standards of protection against solar storms and cosmic rays. The duration of the voyage[1]—20 months for an Earth–Mars round trip—has factored a new, psychological parameter into manned space exploration: astronauts will lose sight of planet Earth. Robots have emerged as the only possible trailblazers.

How important is human presence in space?

After taming almost all the elements on its native planet, it would be galling for humankind to take a back seat while metal upstarts explore space in its stead. Voices are regularly raised asserting that there can be no space exploration without human beings. They argue that it was because the Apollo 9 crew took manual control of the lander module that it touched down and man made history by walking on the moon. They also point to the Apollo 17 mission, where the crew set up experiments and gathered geological samples. And what about the Mars Global Surveyor? In 1997 a solar-panel malfunction prevented the probe's orbit insertion[2], so delaying the main Mars mapping mission by two years. Had an astronaut been on board he or she would have kicked the panel into place, contend the advocates of manned space travel.

(1) Earth is about 50 million kilometers from Mars.

(2) To land on a distant planet a spacecraft must insert itself into an elliptical orbit around the planet before starting its descent orbit.

Four ways in which robots intervene in space exploration

1- As probes that land on a planet and explore its soil. Though stationary, they often have mechanical arms that scoop up soil samples. Their purpose is to prepare the way for human astronauts or autonomous roving robots.

2- As rovers, like *Sojourner*, with veritable built-in mini-laboratories for analyzing a planet's soil and atmosphere. Unlike probes, they are mobile.

3- As troubleshooters. They carry out repair work on spacecraft. Astronauts would otherwise have to venture outside into space at their peril.

4- As assistants to astronauts in routine or hazardous tasks. They also trigger alarms and emergency procedures in the event of danger.

1- The US space probe Odyssey has been mapping Mars since 2002.
© Courtesy of NASA/JPL

2- Robonaut, a droid designed to help the astronauts manning the International Space Station.
© Courtesy of NASA/JPL

3- The lunar rover from the Apollo 15 Moon mission.
© Courtesy of NASA

4- This small free-flying camera is the Autonomous Extravehicular Activity Robotic Camera Sprint.
© Courtesy of NASA

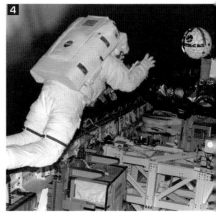

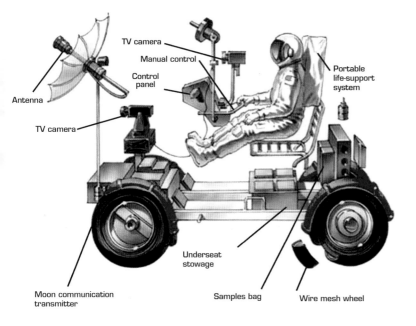

TV camera

Manual control

Control panel

Portable life-support system

Antenna

TV camera

Moon communication transmitter

Underseat stowage

Samples bag

Wire mesh wheel

Rover
A small robotic buggy able to drive over rough terrain to take rock/soil samples and use scientific analysis instruments.

Probe
Unmanned spacecraft whose purpose is to explore space or planets and return photographs and samples to Earth.

Landmarks in robotic space exploration

Year	Country	Mission	Type	Mission Accomplished
1965	USA	Surveyor 5	Probe with mechanical shovel	Yes
1970	Russia	Luna 17 (Moon)	Probe and Lunokhod 1 robot	Yes
1976	USA	Viking 1 and 2	Probe	Yes
1988	Russia	Phobos (Mars)	Probe	
1993	USA	Mars Observer	Probe	
1997	USA	Mars Pathfinder	Probe and Sojourner robot	Yes
1998	Japan	Nozomi – Mars	Probe	
1999	USA	Mars Climate Orbiter	Probe	
1999	USA	Mars Polar Lander	Probe	
2001	Canada	Canadarm 2	Robotic arm mounted on International Space Station	Yes
2003	Europe-Russia	Mars Express	Probe and Beagle Robot	
2004	USA	MER-A and MER-B (Mars)	Probes and the robots *Spirit* and *Opportunity*	Yes

Objective Moon!

1966: The probe Surveyor, and its mobile robot blaze a trail for the great American Space Adventure

Years before the great step for mankind left its footprint on the moon robotic probes had already made their mark. On June 2, 1966 the three-legged module *Surveyor*, effected a soft landing on the moon. It touched down in the Ocean of Storms, a vast expanse of petrified lava and the largest of the lunar 'seas.' Built by the Hughes Aircraft Company, the 280-kilo (617 pounds) *Surveyor* resembled nothing so much as a huge, metal, moonhopping cricket. Its lower stage comprised a mobile robot which, though it did not yet comprise equipment advanced enough for soil analysis, undertook the first exploratory journey of the moon's surface. It stumbled blindly ahead over 1.6 kilometers (one mile), yet managed to send some 11,150 photographs back to Cape Canaveral. There were mixed fortunes for the ensuing Surveyor missions, with the second and fourth failing. But *Surveyor 3* accom-

plished its mission, as did the fifth, sixth and seventh craft. Altogether, they would beam back to earth 107,000 images, which were broadcast on television. By the time of *Surveyor 7*'s moon landing in January 1968, robots were performing to the full their task of preparing the way for a manned lunar landing. They enabled the US to map the moon a year before the fabled *Apollo* flight. *Surveyor*'s 3, 5, and

6 carried clawlike devices that scooped up samples of lunar soil, analyzed and identified as very fine volcanic rock similar to basalt. On November 19, 1969 *Surveyor 3* was paid a visit in the Ocean of Storms by moonwalking Apollo 12 astronauts, Charles Conrad and Alan J. Bean. They dismantled parts of the robot, like its television camera, to examine how it had withstood its lunar sojourn.

Above:
See you on the moon.
In 1969 the Apollo 12 astronauts came to inspect the lunar probe, Surveyor 3, which had landed two years earlier. They took some pictures and retrieved parts for analysis back on Earth.
© Courtesy of NASA/JPL

Page left:
NASA's Surveyor program comprised seven unmanned trips to the Moon between May 1966 and January 1968.
Surveyors 1, 3, 5, 6, and 7 all made successful soft landings.
© Courtesy of NASA

Objective Moon!

1970: *Lunokhod* the first mobile Soviet robot

Soviet ambitions to explore the moon using robots took shape in 1957 in the wake of the launch of the first *Sputnik* satellite. From that time the country began producing films to dramatize the challenge it was undertaking. Its robotic-engineering dream would only come true, however, on November 10, 1970, when *Lunokhod 1*, a fantastical jeep-like robot, rolled out onto the lunar landscape. It had descended the twin ramp lowered from the equally astonishing *Luna 17*, reminiscent of a giant, yet elegant, insect.

The *Lunokhod* rover was mounted on eight wheels measuring 0.5 meters (20 inches) in diameter. Weighing in at 765 kilograms (1,886 pounds), it was 21 meters (69 feet) long and had five cameras, which transmitted a panoramic view of the surrounding moonscape. It was equipped with extendable arms to probe and analyze the soil and a device that measured the distance between Earth and the moon. Although *Lunokhod* could operate in autonomous mode, a five-strong team back at the Baikonur Cosmodrome mission control in

Kazakhstan remote controlled its movements. They soon discovered the maddening frustration of the time lag. By the time they received a picture of the rover, it had moved on several meters. Despite the maneuvering difficulties, though, the operation was a success.

During the 11 lunar days (about 11 Earth months) during which it operated, *Lunokhod* traveled 10.5 kilometers (six and a half miles), conducted 25 soil analyses and carried out bearing-strength and density tests on the soil at 500 points. Furthermore, it beamed back to Baikonur 20,000 photographs and 200 panoramas.

On January 8, 1973 the Baikonur Cosmodrome succeeded in landing *Lunokhod 2* on the moon.

A scaled-down model of its older, remote-controlled brother, *Lunokhod* Junior traveled twice as fast, had six TV cameras and boasted more sophisticated instruments, including one that measured magnetic fields. Yet five days after it was deployed, gloom descended on Baikonur. *Lunokhod 2* had ceased responding. It could no longer be located; there was no longer any point in attempting to control it. But was the mission a failure? Not altogether. The robotic rover struck out on its own, sending 80,000 photographs and 86 panoramas Earthwards. By the time its last image arrived, it had covered 37 kilometers (23 miles) in four months, crossing ravines 50 meters (164 feet) deep and scaling heights of up to 400 meters (1,312 feet)!

An operator at work at the Baikonur Cosmodrome, which remote controlled the lunar robot Lunokhod 1.

Left:
Luna 13 was equipped with a dynamograph, a radiation meter and two arms for obtaining data on the physical properties of the lunar surface.

Below:
Deployed by the *Luna 17* lander in 1970, the Soviet lunar rover *Lunokhod 1*, traveled over 10 kilometers (6 miles) in 11 months.

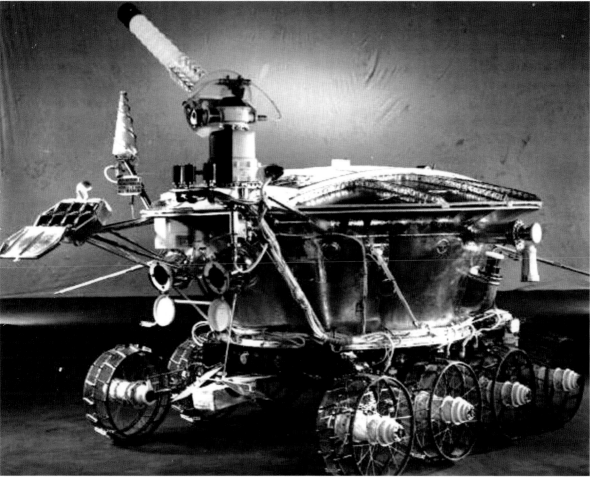

Objective Mars!

Fascinating red planet

If there is one planet that has captured the imagination of Earthlings, it is Mars. Although it is less than half the size of Earth, it boasts an equivalent dry-land surface and is thought to have formed, like our world, some four billion years ago. Yet there seems to be little likelihood of its harboring organic life. It is just too cold and barren.

Missions to Mars have to consider its highly elliptical orbit around the sun and the 780 days it takes for the earth and Mars to align—what is called the synodic period. Because the red planet's orbit is so eccentric, its distance from Earth ranges from 56 million to 100 million kilometers. Any mission plan has to bear in mind its orbital behavior as well as the fuel requirements, cost and duration of a flight.

The first Mars-bound voyage dates back to October 24, 1962, when the Soviet Union launched a craft that fell to earth after breaking up in low orbit.

It was to be the first of many failed attempts, with the US experiencing its share of setbacks. The first of these came on November 5, 1964, with *Mariner 3*. The shroud encasing its rocket did not jettison properly. Weighed down, the craft got caught in the wrong orbit and came no nearer than 68 million kilometers to Mars. That same year the Soviets responded with *Zond 2*, a probe powered by underestimated electrojet plasma engines.

A partial power failure cut communications with the craft, which went on to sail silently within 5,000 kilometers (3,100 miles) of the center of Mars. In 1965, after several near-misses, the US flyby probe *Mariner 4* finally came good, producing the first close-range images of Mars which, in one fell swoop, revealed frost and wiped out hopes of vegetation. The planet's atmosphere turned out to be composed mostly of CO_2, while temperatures at the poles dropped to 118°C (244°F) below zero. Its crater-pocked surface recalled the moon's. Conjecture abounded. What was Mars's atmosphere originally like? How did its geology form and evolve? And was its evolution at any stage conducive to the development of life forms? Would a crew of astronauts alighting there find natural resources?

Mars fascinates and foils. Of all the planets man has sought to explore, it has dashed the most hopes. Only one-third of all Mars-bound spacecraft have ever arrived, and scientists have not always been able to supply explanations. Some have talked of a sort of Bermuda Triangle along the route from Earth to the red planet.

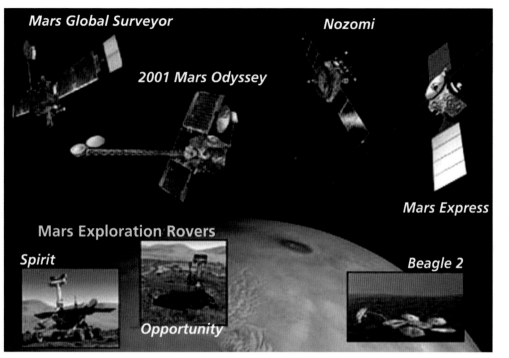

Mars Global Surveyor

Nozomi

2001 Mars Odyssey

Mars Express

Mars Exploration Rovers

Spirit

Opportunity

Beagle 2

Missions to Mars

There were five of them—one British, one European, two American, and one Japanese. All took advantage of the exceptional opportunity afforded by the synodic alignment between Earth and Mars in the summer of 2003 to set off for the red planet.

© *Courtesy of ESA*

Year of launch	Country	Mission	Result
1960	USSR	Marsnik 1	Failure
	USSR	Marsnik 2	Failure
1962	USSR	Sputnik 22	Failure
	USSR	Mars 1	Failure
	USSR	Sputnik 24	Failure
1964	USA	Mariner 3	Failure
	USA	Mariner 4	Success
	USSR	Zond 2	Failure
1965	USSR	Zond 3	Failure
1969	USA	Mariner 6	Success
	USA	Mariner 7	Success
	USSR	Mars 1969 A	Failure
	USSR	Mars 1969 B	Failure
1971	USA	Mariner 8	Failure
	USSR	Kosmos 419	Failure
	USSR	Mars 2	Success
	USSR	Mars 3	Success
	USA	Mariner 9	Success
1973	USSR	Mars 4	Failure
	USSR	Mars 5	Success
	USSR	Mars 6	Failure
	USSR	Mars 7	Failure

Year of launch	Country	Mission	Result
1975	USA	Viking 1	Success
	USA	Viking 2	Success
1988	USSR	Phobos 1	Failure
	USSR	Phobos 2	Success
1992	USA	Mars Observer	Failure
1996	USA	Mars Global Surveyor	Success
	Russia	Mars 96	Failure
	USA	Mars Pathfinder	Success
1998	Japan	Nozomi	Failure
	USA	Mars Climate Orbiter	Failure
1999	USA	Mars Polar Lander	Failure
	USA	Deep Space 2	Failure
2001	USA	Mars Odyssey	Success
2003	Europe	Mars Express	Success
	UK	Beagle 2	Failure
	USA	Mars Exploration Rover	Success

Objective Mars!

The Vikings
strike out
into Martian
territory

The very first close-up pictures of Mars were taken by the flyby orbiter *Mariner 4* in 1965. Their detail radically altered our perception of the planet and enabled scientists to begin mapping its surface. The objective of the Viking mission was to further that revolutionary detail through high-resolution images of Mars's surface, analyses of the chemical make up of its soil and atmosphere, and experiments that, it was hoped, might reveal the presence of simple life forms.

The mission comprised two identical craft, *Viking 1* and *2*, each of which had two modular parts—an orbiter and a lander. The orbiters were to map Mars and relay transmissions between Earth and the insectlike landers busily conducting experiments on the planet's surface.

With their state-of-the-art instrumentation and laboratories, the *Viking*s were the most advanced craft of their time. They cost around one billion dollars and were the fruit of work of 10,000 space scientists in the US.

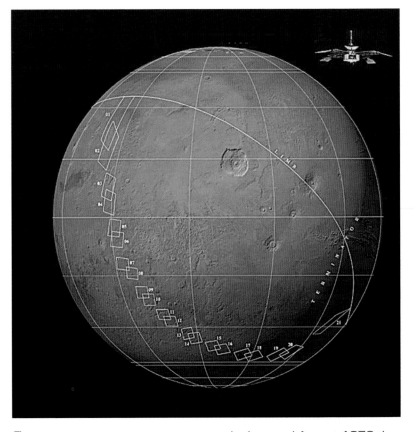

The camera of Mariner 4 *will acquire 21 complete images of Mars.*
© Courtesy of NASA

In June and August 1976 they entered Mars's orbit in search of a suitable landing site. When, a month later, the landers separated and touched down, the orbiters pursued their mapping work. Swooping as low as 300 kilometers (186 miles), they took high-resolution images of the lay of the Martian land. Meanwhile the landers were flexing their digger arms to scoop out shallow trenches in the ground, where they gathered soil samples on which they ran the three life experiments in their integrated-

The landing module of the mission Viking was 3 meters high, for a weight of a half ton. It carried a total of 15 scientific experiments.
© Courtesy of NASA

biology instrument. In one of the experiments nutrients and water were added to the sample. If some simple life form existed within it, it would eat the nutrients and metabolize, so producing organic by-products that could be detected. Though some results suggested life, scientists ruled them out. One argument was that microorganisms brought from Earth had survived and contaminated the samples. There could, therefore, be no certainty that there was life on Mars.

In April 2004, the European *Mars Express* probe detected methane in the red planet's atmosphere, rekindling speculation at the European Space Agency that there might, after all, be 'spiders on Mars.'

Marsokhod,
a robot lost
in space

In 1987 the Soviet Union unveiled plans to fly three missions to the red planet—Mars 92, 94 and 96. The first two would each carry a big wheel-walking rover robot, known as the *Marsokhod*, which would bound about investigating the terrain. *Mars 96* was given a very different assignment: to return to Earth with samples of Martian soil. The overriding objective for all three craft, though, was to prepare the way for manned flight to Mars by early in the 21st century. But the failure of the Phobos missions and the collapse of the Soviet Union set back the grand Russian designs on Mars. Deep cuts were made in the space-

exploration budget, schedules were changed, and the sample-return mission was discarded as unfeasible. The original project was abandoned, but Mars 96 reappeared in the guise of the international space probe of the same name, equipped with a vast array of sophisticated instrumentation to study Mars every which way. The mission's sheer ambition bore the unmistakable stamp of the great Russian tradition of space exploration—numerous scientists had worked for over ten years to prepare it.

The *Mars 96* spacecraft was the most comprehensively equipped ever launched by man and probably the last great Russian exploratory planetary venture.

On November 16, 1996 Proton, the four-stage rocket launcher carrying it, blasted off. Liftoff went smoothly, and when the third stage separated, Proton slipped into low orbit trajectory at an altitude of 145 kilometers (90 miles). At this juncture the fourth stage of the launcher, Block D2, was still carrying its payload. Block D2 was designed to pull the probe clear of Earth's gravity and send it on its way to Mars. *Mars 96* had its own propulsion unit, the ADU, which would boost the craft into solar orbit insertion, the next leg of its Mars-bound journey. But at 10:07 p.m. telemetric communication with the craft was lost and it disappeared from radar screens. Mission Control immediately feared that the rocket had blown up. Not quite.

Block D2 experienced a burn failure and placed its payload in the wrong orbit. So when *Mars 96* separated from the rocket and its own engine cut in to fire it clear of Earth, it was too low when it did so. Although it reached the upper atmosphere, it was slowed by friction and gradually lost momentum. It managed to orbit Earth twice, but could no longer escape the inevitable, burning up on re-entry into the atmosphere.

That is what is thought to have taken place on November 17, 1996, but no part of the probe, not the slightest scrap of buckled metal, ever fell to earth. There was not even a trace of radioactivity. To this day nobody knows where *Mars 96* crashed or what cut short its voyage so prematurely. Lost too was the *Marsokhod* robot, a cutting-edge gem of technology. It fell out of the sky. Mars seemed determined not to yield up its secrets to explorer robots for a long time to come.

*The **Lama** of **LAAS**, successor of the **Marsokhod**.*
© *Courtesy of Simon Lacroix, LAAS*

The JPL mission: faster, cheaper

NASA's Mars missions have become the specialty of JPL[1]. The JPL nestles in the foothills of San Gabriel to the northeast of Los Angeles. It is the hub of NASA's Mars missions. Its beginnings go back to 1936 when six space and rocket-science enthusiasts from the California Institute of Technology sank all their earnings into aeronautical research at the time of the Great Depression. Legend has it that the "rocket boys" used to fool around with experimental projectiles that would explode in showers of sparks and flying metal.

Hounded off the CalTech campus, they set up a makeshift lab in disused workshops in the Southern California foothills, a few miles away. Although sneered at by the top minds of the time, they made a name for themselves for their sheer brilliance in designing rocket engines and interplanetary probes. As part of its "better, faster, cheaper" policy, NASA asked JPL to design a craft for a mission to Mars on one condition dictated by cost cuts—there would be no second chance in the event of failure. JPL's response was the Mars Pathfinder and its small rover *Sojourner*. After the first giant step it took for mankind, the US was now poised to take a further stride.

(1) A division of NASA devoted to unmanned flights to Mars since the mid-1990s.

Objective Mars!

Sojourner:
Triumph of the unassuming robot

In 1997 the Mars Global Surveyor (MGS) and Pathfinder missions brought space exploration back into the limelight. While the MGS orbiter was to image and map Mars and spot likely landing sites, it was *Pathfinder* that grabbed the headlines because of its leading character, *Sojourner*—or Rocky as he came to be affectionately known.

Rocky was a sleek, solar-powered microrover measuring just 65 centimeters in length, 40 in width and 30 in height, clearly built to meet NASA's low-cost directives. For some important parts JPL had simply gone to the open market rather than conducting cutting-edge research. It bought *Sojourner*'s modem from Motorola and fishing-rod winding motors from the Swiss company Maxon to actuate the wheels. The solar array atop the microrover supplied peak power production of 16 watts, barely more than strictly necessary, while a nonrechargeable lithium backup battery provided power for nighttime and low-light experiments.

Mounted on six wheels, each 13 centimeters (5 inches) wide,

and riding on JPL's purpose-built rocker bogie suspension, *Sojourner* could take on obstacles 20 centimeters (8 inches) high and scale gradients of up to 45°. However, for safety's sake, it was programmed to backtrack whenever it encountered slopes in excess of 26°. Its wheels doubled as soil analysis tools. Five of them would lock, while a sixth, the abrasion wheel, would spin in place. Its wear characteristics indicated soil types.

Rocky was built to take a battering from the rough terrain it traversed in the course of

its wanderings. It used laser beams to feel out the lay of the land and, on encountering an obstacle, would try to cross it. If at first it did not succeed it would try again and again. On its third failed attempt it would transmit a help message containing stereo images to the Pathfinder lander, which would relay it earthward.

A Deep Space Network receiver, measuring 70 meters (230 feet) in diameter and one of only three in use, captured the imaged data, from which a digital terrain model was generated.

The Sojourner rover measured 65 centimeters (25 inches) in length, 40 (16 inches) in width and 30 (12 inches) in height, while weighing in at 10.6 kg (23 pounds). The solar panels it carried on its back made it autonomous.
© *Courtesy of NASA/JPL*

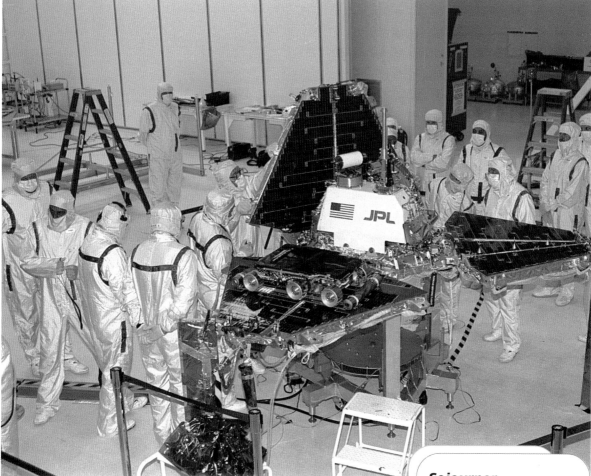

JPL scientists built a scale model of Pathfinder's *landing site to facilitate study of the* Sojourner *robot's behavior during its mission.*
© Courtesy of NASA/JPL

Sojourner

A 12-year-old girl won the competition held by NASA to find a name for the Mars microrover. She called it *Sojourner* in honor of Sojourner Truth, an abolitionist and women's-suffrage campaigner.

Working from the model, JPL would transmit avoidance instructions and directions to the little robot. The lengthy procedure helps explain why Rocky the robot moved no faster than one meter (39 inches) per day. Nevertheless, it returned images and took soil analyses for three months, covering 118 meters (387 feet), taking 550 images and performing 15 chemical analyses.

Although the final count might not seem that much per se, it revealed the presence of quartz and sand and provided valuable meteorological information.

On September 27, 1997—day 83 of its assignment—Rocky sent its last images and soil findings to Earth. Transmissions then ceased. It is generally thought that the little robot had finally succumbed to the icy bite of the bitter cold on Mars.

1- Mission Control on Earth reprogrammed Sojourner *every sol (Martian day). Instructions took 11 minutes to reach Mars.*

2- On July 4, 1997 Sojourner *descended the ramp from the lander and began its voyage.*

3- The Martian horizon as seen from the Pathfinder *landing site.*

© *Courtesy of NASA/JPL*

Sojourner analyses the rock called Yogi with its APXS spectrometer. © Courtesy of NASA/JPL

The first images from Mars show the little robot waiting on one of the station's solar panels. © Courtesy of NASA/JPL

The *Pathfinder* lander camera had meanwhile returned over 16,000 images, which added credence to the belief that Ares Villis, the area explored by the now-silent rover, was a dried-out flood plain.

Mars Pathfinder's success gave NASA fresh hope and, in 1999, the Global Surveyor probe began its primary mission of mapping Mars, during which it gathered an unprecedented wealth of scientific data. Analysis of the images that MGS returned strengthened the case for there having once been underground water sources on Mars. There was only one way to find out—land a robot on Mars. Unfortunately, two subsequent US Mars discovery missions failed. In 1998 the Mars Climate Orbiter was destroyed when it failed to enter orbit at the right altitude, while the following year telemetry with the Mars Polar Lander ceased after it touched down in the south pole, where it was to dig the ice. The loss of the two craft compelled NASA to further squeeze the second term in the "faster, cheaper, better" formula.

The Mars rover Opportunity, *is closed into its aeroshell and padded with airbags.* © Courtesy of NASA/JPL/KSC

Objective Mars!

The robot twins
Spirit and *Opportunity*

By 2004 hope had rekindled at NASA and two new Mars exploratory rovers landed on Mars in January. *Spirit* touched down, or rather, hit the planet's surface, bouncing and rolling to a rest at Gusev Crater. That was on January 3. On the 24th *Spirit* had been followed by its twin, *Opportunity*, which landed on the Martian plains known as Meridiani Planum. By sending two robot geologists to two different locations JPL hoped to double its chances of success.

The roving twins were looking for traces of water in zones that might have supported life. The images returned to Earth over the years had revealed a rugged terrain gouged by valleys and canyons, like Valles Marinensis, which were deeper than anything on Earth. Judging from the relief of the landscape, it was possible that water had once flowed on Mars and that there might even have been seas. The mission had chosen Gusev Crater as a site precisely because it could for-merly have held a great lake, while Meridiani Planum, which *Opportunity* was to explore, lay within an outcropping of gray hematite, an iron oxide that usually forms in the presence of water.

Both rovers were equipped with radiation-hardened micro-processors—operating at speeds of 20 million instructions per second—and augmented memory capacities. *Spirit*'s robotic arm was tooled with a micro-scope imager, which comprised a digital video camera and microscope. It could take pictures to within the nearest few hundredths

Above:
NASA engineers
assemble the Mars rover
Spirit, and its lander.
© Courtesy of NASA/JPL/KSC

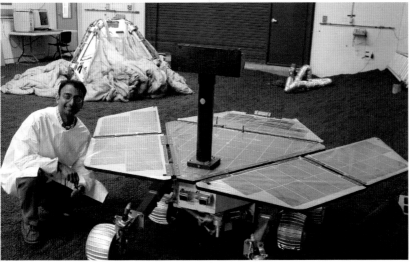

Left:
Prasun Desai, lead engineer
for the Mars Exploration
Rover Mission at **NASA**.
His responsibility is
dynamics and systems
analysis for the entry,
descent, and landing of
the rovers. He has been
with **NASA** for 12 years.
© Courtesy of NASA/JPL

of a micron. The high-precision images of Martian soil were beamed back to mission control at JPL, where scientists studied them in digitized 3D-model form to determine the next step. They then transmitted instructions to the rovers, not forgetting the time lag between Earth and Mars and speeds of transmission that varied according to distances between the two planets.

It was on Tuesday, March 2, 2004, that confirmation of a fabulous discovery became clear—water had once flowed on Mars. Ed Weiler, NASA's associate administrator for space science, said that *Opportunity* had explored an area of the Martian surface that had once been wet.

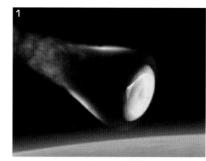

1- The cruise stage is the spacecraft configuration for carrying the rover from Earth to Mars. It is approximately 2.65 meters in diameter, powered by solar arrays, and controlled by its on-board computer.

2- Well protected inside its tetrahedric lander, the rover continues its descent. Six seconds before touchdown the retro-rockets on the aeroshell are fired to slow the lander's speed of descent.

3- The airbags inflate, forming a protective cushion in readiness for landing.

© Courtesy of NASA/JPL/Analytical Mechanics Associates

4- On landing, the petals are deployed and the rover begins its egress down the soft batwing ramplets onto the surface of Mars.

5- The lander is of no further use. It is discarded on the surface of Mars surrounded by a wreath of crumpled airbags.

© Courtesy of NASA/JPL

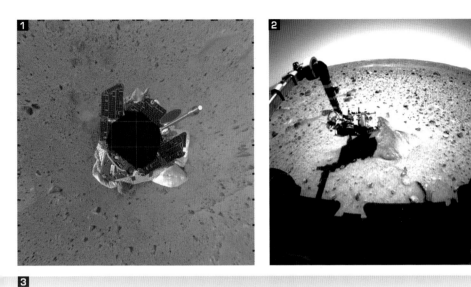

1- *Photograph of the Spirit's lander taken by the 360° Pancam.*

2- *At the end of the robotic arm on the front of the rover is a cross-shaped turret that holds four scientific instruments.*

3- *Mars Exploration Rover.*

© Courtesy of NASA/JPL

A robotized laboratory

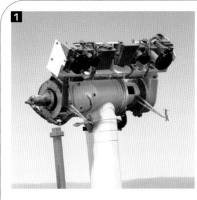

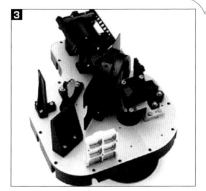

To communicate with Mission Control on Earth, *Spirit* and *Opportunity* use antennae on their equipment deck, or back. They are equipped with six state-of-the-art scientific instruments:

1- Pancam
The Pancam is a panoramic image system consisting of a high-resolution color stereo pair of CCD cameras for imaging the ground and sky, and a navigation camera. The cameras are mounted on a bar that sits on top of the mast of the rover.

2- Microscopic imager
The microscopic imager analyzes the grains in sedimentary rocks. It is the first time in the history of the exploration of Mars that a rover has performed microscopic analysis on the spot.

3- TES
The Thermal Emission Spectrometer determines rock and soil mineralogy by detecting thermal radiation patterns. Like the Pancam, it is mounted on the camera bar.

4- APXS
The Alpha Particle X-Ray Spectrometer determines the elemental chemistry of rocks and soils by measuring the alpha particles and x rays that they emit.

5- Mössbauer Spectrometer
Is an instrument that is designed to study iron-bearing minerals, determining their quantity and composition.

6- RAT
The Rock Abrasion Tool is a small but powerful grinder. It sweeps away surface dust and then grinds its way into the rock below. It thus ensures that the samples that will be analyzed come from the interior of the rock. A problem encountered by *Sojourner*, which was not equipped with the system, was that it gathered samples of the rock's dust-caked exterior.

Sojourner *(left)*
is different from
Spirit *(right)*.
© Courtesy of NASA/JPL

Mars Exploration Rover
telecommunications system

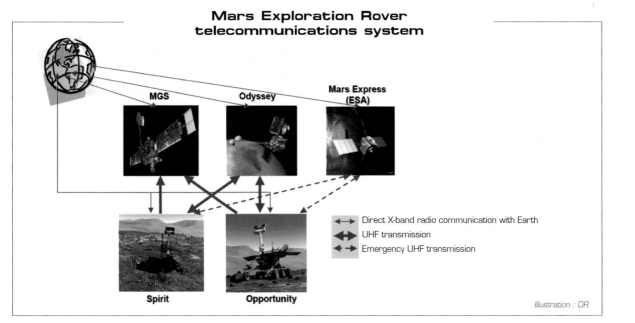

MGS

Odyssey

Mars Express (ESA)

◄──► Direct X-band radio communication with Earth

◄──► UHF transmission

◄─ ─► Emergency UHF transmission

Spirit

Opportunity

Illustration : DR

Mars Express

Mars Express was the European Space Agency's first mission to another planet. Borrowing technology from the failed Soviet probe in March 1996, its prime aim was to explore Mars for traces of water. The *Mars Express* spacecraft consisted of an orbiter and a British-built lander, *Beagle 2*.

The total cost of the mission was estimated at $390 million, the cheapest Mars-bound mission to date. It was planned and implemented in the record lead-time of four years, hence its name, "Express". ESA plans another fast-track mission to Venus in November 2005.

The *Mars Express* spacecraft was assembled in Toulouse by Astrium SA. It carried a high-tech payload of scientific, technical and navigation instruments that had been designed and produced in different parts of Europe. The instrumentation was mounted on a cubic platform measuring 1.5 meters (five feet) long by 1.8 (5 feet 10 inches) wide and 1.4 (4 feet 7 inches) high, on one side of which was the *Beagle 2* lander encased in its protective heat shield.

The craft weighed 1,060 kilograms (1.1 tons) on blast-off, fuel included, and was equipped with a solar array that presented a surface of 11 square meters (118 sq. ft.) exposed to the sun's rays. To communicate with mission control on Earth, *Mars Express* used a high-gain antenna with a diameter of 1.8 meters (5 feet 10 inches) and a smaller low-gain antenna, measuring 40 centimeters (15 inches) in width.

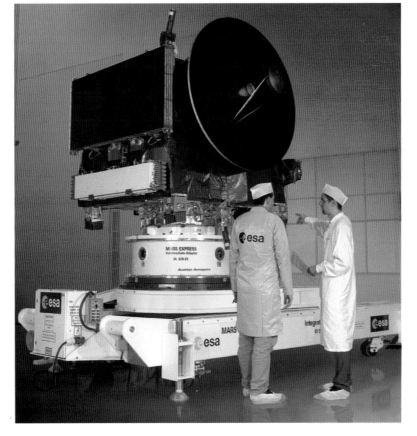

Assembling the spacecraft in France at the National Center of Space Studies (CNES).
© *Courtesy of the CNES*

A navigation system comprising cameras and sensors enabled the craft to find its bearings in outer space, while a set of small motors fired to adjust its trajectory as required. To resist the extreme cold of space, where temperatures drop to 270°C (518°F) below freezing, *Mars Express* was equipped with insulation blankets and small heaters that kept its sensitive scientific instrumentation warm. On inserting itself into orbit around Mars, it headed for the polar cap where it deployed its antennae and instruments.

1- **Mars Express's** platform is 1.5 meters long, 1.8 in diameter, and 1.4 in height (five feet, five feet ten inches, four feet seven inches). The **Beagle 2** was mounted on one of its outside faces.

2- **Mars Express** *is powered by energy from its solar panels.*

3- *To adjust its trajectory, Mars Express was equipped with eight thrusters and four reaction wheels on the underside of the bus.*

© *Courtesy of ESA*

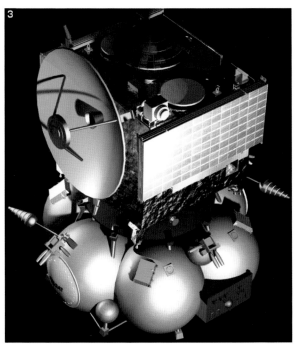

Mars Express's elliptical orbit around Mars is termed a "polar orbit" because the craft passed very close to the planet's north pole, moving away toward its orbital apocenter, then swooping back in towards the epicenter. Close-flying orbit enables its instrumentation to cover the planet's surface, comprehensively viewing it from all angles. On moving out into wider orbit, it communicates with mission control on Earth.

The seven scientific instruments take readings, images and measurement of the surface throughout the two Earth years (one Martian year) of the mission's duration—from March 11, 2004, to November 30, 2005.

Once *Mars Express* has completed the scientific leg of its mission, it will continue to orbit Mars, acting as a data communications relay for future missions, for at least another two years (one Martian year) from December 1, 2005 to November 30, 2007.

The high point of the mission to date was a finding that may be even more important than the discovery by *Opportunity*'s robot of the once-probable existence of water. In April 2004, the European Space Agency officially announced that the orbiting craft had detected traces of methane.

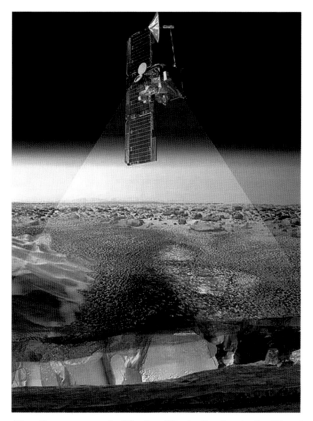

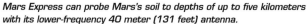

Mars Express can probe Mars's soil to depths of up to five kilometers with its lower-frequency 40 meter (131 feet) antenna.

One possible source of methane is microbes, the implication being that it is emitted by bacteria on Mars. If the methane is indeed produced by microorganisms, the Mars Express mission will have "unearthed" the first indirect evidence of extraterrestrial life on a planet that lies millions of kilometers from Earth.

Beagle 2

The *Mars Express* landing craft, the *Beagle 2*, was one of a kind. It was the smallest craft ever designed to land on Mars, weighing in at a mere 34 kilograms. It resembled a small, flat disk 65 centimeters (over two feet) in diameter and 23 centimeters (nine inches) thick. On landing, *Beagle 2* was designed to deploy its five solar panels, offering an aggregate area of one square meter to capture the sun's rays.

Mars Express *jettisons* **Beagle 2.**
© *Courtesy of ESA/MediaLab*

On the end of its single robotic arm was a wide array of instruments, while its communications systems, batteries, suite of environmental sensors, and gas-analysis probe (GAP) were housed within the main body.

The tiny *Beagle 2* had the highest ratio of instrumentation payload to overall mass of any Mars mission. It was the result of a wildly ambitious project that hatched in the mind of Colin Phillinger, a British scientist with a passionate interest in Mars and the possibility of life in space. The Open University led the project with funding estimated at $52 million, provided by ESA, the British government and private backers.

On June 2, Colin Phillinger was vindicated when the *Beagle 2* took off clamped to the back of the *Mars Express* spacecraft as though it were hitching a ride. When it touched down on Mars, it was supposed to beam back to earth a song especially written for the occasion by the British band Blur. Only...it never landed. The last that was seen of it was an image taken by its orbiting mother ship in December.

Lead scientist Colin Phillinger designed **Beagle 2**. © Courtesy of Beagle 2

Beagle 2's instrument package

Beagle 2 had solar petal panels and rechargeable batteries to power its scientific instruments. On the end of its jointed robotic arm was a rock corer, or paw, and a highly sophisticated instrumentation package. It included a wide-ranging Environmental Sensor Suite (ESS) to measure radiation levels, ultraviolet rays, atmospheric composition and pressure and wind strength on the surface of Mars. The findings that the ESS would have gathered were designed to determine whether conditions on Mars could be life-supporting.

Collecting samples

The *Beagle 2* was equipped with a molelike probe, called Pluto, to gather geological samples. Pluto was designed to inch along the surface of Mars digging up samples. Subsurface samples offer the advantage of not being exposed to radiation and are consequently much more valuable for understanding the true nature of the planet's geology than surface samples. The robotic mole would have proceeded at a speed of one millimeter every six seconds, rummaging intermittently in the ground. It was attached to a wire that drew it back into the lander for analysis of the samples collected.

Beagle 2 *was encased in a golden insulation cladding to prevent heat loss when on the surface of Mars.*

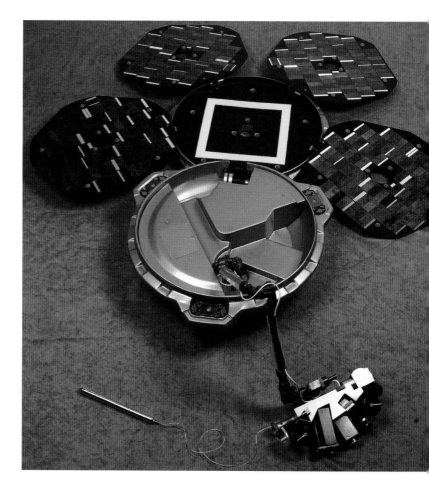

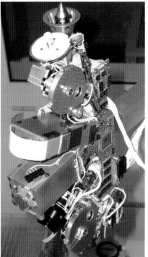

Beagle 2's robotic arm and its impressive array of instruments.
© Courtesy of Beagle 2

The Beagle's arm

Beagle 2's articulated arm, or Paw, fanned out to deploy scientific instrumentation that was the best that electronic miniaturization technology had to offer. Weighing in at only 2.5 kilograms (5.5 pounds), the Paw was equipped with:

• **Two stereoscopic cameras** for 360° and 3D photography. The paired cameras were also designed to identify rocks for analysis. They were built by the Mullard Space Science Laboratory and the University of London.

• **A microscopic imager** for extremely close-up analysis of the soil and rock structures on the surface of Mars. This technological marvel was the first microscopic imager to be included in a mission to Mars. It was developed by the Max Planck Institute for Aeronomy at Lindau in Germany.

• **A gamma-ray Mössbauer spectrometer** for analyzing iron oxide levels in the soil and rocks of Mars. It was made by the University of Mainz in Germany. NASA's Mars Exploration Rovers, *Spirit* and *Opportunity*, were also equipped with Mössbauer spectrometers.

• **An X-ray spectrometer** to analyze rock and soil elemental chemistry, supplied by a team from Leicester University, led by Professor George Fraser.

• **A grinder** that sweeps the surface clear of dust and pebbles and grinds into the surface of the rock, exposing a fresh surface beneath. It was designed and developed by a dentist, Dr. Ng Tze-cheun, from Hong Kong, in collaboration with Leicester University.

• **A small corer** that drills into rock to extract a sample one centimeter (0.4 inches) long by 2 millimeters wide. It is then submitted to gas analysis tests to gauge whether its properties are consistent with existing or extinct life forms.

Europe's *Smart 1* heads for the moon

On September 27, 2003, an Ariane-5 rocket blasted off from the ESA launch pad in Kuru, French Guyana. On board was a small European probe called *Smart-1*, short for Small Missions for Advanced Research in Technology. Its objective was the moon. After crossing space on a journey that lasted just over 11 months, it went into orbit around the moon in mid-November. It will not, however, be landing. Part of its mission, which will last between six months and one year, is to perform scientific experiments in an effort to improve understanding of how the moon was formed.

Smart-1's main mission, however, was to test new propellant technology and instruments. During its cruise to the moon, it successfully ran a battery of tests. The most important of these concerned a revolutionary new solar-electric propulsion system for deep-space missions, which only NASA's Deep Space 1 had previously used. The system was used to power *Smart-1* for part of its journey to the moon and, again, to maneuver it into a stable orbit. The solar-electric propulsion system, also known as an ion engine, works by energizing inert xenon gas with solar electricity. The ions produced accelerate and are driven out through a thruster at very high speed, so propelling the craft. The prime advantage of this technology is that it is very economical—ten times more fuel-efficient than conventional hydrazine propulsion systems. All it needs is a little xenon and electricity supplied courtesy of the sun. The ion engine's efficiency also means that it will run much longer. That, however, is not all. As the ion engine runs, the spacecraft that it powers gathers velocity, which could open up interesting prospects for interplanetary travel.

Even on its experimental lunar mission it was faster than the anticipated 15 to 18 months, claims ESA.

Smart-1

European spacecraft, Smart-1, on its way to the moon. Instead of heading straight for the moon, it was put into an elliptical orbit round the earth, which dragged it gradually towards the moon. At about 200,000 kilometers (124,000 miles) from Earth, Smart-1 was pulled into the moon's gravitational field
© Courtesy of AOES MediaLab, ESA 2002

Smart-1's ion engine.
© Courtesy of ESA/MediaLab

Rosetta

The European spacecraft Rosetta will orbit the comet Churyumov-Gerasimenko on which it will jettison its robot lander Philae.
© Courtesy of ESA/AOES MediaLab

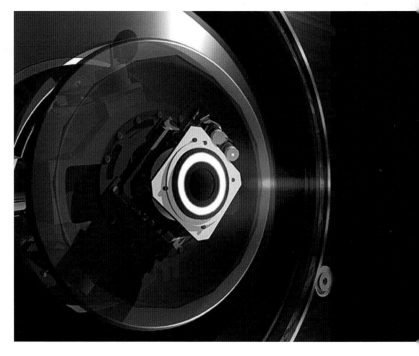

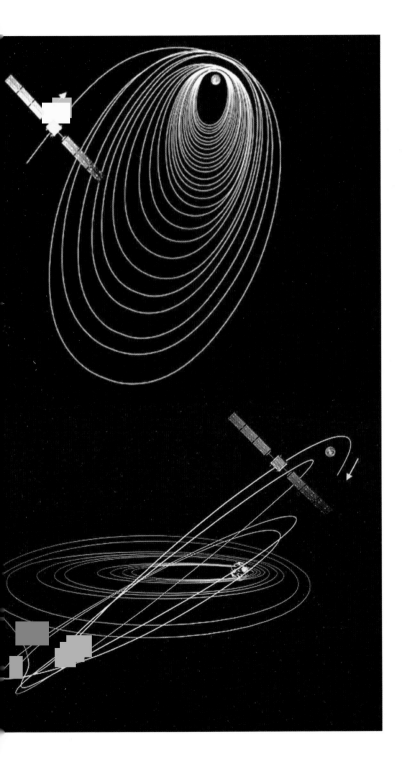

Rosetta send a European robot into a comet's nucleus

ESA's *Rosetta* spacecraft blasted off on February 27, 2004 for the comet Churyumov-Gerasimenko. Its mission is to orbit the comet and maneuver into a position from which the robotic landing module, *Philae*, may self-eject onto the comet's nucleus. Comets carry complex organic mole-cules, some of which may have come into contact with Earth and played a role in the earliest forms of life. If all goes smoothly, *Rosetta* will reach its destination in 2014 after orbiting the sun four times and being bounced about in space like a pinball for ten years.

Rosetta

Rosetta *is named after the Rosetta Stone found in Egypt 200 years ago. The inscriptions on the slab of stone provided the key to deciphering the ancient Egyptian hieroglyphs. It is hoped that the Rosetta mission will provide some keys to the secrets of our solar system.*

© Courtesy of ESA/AOES MediaLab

Other robots in space

Robots designed to help humans in their exploration of space are not just rugged, all-terrain rovers. There is a wide range of devices and machines to meet needs.

Robotic arms for International Space Station

Man has been building space stations since the early 1970s. The former Soviet Union constructed the first series, the *Saliut* stations, followed by the US, which completed its *Skylab* in 1973 to carry out scientific experiments in conditions of microgravity. The last to be built during the Cold War was *Mir*. It orbited Earth

from 1986 to March 2001 and was the first space station designed for long duration stays.

The end of the Cold War spelled the beginning of a new cooperative era in space flight. Fifteen countries (Canada, the US, Russia, Japan, Belgium, Denmark, France, Germany, Italy, the Netherlands, Norway, Spain, Sweden, Switzerland and the UK) agreed to pool their efforts to build the International Space Station (ISS) in what remains the biggest international collaborative venture in peacetime.

Each country has made its contribution to the development of technologies and materials, which will make the ISS the biggest, most highly advanced

orbital laboratory in history. It will provide researchers with the opportunity to work on new materials in an environment that will drastically reduce the effects of gravity on the outcome of experiments. In the field of medicine it will make a valuable contribution to understanding illnesses like osteoporosis.

Work on the ISS got underway in November 1998, when the Russian module, *Zarya* ("dawn" in Russian) blasted off into space to become the first building block. The assembly phase began the following month when the US connecting module, *Unity*, locked successfully on to *Zarya* to form the beginnings of a real space station.

By the time the 450-ton station is fully up and running, 45 assembly missions will have been flown by NASA space shuttles and Russian Proton and Soyuz craft between the inception of the project and scheduled completion of the finished facility in 2005.

Initially there were to be seven astronauts permanently manning the ISS, but in 2001 the Bush administration cut back on NASA's budget when it once again overshot its $12.4 billion allocation. Its visionary administrator, Daniel Goldin, made way for a more pragmatic steward, Sean O'Keefe.

Fifteen countries agreed to pool their efforts to build the International Space Station (ISS).
© Courtesy of NASA

The International Space Station

The ISS can be seen with the naked eye from most countries in the world. It orbits Earth at a speed of 27,000 km/h (17,000 mph) at an altitude of 400 km (250 miles), or 16 times a day. For the astronauts on board, that means 16 dawns and 16 sundowns every 24 hours. The 470-ton station is 108 (355 feet) meters long by 74 (240) wide. It was built to accommodate up to seven astronauts in its 1,200-cubic-meter (42,378-cubic-feet) living quarters and to operate for some 10 years.

Opposite: Autonomous rescue vehicle prototype for the ISS.

© Courtesy of NASA

The ISS was commissioned in 2000, but since April 2003 the size of the permanent crew has not exceeded two. Its scientific yield has been practically zero, because the maintenance work monopolizes the crew members' time. However, they do receive assistance from two robotic arms developed in Canada, whose tasks are mobile maintenance and assembly. Canadarm 1 and Canadarm 2, as they are aptly called, can lift objects of a mass equivalent to that of the space shuttle. A camera surveillance system enables the astronauts inside the ISS to keep an eye on the robots and ensure they are handling and positioning parts correctly.

Each arm has a shoulder, an elbow, and a wrist. These joints give the arms the freedom of movement required to assemble parts, retrieve, and repair damaged satellites, and put new satellites into orbit around Earth.

In 2001 Canadarm 1 and Canadarm 2 shook hands in space. Canadian astronaut Chris Hadfield controlled Canadarm 1, which was installed on a shuttle, while his US crewmate Susan Helms, remote guided Canadarm 2 from the space station.

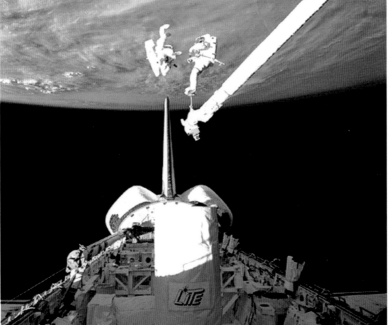

The ISS's Canadian robotics make it possible to telemanipulate and move objects whose mass is equivalent to that of a space shuttle.
© Courtesy of NASA

The live-in droid

NASA's Johnson Space Center in Houston, Texas, is exploring the possibility of forming crews mixing robots with humans in order to make astronauts' extra-vehicular activities (EVA) more productive. Johnson's engineers want to coordinate humans at work inside and outside the space station with teams of auton-omous or remote-controlled ro-bots flitting around it. Two astro-nauts aided by a robotic arm currently perform EVA while keeping a watch on each other. By forming two separate teams of one human and several robots, the six to eight hours of EVA could, thinks NASA, be more productive. In 1997 the AERCam Sprint free-flying robot camera was successfully used in the space shuttle's cargo hold during the STS-87 space-walk.

DARPA and the Johnson Space Center are currently developing Robonaut, a remote-controlled humanoid robot with a head, a torso, hands with five articulated fingers, and arms like those of a human being. Robo-naut, says the Johnson Space Center, "is designed to be used for 'EVA' tasks, i.e., those that were not specifically designed for robots."

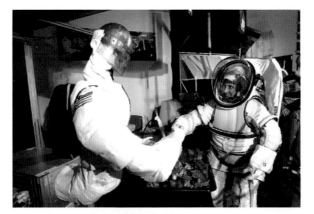

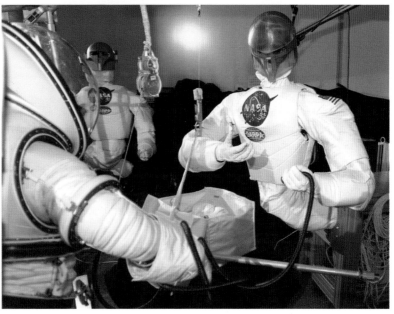

Robonaut is a remote-controlled humanoid robot with a head, a torso, hands with five articulated fingers, and arms like those of a human being.
© Courtesy of NASA/JSC

Its spacewalking tasks could include preparing EVA jobs to be performed by astronauts and cleaning up when they have finished. Robonaut's successor may one day act as a surgeon's assistant and carry out critical spacewalking tasks like assem-bling telescopes or interplane-tary transit vessels. Their anthropomorphic design would be the ideal interface for opera-tors to control them intuitively. Robonaut, it is hoped, will enter service on the ISS within four years.

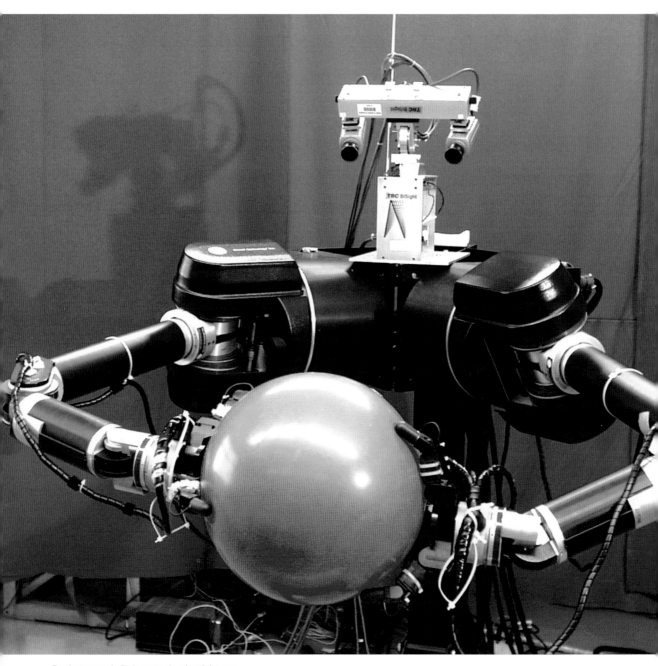

So dexterous is Robonaut that it might one day act as a surgeon's assistant in the event of an operation.

© Courtesy of NASA

Hubble repair robots

The aging Hubble Space Telescope has been orbiting Earth since 1990 and was due for an overhaul and fresh lease of life in 2005. Until January 15, 2004, that is, when Sean O'Keefe announced that NASA was going to write it off. Protest was immediate in the scientific community, and understandably so. A feat of precision engineering in itself, the telescope had helped evaluate the age of the universe and "sight" planets outside our solar system. In March O'Keefe gave ground and agreed that Hubble would be kept on until the arrival of its successor, James Webb, in 2011. NASA would send into space a craft that would dock with the telescope. Onboard would be a robot dexterous enough to change its batteries and carry out other refurbishment tasks.

The automated Hubble Space Telescope can observe the universe using visible, ultraviolet and infrared light spectra.
© Courtesy of NASA

Personal and savior droids

Research is ongoing into experimental robots whose job would not be to service telescopes but to serve astronauts living and working in space. The ball-shaped robots, which NASA calls Personal Satellite Assistants (PSAs), look like devices straight out of *Star Wars*. Their designer, Yuro Gawdiak, readily acknowledges that the inspiration for the PSAs was Luke Skywalker's sparring partner, the lightsaber-training droid in *The Empire Strikes Back*, crossed with the tricorder from *Star Trek*.

PSAs are spherical, flying robots, packed with environmental sensors that detect the pressure, temperature and composition of the air. They are designed for operation on spacecraft and stations, where their task will be to help crews by taking charge of routine, time-consuming chores.

NASA intends to endow the powerfully intelligent machines with a voice-recognition capability, so that they can answer ques-

The Personal Satellite Assistant. This ball-shaped robot measuring 15 centimeters (six inches) in diameter can propel itself using six small ducted fans.
© Courtesy of NASA

tions as well as help crew members by reminding them of things to do, for example, or telling them where to find something they might have misplaced. They sport small LCD displays that could show checklists or numerical readings. The flying, talking ball bots could, ultimately, save lives by taking action in critical situations that are too dangerous for the human crew. They could, equally, give the alert should they sense danger themselves, like rising levels of CO_2.

Robot gardens

Before space agencies send robots into space they put them through their paces in interplanetary conditions simulated on Earth.

The University of Arizona's Lunar and Planetary Laboratory has built a "Mars Garden" from 65 tons of red earth and volcanic basalt rock gathered from the desert. It was here that *Sojourner*, *Spirit*, and *Opportunity* took their first, tentative steps. The garden's sand dunes and lava rock formations bear a startling similarity to the landscapes imaged by *Viking* and *Pathfinder*.

France's Center for Space Studies (CNES) in the city of Toulouse, also built a terrain as part of the Mars 98 project. Covering an area of 600 square meters (6,500 square feet), it reproduces in hyperrealist detail a plot of Martian terrain with its strange little hillocks and orange crevices.

Earth naturally provides harsh environments that are even more like those robots will encounter in other worlds. In the arctic wastes of the Great White North in Canada's Northwestern Territories lies Devon Island. Since the early 1990s NASA has been exploring a crater on the island

to test survival modules and space rovers.

At the other end of the earth, a robot developed by Carnegie Mellon roamed an icy expanse of land, known as Elephant Moraine, in Antarctica in 2000. Until this foray, robots were either controlled by non-reconfigurable programs or remote handlers. *Nomad*, however, undertook its mission to find and classify meteoric rock samples autonomously. It was in search of specific types of rock, which its machine vision enabled it to spot by distinguishing their dark color against the white ice background. *Nomad*'s high-resolution camera then zoomed in on the rock, determining size and color as evidence that it was a meteorite. On January 25, *Nomad* made its first autonomous discovery of meteoric rock.

Above:
At Arizona University: testing the flexible ramplet down which the rover Spirit, *rolls out of its lander onto the surface of Mars.*
© *Courtesy NASA/JPL*

Page right:
Laboratories are ever more inventive in their efforts to create prototypes of robots that are best adapted to the hostile conditions encountered when exploring planets. Robots are tested on terrain similar in nature to the environments in which they will have to operate.

1- Athena
2- Dante I
3- GoFor
4- Hazbot
5- Marsokhod
6- Nanorover
7- Ratler
8- Robby
9- Rocky I
10- Rocky III
11- Rocky IV
12- Rocky V

© *Courtesy of NASA/JPL*

NASA's Advanced Spacesuit Lab prototype robots for collaborative work between man and machine in space.
© Courtesy of NASA/JPL

The Extravehicular Robotic Assistant (ERA) is the perfect work companion for a space-suited human astronaut. Developed by NASA's Advanced Spacesuit Laboratory, ERA has taken part in simulations of spacewalking tasks that involve working side by side with a human. The results of the simulations have been used to further improve the robot and graduate to more complex collaborative work between robot and human.

The aim of the ERA project is not to develop robots for real-life missions, but to continue to use them as operational test beds for scientists. Researchers at the Johnson Space Center use it to try out advanced technology in the field of natural language and gesture recognition. The outcomes of their experiments will help them to design similar robotic assistants for tasks on Earth, in orbit around Earth, and in deep space. An ERA that can carry astronauts is currently under development.

Fido, *precursor of* Spirit *and* Opportunity

Fido stands for Field Integrated Design Operations and is the name given to a prototype rover from the Jet Propulsion Laboratory in Pasadena. It has a dog's

ERA in the Desert Research Station, Utah.

Lama, developed by the CNES mobility testing team at the Space Center in Toulouse, France.

name and a dog's size—that of a St. Bernard. It is 75 centimeters wide (29 inches), one meter long (3 feet 3 inches), 55 high (1 foot 9 inches) and weighs 70 kilograms (155 pounds). It moves at speeds of 200 meters per hour (650 feet per hour), using its own vision system. Its head and panoramic eyes are perched on top of a long neck, which gives it an all-around view of surroundings. It uses solar energy to power its batteries and can operate without human assistance. *Fido* was first put through its wheeled paces on Nevada desert terrain, similar to the surface of Mars, on May 20, 2000. The JPL control team in Nevada kept track of it from the laboratory in Pasadena. *Fido*'s purpose was to serve as a test bench for the rovers *Spirit* and *Opportunity*.

Fido. © Courtesy of LAAS

Robot laboratories

Research is exploring numerous avenues in attempts to optimize the use and control of space robots and to improve rovers' autonomy.

Faster remote control

Experience has shown that controlling a Mars rover from Earth is anything but ideal. Instructions from mission control take between 40 minutes and several hours to reach Mars, and data transmission in the other direction is just as long. As a result, rovers' responses and movements are maddeningly slow. *Spirit*, which follows a partially preprogrammed route, is a brisk walker compared to others, covering between 27.5 meters (90 feet) and 40 meters (130 feet) per day, depending on conditions.

To offset slow rover locomotion NASA plans to improve data flow by putting in place a Mars network. Creating the network would involve positioning satellite communication relays between vessels orbiting Mars and Earth, in order to use the denser network of 37-meter-wide (120 feet) antennae, which are four times more numerous than their much bigger 70-meter (230 feet) counterparts.

Spiderbots are an area of research on which NASA is currently focusing. Designed to operate in teams, they can reach and move in terrain which is inaccessible to wheeled rovers.

Autonomous robots

An additional objective of ongoing research is to make robots more independent of their tele-present handlers. After Mars 98 was abandoned, a new collaborative program was put in place under the aegis of the CNES, bringing together Russia, Hungary and Spain. Known by its French language acronym, IARES, which stands for Mobile Robotic Autonomous Illustrator for Space Exploration, the project aims to create an autonomous robot with an integrated computer that can map hazards and determine the best route by itself.

Hyperion is a prototype robot that has demonstrated its ability to cross the Canadian Arctic by itself. Developed by Carnegie Mellon University's Robotics Institute, Hyperion carefully chooses its navigational route to avoid dark and shade, so that its solar panels receive as much light as possible for as long as possible. Furthermore, the robot is smart enough to know when it is lost or in trouble and to take corrective action.

Spider robots

Spiderbots were developed by JPL and, as the term suggests, they do not use wheels but a spiderlike style of locomotion, albeit on six legs. They were designed by robotics engineer, Robert Hogg, who makes the important point that "rovers have very efficient wheels, but there are things we'd like to explore with legs that you can't do with wheels."

One of spiderbots' advantages is their diminutive size—the first prototypes can nestle in the palm of your hand. Their antennae detect obstacles, while they use their cameras to explore their environment. They belong to the collaborative school of robotics, which currently has the wind in its sails. Spiderbots are designed to work in large groups. Their legs are already instrumented, but to endow them with greater capabilities like running, climbing, digging, and repairing, they could be given even more legs—between 12 and 50. Their resulting versatility would enable them to adapt effectively to different tasks in different environments.

Once fully developed, spiderbots should be able to exchange information, constantly "aware" of each others' positions. Although primarily intended to explore planets' surfaces, they could well be useful for other tasks, like maintenance work at the International Space Station.

Floating robots

To date, robot rovers' exploratory missions have seldom exceeded a few dozen meters per day, because their energy is intended to focus more on collecting data than on moving. To save energy and travel farther, robots could be carried by hot-air balloons. That is the idea of a French researcher, Jacques Blamont. His balloons would require no energy to power them, while their range would be much greater than that of a wheeled or crawling robot. There have also been proposals for solar balloons that would simply float over ground-level hazards, carrying robots long distances between exploration sites.

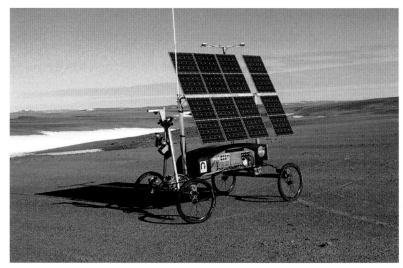

Left:
Many-legged space robots would be energy efficient, covering long distances and carrying heavy loads.

Below:
The prototype robot Hyperion.

Entomopters

Is it an insect or is it a bird? The answer is "yes", because the entomopter is a little of both. Great hope has been placed in them for exploration of Mars because of their ability to fly and land. Once again, NASA's concern is to increase the range of robots that explore the surface of the red planet. Locomotion is their weak point.

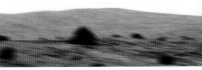

Fixed-wing flying rovers have been dismissed as a viable solution because they would have to fly at speeds of over 420 km/h (250 mph) to stay in the air in the rarefied atmosphere of Mars. At that kind of speed, they would also have to be acrobats to land.

Insects were the answer, thought the researchers who developed the entomopter. Robert Michelson's team at the Georgia Tech Research Institute collaborated with researchers from Cambridge University and ETS Laboratories in work on the flying robots. The flapping motion of their wings is driven by a mechanism known as Reciprocating Chemical Muscle (RCM). Variations in the electrical charges that actuate the RCM motion cause variations in wing beats, which makes directional flight possible.

Originally under study for military reconnaissance use, entomopters came to the attention of NASA because their attributes would enable them to fly over the surface of Mars. To that end the surface area of their wings and their flap rate would have to be greatly increased because gravity on Mars is 37 percent less than on Earth.

The flying robot could fly slowly, even hover, over Martian terrain, landing to gather samples and returning to its launch point to recharge its batteries, transmit data, then take off again. An advanced version of the entomopter will soon, it is thought, be part of missions to Mars.

The operating range of an entomopter is potentially very wide, not only because of its flying capability, but because the rover on which it docks can also move.

The objective of Garnet Hertz's project,
"Control and Communication in the
Animal" is to develop a cockroach-
controlled mobile robot. The insect-driven
system fits into a context relating to
notions of intelligence, hybridity and
post-human embodiment. Garnet Hertz
is a Canadian media artist, Fulbright
scholar, and research fellow at the
California Institute for Telecommunications
and Information Technology.
He is currently developing his work under
the auspices of the interdisciplinary
Arts/Computation/Engineering Graduate
Program at the University of California
Irvine.

© Courtesy of Garnet Hertz

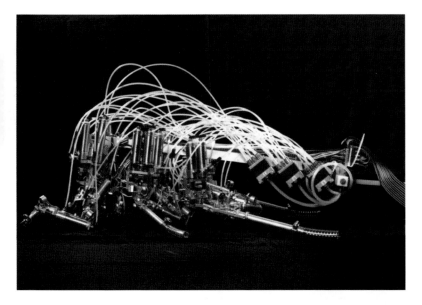

Robot III, a hexapod with 24 degrees of freedom, is based on studies of the cockroach Blaberus discoidalis.
© Courtesy of Roger D. Quinn - Case Western Reserve University Cleveland

Robots from life

Tourists who climb valiantly up the steps of the Eiffel Tower are always admiring of the stability of the structure, which was originally designed as a temporary exhibit. Another thing that they may not know is that its architecture was based on the human skeleton.

Forty years before the Eiffel Tower was built, Swiss professor of anatomy Hermann von Meyer, from Zurich, grew intrigued by the way in which the hip bone is connected to the thigh bone, etc. He published a paper describing how the structure of the skeleton pushed the body away from its center. Building on Von

Meyer's description, engineer Karl Cullman showed that our bones spread weight away from the center of the body, redirecting it into the long bones of the legs. Gustave Eiffel designed his famous tower by distributing lines of force in the same way as the human body does.

Closer to us, all-pervasive Velcro was also born from the observation of nature. It was invented by the Swiss engineer Georges de Mestral, who noticed, on returning from walking his dog, that burdock had snagged in its fur. His clothes, too, were covered in burrs. Intrigued, Mestral thought that he could make a material that would fasten to other materials with as much grip as burrs. After years of trying, he finally invented Velcro

from two strips of nylon, one of which had thousands of tiny plastic hooks, the other thousands of tiny loops.

The past is full of examples of ways in which nature has been the scientist's muse. What is new today is the deliberate study of natural mechanisms to find ways in which they could serve as models for machines whose motion and behavior would mimic living creatures. This study is known as "biomimetics", or "biomimicry".

Jack Steven is generally credited with originating biomimetics. Steven was a researcher with the US Air Force. Addressing a meeting at the Wright-Patterson airbase in Ohio, he defined the science of copying nature's attributes and applying them to machines as "bionics". Since then, however, science fiction's liberal use of the term has led to its coming to mean the replacement of biological body parts with artificial—mechanical and electronic—ones. To redefine the focus of their studies, chemists proposed the term "biomimetics," or "biomimicry."

The general philosophy behind the discipline is as follows. In order to ensure their survival as species, animals and plants have evolved, developing ideal solutions to the dangers and problems posed by the nature of their environment. They have thus arrived at optimal solutions, which incorporate factors like size, weight, and energy needs.

One astonishing example is the survival strategy evolved by ants living in icy conditions. They dig underground shelters in which they huddle together to keep warm. Other insects produce great quantities of antifreeze. A species of fly, 70 percent of whose body weight is water, is able to live at minus 60° without freezing. "The solutions evolved by some insects could revolutionize robotics," says Nicolas Franceschini, who leads the CNRS research unit in Marseilles. He has been studying the vision systems of flies for thirty years for application to robotics.

The fabulous creatures of MIT

One of the world's foremost research and development organizations, MIT is involved in numerous biomimetics projects applied to the field of robotics. Robostrider is one such project and has produced a result that would be described as a miracle if performed by a human: walking on water.

Robostrider resembles the aquatic, foraging insect after which it is named, the water strider. Indeed, at first sight, it looks nothing like a robot. The water strider can walk on water, even though the weight of its body should cause it to sink. Patient observation of the tiny vortexlike waves it produced as it propelled itself over the water revealed to scientists that it moved by rowing, at a speed of 25 centimeters (10 inches) per second, without breaking the surface. Robostrider has proved researchers right, creating minute whirlpools in the surface tension of the water to stay afloat and upright.

As part of research into how liquids behave at a very small scale, another MIT team has developed a snail robot, which measures 30 centimeters (one foot) in length. Mounted on a rubber foot, it can move through muddy terrain.

The RoboLobster is the work of Thomas Consi, who has applied his observations of the crustacean to robotics. Lobsters have feelers that detect the clouds of excrement discharged by the snailfish on which it feeds. Similarly RoboLobster's sensors could sense chemical pollutants, such as oil, in water.

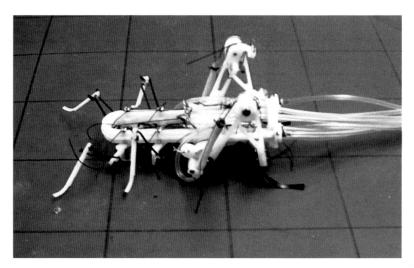

The rear leg of an autonomous prototype robot based on the cricket. It can walk and hop.
© Courtesy of Roger D. Quinn - Case Western Reserve University Cleveland

The spider robot with its creator, Amir Shapira.
© Courtesy of Technion

Cloning lizards and snakes

The Poly-PEDAL Laboratory at Berkeley studies the mechanisms that enable insects, reptiles and amphibians to climb. One animal that has greatly intrigued researchers is the gecko. It can clamber about on any kind of surface by positioning its five "fingers" flat against that surface. Their ability to grip comes from the microscopic hooks and suckers covering them. If the gecko's natural biotechnique for climbing could be translated to a robot, its portability and "upward mobility" could have some useful military reconnaissance applications.

Researchers at the Tokyo Institute of Technology have been exploring ways of developing a robot that can crawl like a snake. Souryu-I has now evolved from numerous prototypes. It weighs 10.2 kilograms (22.5 pounds) and is 1.2 meters (4 feet) long. Like its real-life model, it can slip into nooks and crannies. If it loses its balance, it can right itself and continue on its way. Souryu-I incorporates a camera and microphone and is intended for use in rescue missions to find survivors of earthquakes.

Carnegie Mellon University is also working on snake-shaped search robots.

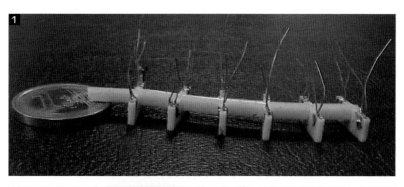

1- The robotic worm from the Center for Research in Microengineering.
© Courtesy of Lucia Beccai, Scuola Superiore Sant'Anna in Pise

2- JPL's visual-inspection snake robot.
© Courtesy of NASA/JPL

understand the fluid mechanics they use to swim and turn so fast without being hindered by drag. On a strictly mathematical level, the muscles of fish and water mammals are not strong enough to produce such speeds. The dolphin, however, travels seven times faster than predicted by equations! So much for equations.

With its six legs, iRobot's experimental Ariel looks like a crab. It can withstand collisions, move forwards and backwards, carry weights of six kilograms, scale obstacles and cross fissures that are too wide for wheeled robots. Ariel was developed under the aegis of the US Army and Navy to detect mines and dispose of them by squeezing them in its pincers until they explode.

iRobot's fish and crab

Although it withheld details of its work because of its confidential, military nature, iRobot began some biomimetics research in (it is thought) late 1999 or early 2000. Among the Boston-based company's projects was Darts, the aim of which was to develop the efficiency, speed and agility of fish in autonomous submarine robots, and Ariel, a mine-detecting crab.

Darts draws on research at MIT's Oceanographic Laboratory by John Kumph, which led to the prototype RoboPike. The artificial fish is 80 centimeters (2 feet 7 inches) long, has a fiberglass skeleton housing a motor, and a flexible Lycra skin. It propels itself by side-to-side movements of its tail fin, a motion that mimics the way in which the real pike slides so smoothly through the water. Kumph worked on pike to try to

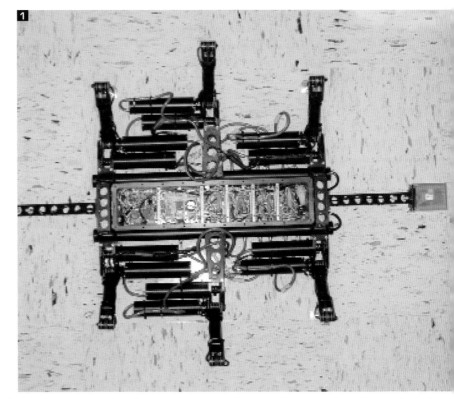

1- Ariel is a robot whose biological model is the crab. Its legs enable it to scale high, wide obstacles. Working together, a group of Ariel robots can capture and defuse mines in breaking waves.
© Courtesy of iRobot

2- Genghis the robot.
© Courtesy of iRobot

Cloning and mixing with nature

An effective way of observing the behavior of wild animals can be to introduce an artificial replica that runs with the pack or swims with the school. In July 2003 the National Marine Aquarium in Plymouth, UK, placed an artificial shark in a tank. Roboshark 2 was modeled on the gray shark and behaved like one, trying to push the real sand tiger sharks around. Roboshark 2 moves autonomously at a speed of 4.8 km/h (3 miles per hour), is two meters (over 6 feet) long, and weighs 35 kilograms (77 pounds). Its sole practical limit is that its batteries have an operating time of only four hours.

The following pages describe further examples of biomimetics in which robotic creatures are put to a wide range of uses.

EOD PackBot.
© Courtesy of iRobot

Robots to the rescue

The tragic attack on the World Trade Center on September 11, 2001 has etched itself indelibly on our minds. It immediately became evident from the sheer scale of the damage inflicted and the difficulty rescue teams faced in reaching possible survivors that human resources were not enough. Highly specialized equipment was needed. Dozens of robots were swiftly put to work.

A retired marine, Lieutenant-Colonel John Blitch, was put in charge of a rescue team from Colorado, that used reconnaissance robots that the army released to help in the emergency. They included shape-changing robots, which flatten themselves to crawl through narrow openings, and rear up so as to scale walls and see what is happening on the other side. An expert from Carnegie Mellon University says that the shape-changing scout robots were based on Urbie, a military robot used for reconnaissance missions that operates semiautonomously: it can be remote controlled, but it is also capable of locating its bearings and knowing whether to raise or flatten itself.

Another team at work at Ground Zero was led by Professor Robin Murphy of the University of South Florida. She had lead-designed seven experimental "mother" marsupial robots. These carried smaller child robots, which they released into holes and crevices in the mountains of rubble which were too small for the mothers to reach. Mothers and "daughters" communicated through tethers, but also through wireless technology. When signs of life or possible survival pockets were detected, the mothers informed rescue crews.

Inuktun, a Canadian British Columbia-based company, which supplied the US Navy, joined in the effort. It provided its miniature robots, usually used to inspect nuclear power station pipes and underwater constructions, to comb through the rubble of the World Trade Center. Their ability to push themselves into the tiniest voids was vital.

On hearing of the tragedy, Inuktun's engineers immediately reconfigured the robots: they could be tethered, remote controlled, and carry spare batteries. One week later, they were ready and were dispatched to Ground Zero where they searched in rubble that would otherwise have had to be cleared by huge bulldozers and excavators, so risking the lives of any survivors.

The September 11 rescue robots

The Center for Robot Assisted Search and Rescue (CRASAR) is a nonprofit organization run since 2002 by Robin R. Murphy, who teaches robotics and cognitive sciences at the University of South Florida. Immediately on learning of the terrorist attack on the World Trade Center, Murphy and her team headed for New York with their search-and-rescue robots. Their determination to save lives led them to carry out five search-and-rescue missions with their robots in 11 days. Immune to the dust and toxic air, the devices located five survivors. To enhance her robots' life-saving capabilities, Murphy is working on algorithms enabling them to recognize shapes and colors.

© Courtesy of Crasar

1- InuktunCopter. This unmanned aerial vehicle flies reconnaissance missions over disaster areas. The information they collect facilitates the tasks of robots searching on the ground.
© Courtesy of Crasar

2- Inuktun is equipped with an infrared, heat-sensitive camera. For search-and-rescue operations vision systems and sensors are crucial.
© Courtesy of Crasar

3- Medical instruments on robots enable the injured to be diagnosed rapidly.
© Courtesy of Crasar

4- Enryu is one of the world's largest robots. It was designed to clear debris with its two arms and lift people trapped in debris to safety. It can be controlled by a human riding onboard, or, if a location is particularly dangerous, remote controlled.
© Courtesy of Tomotaka Takahashi

Caterpillar robot

If there is one country where search-and-rescue resources is a priority concern, that country is Japan, prone as it is to earthquakes. Accordingly, the government sets great value by robotic rescue applications. One such system, the brainchild of a Tokyo roboticist, Kazuo Yamafuji, is an exoskeletal arm that, when donned, increases the strength of the human arm eight fold. That kind of strength can make the difference between life and death when trying to pull a survivor clear of the debris. Nevertheless in Japan and elsewhere, rescue robots are essentially automated exploration machines.

Biomimetics could be a promising new area of rescue robotics research. Tokyo University and the Kyoto Science University have both designed robots based on caterpillars for searching for survivors. Kohga is two meters (6 feet 8 inches) in length and is remote controlled. Its batteries' operating time is only 30 minutes. Its designer, Fumitoshi Matsuno, explains that he had to find a trade-off between longer operating time, and therefore bigger batteries, and weight. The caterpillar's body should be as light as

possible to avoid the risk of disturbing rubble and causing a building to further collapse. Moira is shorter than Kohga and can wend through debris autonomously.

Prevention is better than cure

When lives are in peril they must be saved. Better still is to prevent accidents in the first place. Once again, robots can play a crucial part.

A good example is Poseidon, the swimming accident-prevention robot, designed by Vision IQ from a Frédéric Guichard original idea. An outstanding feature is its sophisticated computer-assisted vision system. Guichard describes it as "a veritable accident-prevention aid" for swimming pools. Its strength resides in the fact that it can gain seconds, so making the difference between life and death. If a drowning person loses consciousness and is revived more than three minutes later, then lesions can be fatal, or permanently life impairing. Less than three minutes later, and he or she is able to recover. To prevent an accident from getting to that stage, Poseidon alerts lifeguards as soon as it detects someone in difficulty. It provides the exact location in the pool, enabling the lifeguards to act immediately. Poseidon is able to respond so quickly thanks to its combined aerial and underwater cameras, which transmit images to a computer. The computer analyzes them and detects potential danger.

Sea exploration robots

On deep-sea ocean floors visibility is nil, while pressure increases to beyond levels that a human can withstand. Robots are ideally suited to deep-sea beds, diving as far down as 10,000 meters (over 30,000 feet). Their tasks include laying and maintaining pipelines, servicing oil rigs, searching for wrecks, and scientific exploration missions.

As early as the late 19th century, scientists sought to design and build remote-controlled unmanned undersea vehicles that could dive deep. The first marine remotely operated vehicle (ROV) was developed by the underwater photographer Dimitri Rebikoff, undertaking its first mission in 1953.

Throughout the 1960s and 1970s, the US Navy worked extensively on developing robotic ROVs, principally to retrieve craft and equipment lost during submarine experiments. In 1966 an underwater ROV helped to find an atomic bomb that had gone astray off the coast of Spain. In 1973 ROVs played an important part in saving the lives of a submarine crew off the coast of County Cork in Ireland.

In the last 20 years, the number of robotic submarine vehicles in operation has grown fast, particularly in the oil industry. It was just such a craft that helped to find the wreckage of the *Titanic* in 1986. The blaze of media attention that followed threw the spotlight on these deep-sea diving machines. Robert Ballard's blow-by-blow account of his discovery and retrieval of the *Titanic*'s wreck became a global best-seller, while James Cameron's 1998 film, *Titanic*, revealed the underwater robot to the filmgoing public. One scene showed a submersible being released into the lower depths to explore the insides of the great liner's broken hull.

Some 3,000 robotic submersibles are now in operation worldwide and their capabilities have been considerably upgraded. Their vision systems use halogen lighting to illuminate the spots they film, sending live pictures back to the surface. To navigate they use sonar, a device that detects distance and location by emitting sound-frequency waves and registering their vibrations when they bounce back from objects. Dolphins use highly sophisticated sonar systems.

Gregory Dudek, director of the Mobile Robotics Lab at Canada's McGill University's Center for Intelligent Machines, asserts that underwater terrain is much trickier to negotiate than the surface of Mars, where wheeled rover technology has been mastered. Deep down, robots have to cope with extreme, shifting environments made up of high pressures, swift currents, low or no visibility, and the constant risk of collision. Experiments with Dudek's robot, AQUA, in coral reefs off the coast of Barbados confirmed the ability of the small robot (it is only 65 centimeters, or just over two feet, long) to use its six flippers to swim, walk and stand motionless underwater. It can navigate by itself and build imaged models of what it sees.

The underwater robots of JAMSTEC
(Japan Marine Science and Technology Center)

1- *Hyper Dolphin.*
2- *Kaiko, a remote-controlled ROV capable of diving as deep as 10,000 meters (32,000 feet) to study marine microlife.*
3- *Urashima is an autonomous underwater vehicle developed by JAMSTEC. It can range over 300 kilometers (186 miles).*

© Courtesy of Jamstec

One of the best-known robotic ROVs is the Triton XL, which came to the fore in early 2000 when it was used to prevent seepage from the oil tanker *Erika*, which was sinking off the coast of Brittany, France. The three-ton robot boasts additional capabilities, like imaging, building, digging and equipment maintenance.

In January 2004 a robot was also used to retrieve the flight recorder and fuselage of the Flash Airlines Boeing 737-300 that had come down in the Red Sea and lay at a depth of between 600 and 800 meters (2000 and 2600 feet). The French authorities contracted the underwater engineering group Comex to retrieve the wreck. The company used Super Achille, a robot capable of working at depths of 1000 meters (3,000 feet) assisted by a similar ROV, Scorpio.

IFREMER

Submersible robots have long been used in marine exploration, engineering and rescue operations. Increasingly they are now being used for surveillance tasks. IFREMER has drawn on its wealth of deep-sea technology to design autonomous and remote-controlled underwater vehicles (AUVs and ROVs) for scientific missions such as mapping and offshore inspection.
1- Asterix the AUV.
2- Robin the ROV, here inspecting a wreck, is remote controlled from a submarine.
3- Victor 6000. Photo: S. Lesbats
4- Victor 6000 and Atalante.
Photo: O. Dugornay
© Courtesy of IFREMER

Mapping corals and studying whales

France's marine research agency, IFREMER, has developed a robot dubbed *Victor 6000*, built for deep-sea exploration. The equipment bay on its underside is modular, enabling its instrumentation to be configured to its mission requirements.

Between June 1, and August 7, 2003 *Victor 6000* took part in two scientific missions to map cold-water coral growths as part of a German polar research program. On board the research and supply vessel, *Polarstern*, 150 researchers operated *Victor 6000* to explore the mysterious; cold-water coral, *Lophelia pertusa*, off the southwest coast of Ireland and on the Haakon Mosby "mud volcano" off Norway.

Japan's autonomous undersea robot, *Aqua Explorer 2000*, is made to track shoals of fish. Developed by Tokyo University and the KDDI research center, the robot is three meters (10 feet) long and weighs 300 kilograms (660 pounds). It can dive to depths of 2,000 meters (6,000 feet) and, with an acoustics system comprising ultrasensitive sonar, can detect whales from distances of 5 kilometers (3 miles).

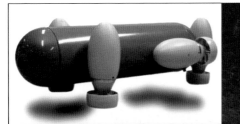

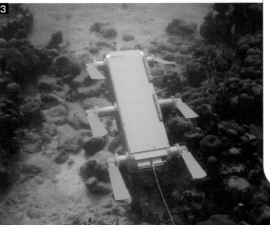

1- The small 40-centimeter (15-inch) Serafina, *designed and developed by Uwe R. Zimmer's team, can move at speeds of up to one meter (39 inches) per second.*
© Courtesy of Uwe R. Zimmer, Australian National University

2- Deep Sea Crawler *is controlled via the Internet. Its great flexibility makes it ideal for underwater observation and in situ experiments.*
© Courtesy of Dr. Volker Karpen, International University Bremen and Meerestechnik Bremen GmbH

3- The autonomous underwater robot AQUA, *was developed from the terrestrial RHex, a cockroachlike, hexapod robot.*
© Courtesy of Michael Jenkin, Computer Science and Engineering, York University

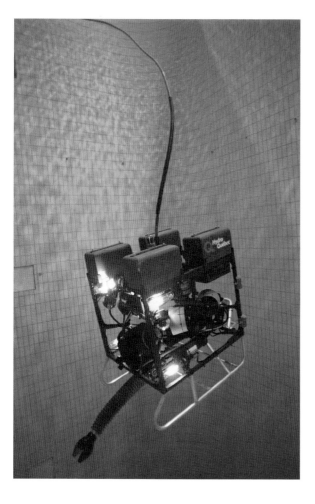

Hydro-Quebec's robots are tailored to the needs of hydroelectric utilities operators. One of the strengths of their technology is that it integrates a robot for carrying and guiding the robot.
© Courtesy of Hydro-Quebec

tems for high-voltage transportation—levitation, supply of current, lowering of voltage. Six 500-watt projectors shed light on the murky waters, while footage from the four video cameras that build 3D pictures of the robot's environment enables its handlers to navigate.

Using the robot has reduced the numbers of inspection missions by human divers, and enabling some tasks to be carried out from a distance. Between 2000 and 2003 the robot brought Hydro-Quebec savings of $5.1 million. In February 2003, when the third version of the robot was already in operation, a project to develop an industrial prototype got underway with an investment budget of $4.5 million.

The dam unbusters

The technological expertise of the Hydro-Quebec Research Institute in Canada includes the maintenance of some 500 dams, of which 100 were built over fifty years ago. To ensure that they are safe, they have to be constantly monitored and any cracks instantly reported and repaired.

Inspection of the immersed part of dam walls, which can be up to 200 meters (over 600 feet) deep, was traditionally entrusted to teams of divers. Because they had to contend with hazardous working conditions due to very low visibility and floating debris, a robot seemed like a very useful addition to the team. In the mid-1990s one was introduced. It was equipped with seven 1.5 hp motors, which operated along the lines of power sys-

Subterranean robots

Venturing into the bowels of the earth is an undertaking that is fraught with risk for humans. We cannot withstand the high temperatures or toxic substances, and are the wrong shape and size for narrow passages and tunnels.

French railroad operator SNCF found itself facing a problem. Its tracks ran over bridges that were too low for humans to worm themselves into. To explore and inspect beneath them, it contracted a company called ROV Développement to design robots that could slip into spaces that measured less than 70 centimeters (27 inches) and carry out inspections. Until then, engineers could only perform visual checks.

Missions impossible

Robots are the obvious answer when extremely high temperatures or radioactive toxicity preclude physical human presence. The Hanford Nuclear Site in Washington State is the most heavily contaminated site in the US. During the Cold War, it became a dumping ground for much of the military's spent plutonium. To dismantle and clean out the underground storage facil-

ities, which held 250 million liters of toxic substances, remote-controlled robots were designed. Their task was to go three meters underground to analyze waste, remove it, and dismantle the installations.

Collapsed mine shafts are another example of places that are much too dangerous for human teams. Whether to rescue miners trapped underground, to analyze the air, or to reconnoiter, robotic scouts are the answer. But that is easier said than done.

Most of the robots that trundle through mineshafts are rover-like vehicles that are remote-controlled from the surface. They are equipped with powerful sensing technology, such as infrared vision, and can communicate over long distances by radio frequency. In addition they must be able to proceed along extremely rugged, uneven terrain with abrupt changes in level.

On November 24, 1998 Rattler was dispatched to the rescue after fire had broken out in the Willow Creek coal mine in the US state of Utah. Rattler, a robot made by Sandier, descended into the hell-hole, and successfully managed to make its way up and

down steep inclines and across gaping cracks, while mine personnel watched it go from a safe distance.

Climbing spider

Biology and electronics have come together in biomimetics to yield some unlikely creatures. Roboclimber resembles a spider. Yet it has only four legs and weighs over 3,000 kilograms (about 3 tons). Still under development as part of the CRAFT technology program funded by the European Union, it is set to be one of the world's largest robots. Its role is to prevent landslides, a safety operation that is currently hazardous because it involves installing and mounting scaffolding on unstable ground. The remote-controlled Roboclimber can work horizontally and vertically. If analysis of a slope reveals that it is likely to slip, the robot drills holes and inserts into them 20-meter-long (60-foot) stabilizing poles to hold the loose terrain.

The spider robot designed by NASA will be used to inspect fuel storage tanks and the surface of distant planets. They work in collaborative groups on their allotted task.

In the land
of the little

Tiny robots have a huge future. In 1998 Toshiba unveiled a robot so minute that it can be sent into pipes as narrow as 25 millimeters to retrieve unwanted foreign bodies. The metallic insect weighs 16 grams (half of one ounce) and is equipped with a camera measuring 7-by-12 millimeters to enable operators to see what it is doing. Large-scale research work into micro-scale robots is ongoing in Japan for applications in a number of fields, most notably medicine (*see Chapters 8 and 11*).

Cryobot, a robot that can drill through ice to depths of 30 meters (over 90 feet).

Inside the pyramids

Robots have also proved valuable aids in missions to uncover our cultural heritage. The exploration of the pyramids is a graphic example.

For centuries the Queen's Chamber in the Great Pyramid of Cheops that lies on the Gizeh Plain near Cairo remained impenetrable. In March 1993, 4,500 years after it had been built, came a major breakthrough (literally). German engineer Rudolf Gantenbrink sent a miniature robot, Upuaut II, into passages that led to the Queen's Chamber. To the south of the chamber, it entered a mysterious passage that was 20 centimeters (8 inches) wide and sloped at an angle of 40°. The shaft stretched over 60 meters (over 200 feet) into the heart of the pyramid. At the end of the shaft, the robot came to a limestone door with copper handles that had been sealed to the door. The door intrigued Egyptologists the world over. What could be behind it? Papyrus parchments? Tools? A statue of the Pharaoh Khufu, who built the pyramid? The beginnings of an explanation would not be known for ten years.

The Egyptian authorities, who had taken charge of the project, gave the go-ahead to a National Geographic expedition, which used a robot built by iRobot. Dubbed the Pyramid Rover it had cost more than $250,000 to build.

Under the excited supervision of Egyptologist Zahi Hawass, Pyramid Rover began its journey into the past on September 17. The National Geographic Channel broadcast it live to over 140

Pyramid Rover

National Geographic scientists and robotics engineers from iRobot designed the little robot to explore mysterious shafts in the Great Pyramid.

© Courtesy of iRobot

countries. Pyramid Rover took two hours to proceed the full length of the shaft. On reaching the door with the copper handles, it drilled a small hole into it. It introduced a miniature camera into the hole and started its visual reconnaissance. Suspense was at a breaking point. The Queen's Chamber was about to yield up its secret! Or was it? No, because behind the door was another door.

Hawass, who had guided the robot, was exultant, even though the mystery had now deepened. He explained that the discovery of the door was the most important one in the Great Pyramid for 130 years and that it could radically change our understanding of its interior geography. He felt it was significant that the door stood at the same distance from the Queen's Chamber as another door discovered at an earlier date. Other Egyptologists, however, voiced disappointment, saying the robot's journey down the shaft had received lavish media coverage for a meager result. Hawass also came in for criticism for what some felt were extravagant statements.

There are plans for another robot to probe the secret of the second door in 2005. Whatever lies behind it, the tiny robotic archeologists will have played a crucial part in the discovery.

This flying robot, Blimp 2, mimics biological mechanisms, particularly those of the fly, to navigate autonomously.
© Courtesy of Jean-Christophe Zufferey, Swiss Federal Institute of Technology, EPFL

A human extension

Eventually fully autonomous robots will be more effective explorers and adventurers. For the time being, however—a time that will probably extend over many years—a robot's software will continue to be inseparable from the human being controlling the device. Robots that go where humans cannot or will not tread thus act as powerful interfaces that extend our range of perception and action.

The priority in the short term is to augment and refine machines' sensing capabilities— e.g., 360-degree sound fields, enhanced articulation, heat, air, and taste sensors, infrared vision systems and their ability to combine them to build hi-fidelity pictures of what they sense.

With such sophisticatedly sentient robots, telepresence would be as good as being there, in environments that might be life-threatening because of high radioactive levels, low oxygen, intense heat or low gravity.

Until such time as they can go it alone, then, robots will act as extensions of the much more fragile human body, breaking down barriers that once physically curbed human curiosity.

Interview
Francesco Mondada

Author and coauthor of a number of books on robots, Francesco Mondada began his research into mobile robotics in 1991 at the internationally renowned Swiss engineering university Ecole Polytechnique Fédérale de Lausanne (EPFL). He spent five years in industry as chief executive of the K-Team company, before deciding to return to research to develop a new concept in robotic explorers.

Could you describe the EPFL's work in robotics research?
EPFL pioneered research into biobased and evolutionary robotics. It was here at the EPFL that the first genetic algorithms were implemented on an actual robot. Before that researchers had always used simulations. In 1992 the EPFL began developing the robot Khepera which has become widely used as an educational and research platform. There are several thousand of them in use worldwide. There are some fifty mobile robotics researchers at EPFL. They work in a wide range of related branches, like biomimetics, all-terrain exploration and interior mapping.

Francesco Mondada.
© Courtesy of Francesco Mondada

What are EPFL's greatest achievements in terms of explorer robots?
There's one that maybe stands out from the rest. It's Shrimp. It's an all-terrain, six-wheeled robot with passive mechanisms that is capable of negotiating large obstacles. It automatically adapts to the terrain it encounters. It can climb stars, walk on flat ground, configuring itself to terrain as it comes to it. Its joints are passive, which means that it doesn't lift its wheels or change their shape when it comes to an obstacle. Motors are located only in the wheels; it's the articulation system, designed from axles and levers, that adapts to surfaces.

You yourself work on robots that come together in groups to explore locations that are difficult to access. How do they work?

Swarmbots are relatively simple robots that do not individually have any special capacity to scale obstacles. The can hook up to each other and so form super-structures consisting of as many as thirty or forty robots. They assemble into these collective structures by using hooks and loops. They have two each, so every robot can hook into another's loop, and so on. They can form caterpillars, robot chains, circles, squares, and more.

Can Swarmbots stack one on top of each other?
No. For the time being they operate in "two-and-a-half" dimensions. They can bend and are strong enough to lift another robot. But we haven't made much progress in 3D operation and perception because there have been enough challenges in two dimensions. For example, if a robot comes to a hole that it cannot cross by itself, it calls its friends and together they build themselves into a structure that will enable them to get across. Our work in this field draws directly on the way ants behave.

What kind of practical applications are there for such robots?
They would be useful in sudden, unexpected situations for which they have no map of the environment. They could be

Swarmbots can work in antlike colonies.
© Courtesy of LSA, EPFL, Swarm-bots.org

useful for search-and-rescue operations or for exploring the surface of a planet or the ocean bed. Anywhere, in fact, where an unexpected situation calls for flexible robots to respond.

When will we see Swarmbots in routine use?

I don't want to be overoptimistic, because I think the kind of development required takes time. Ten years ago it was commonplace to think that some applications would materialize quickly, like the robot postman, for example. That was all the rage. But it didn't happen. All we've got at the moment are vacuum-cleaning robots, which are less efficient than the hand-held devices most of us still use. I say all that because I want to try to get across why I don't think explorer Swarmbots will play a part in helping human explorers for another fifteen years at the earliest. The fact is that there are still lots of problems related to robots' sensing capabilities. Many robots became useless for search-and-rescue operations at Ground Zero after the September 11 attacks. Basic factors, like dust, clogged up their sensors. They also had problems navigating in the rubble.

... Interview cont.

Resolving all those problems is a complex matter and people have a tendency to simplify and get ahead of themselves. Speed doesn't necessarily prevent researchers from getting to the bottom of problems, but when they want to step up the pace, they naturally overlook the complexity of real-life environments. For example, just take a child's bedroom and try to model it, deciding what's tidy, what can be thrown out, and what goes in the toy box.

How do you assess the exploration robotics sector globally at the current time?
Space and underwater exploration are the fields where robotics has proved its use. The same can be said for pipes and ventilation systems. But that's where things stop, because they are specific applications for which robots can be configured. In a building that has collapsed, constraints are totally different and there's still a lot of research to be done. That said, such applications are typically those where robotics comes into its own, either because they're too dangerous for humans or because humans are too big.

To what extent can robots make it possible for humans to explore fields where they could not go before?
The robot is par excellence an example of how humans make tools that enable them to break new ground.
In space exploration that's obvious. Microexploration, the exploration of the very small, still belongs to science fiction, but we're getting there. Limitations are primarily of a mechanical nature.

What technological breakthroughs are to be expected in coming years, which could bring major advances in robotics?
We can now build good motors and there is a lot of focus on improving batteries and operating times. We're looking for a breakthrough in sensing technology. Distance-measuring laser sensors, which map the environment, are still very costly— between $6,500 and $13,000. It's at this level that major technological progress is needed.

Solero *(in brown),* the Shrimp *(in blue) and their little brother,* U-Shrimp.
© *Courtesy of ASL, LSA, EPFL, BlueBotics*

Interview
Raja Chatila

Raja Chatila is director of research at France's information technology research center, Laboratoire d'Analyse et d'Architecture des Systèmes (LAAS), based in Toulouse in southwest France. He leads research into robotics and artificial intelligence.

Raja Chatila is director
of research at LAAS.
© Courtesy of Raja Chatila

The Eden project was designed to produce a robotic explorer. What were its practical applications?
Our challenge was to develop a robot that was able to move and explore an environment that it did not know, the kind of functions that are required today for exploring Mars. By addressing the question of autonomous decision-taking, we developed an architecture that helped to structure a robot's capabilities. The robot we developed was called Lama. It has never taken part in any real missions, but that's because Europe has abandoned planetary exploration. Nevertheless our research did lead to applications such as spacecraft and satellites, where we optimized robotic

tasks, which kicked in at the right moment once the situation had been identified and assessed. Another area of progress has been hot-air balloons, which are like autonomous helicopters and pose the same problems in a different form. The Karma balloon was used within the framework of the European project Comete, the aim of which was the coordinated use of several aerial vehicles to detect forest fires. Karma was tested in Portugal in an experiment where a forest fire was simulated.

What is the aim of the group that works on predicting, tolerating and eliminating defects in robotic systems?
The aim is to try to factor in the unforeseeable, given that the robot's environment is one that evolves and that there can be breakdowns and sensors that return data not precise enough. The aim is to make the decision-making capability robust and able to tolerate imprecision, so that a robot can adapt to its environment and any internal inadequacies.

What other areas of research is the robotics and artificial intelligence group working on?
The main thrust is to provide computer-related entities with intelligence—that spans a robot's environmental perception ability, locomotion, obstacle avoidance, and interaction with other robots and human beings. In 2000 we started up a company called Kineo that is dedicated to marketing motion planning software.

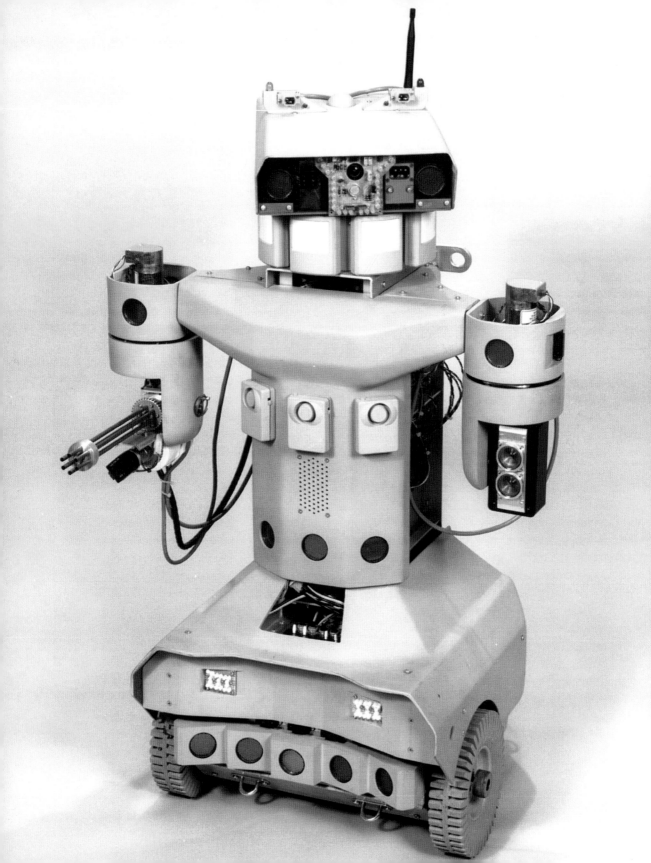

SECURITY ROBOTS

"Somebody's got to do the dirty work," could well be the motto of the rugged robots that the military uses. Whether their mission is furtive reconnaissance, the disposal of explosive ordnance devices, or to shoot on sight, the all-action robots that intervene on the battlefields "ain't no softies".

Tough, rough and rugged they might be, but, apart from the unmanned vehicles and drone planes that are now a routine part of warfare, they have not come into their own because they are not yet failsafe.

Autonomous security robot prototype Robart was developed as part of the US Army-Navy Mobile Detection Assessment and Response System (MDARS). Robart I could only sense suspected intruders, while its follow-up, Robart II, sensed and assessed the threat they posed. Machine-gun toting Robart III (see picture) can take them on, though nonlethally. Its weapon is a six-barrel airgun that simultaneously fires tranquilizer darts and rubber bullets.

© Courtesy of SPAWAR.

Robot war

The mission of the robot soldier is to save lives—those of the human troops on its side, at least. A praiseworthy mission, indeed, but one that is not dictated by purely humanitarian concerns. However just they may claim a *casus belli* to be, heads of state are loath to have to announce loss of life on the battlefield or that troops have gone missing in action. Casualties in war cause casualties in politics. But other factors, too, warrant the use of automated combatants—ones of a practical nature. Most Western countries have

abolished compulsory military service, so there are fewer troops to be called upon in the event of conflict. Furthermore, there are missions where humans are ineffective or at risk, particularly if chemical or nuclear weapons are deployed.

So robots have been enlisted as players in the theater of war. They look set to eventually take over from soldiers for missions like mine sweeping, stealth reconnaissance or just wreaking havoc in their wake. At times handlers will remote-control them, at others they will act alone.

They can already claim to have possibly saved lives. In April

2004 the chief executive of iRobot, Colin Angle, cheered publicly when one of his PackBot mine-clearing robots was blown to smithereens on the Iraqi battlefield. "It was a special moment— a robot got blown up instead of a person," said Angle. His cry came from the heart, though it was unclear whether it was the heart of a humanist or a businessman, who could not have looked askance at the opportunity to supply the US army with a replacement robot costing $500,000.

Warbot vendors who tout their wares' life-saving capability might not in the future be marketing them to the military alone. The media are increasingly worried about the safety of their war-reporting staff in zones that have become particularly dangerous. MIT and University of California have both developed prototypes of unmanned vehicles that could beam back pictures of high-risk war zones in place of photographers and camera teams.

The Novel Unmanned Ground Vehicle (NUGV) was designed and developed by ACEi for the Space and Naval Warfare (SPAWAR) Systems Center. It can change its conformation when in motion depending on the terrain and obstacles it encounters.
Photos: Courtesy of SPAWAR.

1

The company iRobot launched its PackBot development program to fulfill an order from DARPA for a rugged portable robot to be used in urban guerrilla warfare. It can climb stairs, fall from heights and be thrown through windows without sustaining damage.

1- The US army's PackBot robot looks for booby traps around a truck at the Najaf airbase in Iraq on March 31, 2004.

2- 3- The PackBot was first used in Afghanistan to detect mines and explosives.

4- The PackBot EOD is a portable ordnance-disposal device that can be deployed within two minutes.

© Courtesy of the U.S. Army

Robotic reinforcements on the way... but when?

Projects to fully automate, if not roboticize, the US army have been on the drawing board for twenty years, yet have still not come to fruition.

In the early 1980s the US had to face up to the uncomfortable fact that the Soviet Union was not only its nuclear equal, but outstripped it two or three times over in conventional weaponry. To regain the edge, the US had to look to computer technology, a field in which it was indisputably stronger. The Pentagon decided to steer research in this direction and draw on hi-tech expertise far superior to that of the Soviet Union as part of an effort to develop a wide spectrum of intelligent weaponry. It launched the Strategic Computing Initiative (SCI), which enjoyed a budget in the region of a billion dollars and aimed, said Robert Cooper, who headed DARPA at the time, to create machines that could mimic human beings in different ways. Research first focused on unmanned vehicles able to conduct reconnaissance, some of which used real-time vision systems and a suite of sensors to scout in unknown

terrain. Work was also undertaken on aircraft copilots that could engage critical action like missile avoidance maneuvers. McDonnell Douglas explored monitoring systems able to detect situations in which pilots failed to take required action.

Below:
Robotic submarine.
© Courtesy of the Boeing Company

Page right:
1- The Boeing Joint Unmanned Combat Air System X-45. The robotic aircraft is pictured here dropping an automated, guided bomb.
Photo: Jim Ross, NASA
© Courtesy of the Boeing Company

2- Fire Scout 379.
Courtesy of Northrop-Grumman

3- The X47B.
Courtesy of the Boeing Company

By the time of the first Gulf War in 1991, new automated weapons had demonstrated their power. Accordingly they went to work. High-precision missiles able to home in on their targets from thousands of kilometers away displayed their deadly capabilities, while Lockheed drones, built in the top-secret Burbank facility, foiled radar to carry out flawless, soundless reconnaissance. These aerial scouts were supported underwater by crewless, self-propelled robotic submarines that the US navy officials described as small, economical and practically undetectable.

The success of its technology prompted the Pentagon to step up its development effort with numerous robot projects which, by 1995, accounted for 18 percent of army expenditure.

Yet in spite of the Pentagon's clearly stated commitment to autonomous war machines, they have been little in evidence in more recent conflicts. Although military officialdom has obviously drawn a thick veil of secrecy over its most advanced weapon technology, TV pictures of the Iraqi war front only show soldiers in battle fatigue astride tank turrets and at the wheels of jeeps. Conspicuous by their absence are the automated devices that were supposed to

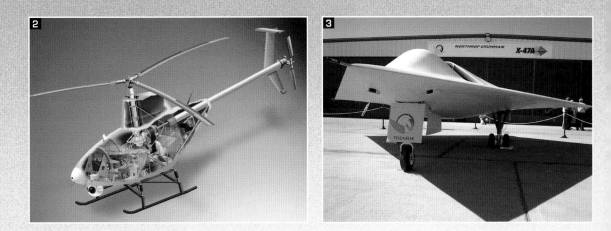

protect their lives. True, the much-heralded network-centered assault on Iraq was technically awesome, with its real-time vision-guided missiles, virtual reality simulations, and wireless connections between troops and command stations. But where were the robots?

The ScanEagle UAV was produced by Boeing and the Insitu Group. In June 2004, two were supplied to a Marine expeditionary force. Equipped with electro-optical and infrared cameras, the ScanEagle can fly for over 15 hours on intelligence surveillance and reconnaissance (ISR) missions.

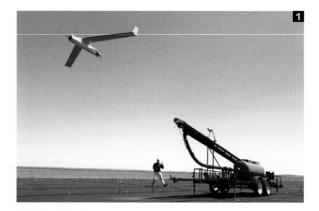

1- The ScanEagle UAV takes-off from a catapult launcher.

2, 4- After its surveillance flight, the ScanEagle flies low as it approaches a boat to hook on to a rope so that it can be retrieved.

3- The ScanEagle has a removable avionics bay for the integration of new payloads.

Photos: Courtesy of
The Boeing Company

5- The A160 UAV from Boeing Phantom Works.

© Anthony Romero/Boeing
Courtesy of the Boeing Company

Glossary

MEMS: Micro Electro-Mechanical System
MAV: Micro/Mini Aerial Vehicle
OAV: Organic Aerial Vehicle
TUAV: Tactical Unmanned Aerial Vehicle
RPV: Remotely Piloted Vehicle
UAV: Unmanned Aerial Vehicle
UCAV: Unmanned Combat Aerial Vehicle
URAV: Unmanned Reconnaissance Aerial Vehicle
VTUAV/VTOL UAV: Vertical Takeoff and Landing Unmanned Aerial Vehicle

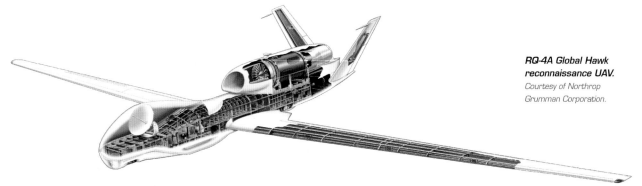

RQ-4A Global Hawk reconnaissance UAV.
Courtesy of Northrop Grumman Corporation.

The Cypher UAV, built by the Sikorsky Aircraft Corp., stages a surveillance demonstration at a military base in Fort Benning, Georgia. It patrols delineated surveillance zones looking out for any intruder, which it spots through its video camera. It can fly at altitudes of 2,500 meters (8,200 feet) and cruise at 150 km/h (93 mph) for two and a half hours.

The unmanned Shadow 200, used by the US in Iraq.

The Pentagon is looking to the future, painting enticing pictures of smart, lightweight equipment for infantry troops designed by organizations like MIT and Dupont de Nemours. Similarly, numerous websites dedicated to US military research—drawn from an annual budget of over $400 billion!—showcase formidable arms like the Javelin, a portable anti-tank missile, and weapons applications from companies such as Sciences Applications International, Boeing, Lockheed, Northrop and General Electric. In the future, the US military's Future Combat System will be deployed and networked. Battlebots will assist troops on the ground, backed by unmanned aircraft that can be deployed in just four days. In the future, but when, to be precise?

At the Paris military hardware trade fair, Salon Eurosatory 2004, advanced-warfare systems were on show. One was electronic equipment for the foot soldier of the future, designed by the French electronics company Sagem. It included multiple personal protection systems and devices for seeing through walls. But once again, no real-life remakes of Terminator. The French military program BOA[1] does incorporate autonomous robots, but they are to be eased into active service between 2012

and 2030. Projects are under development, nevertheless. In Alabama MESA Associates have received an order for a small robot, designed to undertake kamikaze scouting missions. To reconnoiter terrain, it will be able to scale obstacles, take photographs and video footage, detect mines and handle explosives.

The TACOM group is developing a doglike robotic beast of burden, intended to relieve soldiers on the battlefield and able to carry up to 50 kilograms (110.23 pounds) of munitions, supplies and equipment. The four-legged

friend will trustily follow its master through thick and thin, running or walking, carrying its heavy load and mapping the lay of the land, all at the same time.

One robot that has created a sensation is DARPA's RoboFly. It is a robotic spy fly with four wings and a tiny camera in its eye.

---

(1) BOA: Bulle Opérationnelle aéroterrestre (aero-terrestrial operational bubble).

Firing a Javelin antitank missile.
Courtesy of the US Army.

Minesweeping robots

Although advanced robotic weaponry belongs to an unspecified future, robots currently operate on the battlefield performing tasks that would put human life pointlessly at risk. Mine clearing, or explosive-ordnance disposal (EOD), robots, like the PackBot mentioned above, are one example. The 19-kilogram (42-pound) machines, designed to rove hostile terrain and even cross waterways, are in use in Iraq and Afghanistan. Similarly, DARPA's tiny tactical mobile robotics (TMRs) are semi-autonomous platforms that can be deployed by troops on the ground. Once deployed they can climb up and down stairs and open doors unaided.

Canada's defense R&D program, DRDC, has a robotics focus that has led to the development of a prototype for remote-controlled landmine detectors. Another Canadian robot project addresses wider security concerns with the RAFE system. Designed by a Toronto-based firm, law enforcement forces use it to safely screen suspect luggage and parcels. An autonomous EOD is also under development in France as part of the BOA program. Called Syrano, it is a compendium of technologies designed and developed by the research laboratories and contractors Thales, Sagem, Giat and Cap Gemini. Once completed, the four-ton robot will be able to proceed to locations it has been shown on a map in order to destroy any mines and perform enemy surveillance missions in poor visibility.

Mine-clearing robot at the Villacoublay airbase in France.
© FYP

Page right:
1- Mine-clearing robot from Cybernetix. Cybernetix has designed the robots Castor and TSR to dispose of ordnance devices.
Courtesy of Cybernetix

2- The robot cop, Wolverine, made by Remotec.
Photo by Randy Montoyoa. Courtesy of Sandia National Laboratories.

3- Detail of Cybernetix's mine-clearing robot.
© FYP

Centibots is another DARPA-initiated project that aims to show how effective teams of distributed robots can be for search-and-rescue missions. They operate like the sort of video game commando unit of which author Tom Clancy is so fond, locating and getting into a building to find and guard "objects of value". The Centibots work in phases and tightly organized groups. One group explores the premises, precision-mapping them to the nearest inch. The second group uses the map to seek out the object. They then keep watch and track any movement within the building. Three companies came up with proposals for Centibots systems, one of which was SRI International.

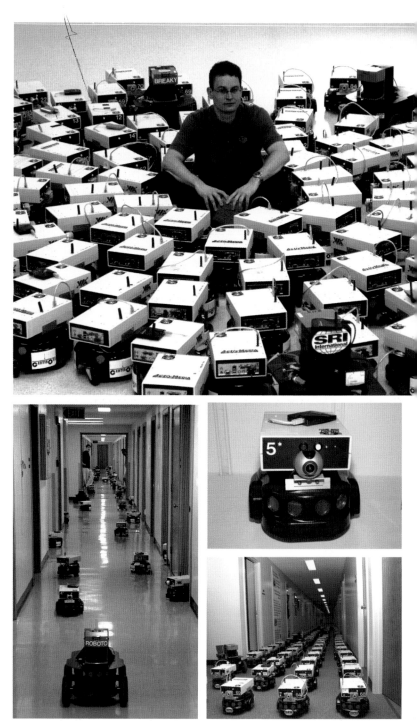

SRI's Centibots
Courtesy of Régis Vincent, SRI

Interview
Régis Vincent

Following his thesis on artificial intelligence at the IT research center of Sophia Antipolis in southern France, engineer Régis Vincent received an offer from the University of Massachusetts in 1997. Four years later the former Stanford Research Institute, SRI International, asked him to join it to work on advanced robotics projects such as Centibots.

Régis Vincent, head of robotics projects at SRI International.
Courtesy of Régis Vincent, SRI

How did the Centibots project come into being?
Centibots was a challenge that DARPA decided to address in 2002. One of the special focuses of its research-and-development work for the army was robotics. It had been funding projects for some ten years. Yet with one or two exceptions, nothing had yet been developed that the military could use in the field. With Centibots its concern was to use robots to multiply troop capabilities. Take the long military convoys the army deploys in Iraq. They comprise between 1,000 and 1,200 vehicles and can be 60 kilometers (37 miles) long. At the very beginning of the American-led invasion of Iraq, in 2003, one of these convoys was attacked. The attack got the military thinking and they

wondered what the point was in having a driver in every vehicle, when they could be attacked at any time. Why not have just one and the rest would follow? Centibots, a robot-assisted search and rescue system was designed to respond to needs that would multiply capabilities without putting human life at risk.

Where does the name Centibots come from?
DARPA looked at past experiments and found out that teams that had worked on similar issues before couldn't handle more than ten robots at a time. DARPA wanted a system that would make it possible to control a hundred. The objective of

the original project was for two humans to control 100 robots on 24-hour missions. It drew up its program specifications, issued an invitation to tender and short-listed three bids—from SAIC in San Diego, California, iRobot from Boston, Massachusetts, and our company, SRI. We're a not-for-profit R&D company based in Menlo Park, California. DARPA then got us to compete with each other. All three companies were given a building whose layout they knew nothing about, the same unspecified mission duration of up to 24 hours, and no network infrastructure.

Why do you think SRI won?
SRI has a tremendous track record in computer science, broadcasting technology and robotics. It created the first mobile robot, Shakey, in 1965. Then in 1992 Flakey was the first winner of the autonomous navigation competition organized by the American Association for Artificial Intelligence. SRI has also been involved in other projects like remote surgery and Lego Mindstorms. Before that it invented high definition TVs and the computer mouse. In 1969, SRI joined the Arpanet project, the forerunner of the Internet and the first email ever sent arrived in the inbox of an SRI computer.

What was your approach to developing the Centibots?

We decided to make each robot intelligent and completely autonomous. They all had to have their own computers. Each robot had to try to collaborate efficiently and effectively with the others but had to be able to carry out the mission by itself if need be.

A Centibots mission comprises three phases. Can you tell us about them?

The first phase entails getting the robots into a building so that they can draw up reconnaissance maps.

It's crucial because, in the event of an earthquake, for example, when a robot gets into a building that has been damaged, any existing plan is no longer valid. It has to remap the layout of the house to work out how to find and reach any victims. Phase two involves finding a specific object. In reality, they would be people who had been injured or at risk. But DARPA was just interested in finding out how the robots would go about locating the object and finding it among the debris. There were several possible solutions, such as heat sensors or cameras. We chose cameras. Once the robots had found the object, the next phase was to guard it and keep control informed. Our idea was to relay information and video footage,

with the robots networking with each other and the control room. That involved the robots taking up position in relation to each other so that transmission of updated information was constant. This phase had no set duration so the robots worked to rote, with some going off to recharge while others stood in for them.

Did you test the Centibots in the field?

Absolutely. DARPA really wanted to know how the systems worked and they made it impossible for any of the teams to find out where the final demo would take place. In the meantime we tested our robots inside three huge buildings. Their surface area was between 1,000 and 2,400 square meters (1,200 and 2,800 square yards). But DARPA conducted their last test in a smaller place, which was about 650 square meters (780 square yards).

Can you tell us about the DARPA trial?

It was in January 2004 in a building we'd never seen before. The robots were sent to the East Coast. We didn't know anything about the building, its size or its surroundings. Only the robots were allowed inside. Since we'd designed them to be autonomous they were able to fend for themselves. The commanding

robot oversaw the mission to make sure it was progressing smoothly. The first robots were sent in and had to draw up a map. It didn't matter how they did it. We didn't intervene at all, they operated by themselves. There were several phases to the test. In the first one, a single robot mapped the building in 21 minutes, then two robots did it in 17 minutes, while three managed it in16. More robots didn't make any difference. In the building you could get into all the rooms from any other. We found out that the other two teams had taken longer than us, about an hour longer, because they relied on remote-controlling their robots.

What happened in the second phase?

Once we had the map, we transferred it to robots that were smaller and less well equipped for reasons of cost. We used between 30 and 40 robots and gave them a valuable object to seek out. They used the map to navigate through the building themselves, working out their position and how to get to the object. A central computer then decided on the best route and checked that no stone had been left unturned by at least two robots. There were four object-location exercises, and since DARPA was assessing us we were able to get reliable feedback on the robots' performance. On average the robots took 33 minutes to find their objects with a precision margin of 20 centimeters (eight inches). The fastest took just 16 minutes.

Would that be fast enough for a rescue mission or to locate a suspect device?

That's pretty fast given that our robots only move at thirty centimeters (one foot) per second. But we're talking about a research project. Faster robots could be used in real life.

What about the surveillance phase?

For the third, the surveillance, phase we sent in a second group of robots to keep watch over the object and send back video footage of it to the control room, which was 35 meters (115 feet) away.

How would the third phase come in useful in real life?

If it's a suspect device, it's important to make sure nobody comes too near. If it's an injured person, the control room can talk to them through a robot's microphone and see them thanks to the robots' video cameras. The robots could also help to reassure the injured people.

Would multi-articulated snake robots be used for the last phase?

Probably. The SRI makes spider robots that could climb over any debris; they could be useful, too. But this kind of base robot was not the point of the test. What was more important was that the robots should map a building they knew nothing about. As far as the actual robots are concerned, DARPA wanted us to use the cheapest ones.

Will there be a follow-up to Centibots?

The project came to an end in March 2004. We'd had eighteen months to do it in. But we're now adapting the software to flying drones so they can carry out surveillance missions.

SRI laboratories.
Courtesy of Régis Vincent, SRI

Surveillance and keeping the peace

Robots help the police. The first real-life RoboCop went into action in 1993 in Maryland in the US to apprehend a murderer who was holed up in an apartment. On spotting the man hiding behind a pile of clothes, the remote-controlled, wheeled RMI-9 squirted him out with a high-pressure jet of water.

Robots now intervene routinely. When they believe a suspect is carrying explosives, is mentally unhinged or otherwise dangerous—indeed, whenever a job looks fraught with risk—the Los Angeles police have no hesitation in sending in Andros from Remotec to take the pulse of the situation. Come rain, shine, snow or sweltering heat, nothing puts Andros off. Armed fugitives sometimes prefer just to give themselves up when the robot rolls in.

Wolverine can use its claw to grip and open car doors.

Phil Bennet (right) of Sandia National Laboratories talks with Albuquerque police about Wolverine's capabilities.

Photos: Randy Montoya.
Courtesy of Sandia National Laboratories.

Smart systems not foolproof

Despite the colossal sums the military have spent on researching and developing robots, the day when they march onto battlefields in place of flesh-and-blood troops is a long way hence. The reasons for not using them outweigh arguments in their favor.

In the first place, the cost of developing robots that can maneuver on all terrain in all weather is prohibitive. One senior DARPA official even admitted in private that he did not believe it was worth spending so many millions of dollars just to avoid breaking some mother's heart. That is one way of looking at it.

Furthermore, the software and computer technology that make robots intelligent are prone to error. It is not uncommon to hear on the news that a

missile scored a direct hit on the wrong target, so claiming innocent lives.

One of the most gruesomely infamous of such mishaps in recent times took place on July 3, 1998. Aegis was a US army automated radar system designed to spot aircraft and assess whether they were friendly or hostile, civil or military. On the fateful day Aegis detected what it mistakenly identified as an enemy craft. The result? An Iranian Airbus carrying 290 civilian passengers was shot out of the sky.

Both Iraqi wars have been punctuated by automatic weaponry errors that have cost innocent lives. They have prompted the Pentagon to design further new systems that yield more accurate, reliable readings of radar screens, acoustic signals, etc., so that friend is not mistaken for foe. In a similar vein

Carnegie Mellon University and Boeing are designing an autonomous combat vehicle called Spinner that can operate on all terrains.
Courtesy of the Boeing Company

three projects—Forester, Jigsaw and SPI 3D—seek to locate enemy movement in forest land without having to use any sinister defoliating substance.

But ultimately, is there any foolproof way of being sure that a smart, autonomous robot can distinguish unerringly between what is a target and what is not? Conversely, how will a robot tell that the foe in disguise is not a friend? And finally, how can the enemy be prevented from hacking into a robot's software and, if he does, how to tell? Military robots raise too many questions to yield military solutions.

License to kill

Last, but not by any means least, are the important ethical issues raised by the decision to design and develop automated killing machines. Killer robots are, in fact, with us already in laboratories across the world. Thai researchers have developed Roboguard, a gun-toting mean machine programmed to stand guard outside museums and corporate or government buildings and to shoot if necessary.

As part of the US Army's Future Combat Systems program, iRobot has signed a deal estimated to be worth $32 million to develop a more compact, smarter PackBot, which is scheduled to be operative by 2012. Like other small unmanned ground vehicles under development, its capabilities will include reconnaissance and other tasks. What other tasks? The Raytheon company, which is developing the different devices on the end of the robot's arm, has incorporated a shooting system with viewfinder. What limits will the PackBot be set to? Will it be able to tell clearly between civilians and soldiers? In the event of a cease-fire or truce, will it be possible to immediately disable all the robotic soldiers that have been programmed to destroy? And how can such creations, tailored as they are along the lines of Terminator, be configured to stop attacking an enemy who wants to surrender? The answers to such questions are doubtless to be sought in highly advanced programming and artificial intelligence concepts and experimentation. But they have not, as yet, been found.

Robart -1.
© Courtesy of
Spawar.

Lieutenant-colonel Cyril Carcy

The French airbase of Mont-de-Marsan is the scene of advanced experimental research into drone aircraft. Lieutenant-Colonel Cyril Carcy has been second in command there of the Drone Experimental Squadron (EED) since September 2004. He talks about the technology.

What exactly does the Drone Experimental Squadron do?
In terms of robotics the EED's job is to run experimental research into the SIDM[1] unmanned aerial vehicle, which we have dubbed Eagle and which was developed by the company EADS. It is an autonomous airborne system that can fly long-endurance missions and relay data to command in real time.

How long has the EED been working on drones?
For eight years now. The French joint chiefs of staff decided to acquire the technology in 1996 in the wake of an air force mission. Four Hunter drones were bought to enable us to familiarize ourselves with what will be a key system in the future.

How do you think drones can help the military?

Currently operating drones have two core capabilities—long endurance and real-time data transmission. They provide enhanced knowledge of the battlefield with constantly updated data, allowing us to understand enemy intentions and anticipate. There are other strengths, too, like low observability and variable payload bays that offer a wide range of potential mission configurations.

What solutions have drones provided that manned craft can offer only imperfectly, or not at all?

Unlike manned craft, the unmanned drones have long-endurance capabilities and can fly over heavily defended territory. We can thus use them to carry out long periods of surveillance of strategic targets without having to halt missions because crews are suffering from fatigue. In a similar vein, a drone can fly low over areas defended by ground-to-air missiles. It is stealthier than a manned craft and if it is shot down there is no loss of life.

How does the use of robotic planes help to save soldiers' lives?

In the future unmanned robotic air vehicles will supply us with enough battlefield intelligence to anticipate enemy maneuvers. Soldiers will use them to spot ambushes or highly mobile enemy units. And with its sensors, a drone can act as troops' eyes and even as a strike force if it is equipped with weapons.

Will drones also help to save civilian lives?

Drones look likely to be used routinely in the future in critical situations like the surveillance of forest fires, to track and arrest criminals on the run, or to look out for survivors of natural disasters like floods and earthquakes.

What other applications will there be for drones in the future in your opinion?

The air force is interested in systems to replace SIDM craft, such as MALE[2] drones that will eventually have flying times of over 24 hours and communicate intelligence in real time over distances as great as 1,000 kilometers (620 miles). But the army is interested in other kinds of drone. The HALE[3] drone could in the future offer the ideal strategic system by collecting intelligence and transmitting it in real time to ground control over very long distances across continents. I should also mention minisize unmanned aerial vehicles, tiny drones that can fly into buildings and transmit valuable intelligence to units preparing an assault. As for tactical drones, they will be used by the army. They have a range of about 100 kilometers (60 miles) and can inform infantry and artillery units of enemy positions and movements. The Dassault company has received a large state subsidy to research drone technology using three DUC unmanned craft—one small, one medium-size, and one large. They will have the firepower to respond to attacks from air-to-air and ground-to-air missiles.

---

(1) SIDM: (French acronym): Intermediate MALE Drone System.

(2) MALE: Medium Altitude Long Endurance.

(3) HALE : High Altitude Long Endurance.

ROBOTS AND MEDICINE

The idea of being operated on by a robot is a daunting one. Would you dare to entrust your life to a machine that, smart though it may be, has no awareness or conscience?

Yet medicine has happily embraced robots and they have swiftly become part of the surgeon's panoply of instruments. Patients may rest assured: most robotics operators are remote-controlled by specialist doctors. They bring greater precision and allow surgeons to operate from thousands of kilometers away. Medical-assistance robots have unobtrusively taken their place in surprisingly large numbers in hospitals and even in homes.

The ZEUS Robotic Surgical System. The use of robots in minimally invasive surgical operations helps reduce surgeons' fatigue during long operations.
© Courtesy of NASA/JPL

Robot assisted surgery

The operating room has undergone radical change in just twenty years. Until the mid-1980s surgery was a hands-on skill. Using basic manual tools such as scalpels, tweezers and needles, doctors operated directly on their patients.

Two decades later, they sit down at a control console and peer at 3D pictures of inner organs. Twiddling joysticklike control levers sends electronic signals from a computer to a robotic arm, which manipulates surgical instruments, cutting into the flesh and bone in synchronization with the surgeon's movements.

So successful has robot-assisted surgery proved that it has cleared the way for telesurgery, whereby the doctor can be thousands of miles from the patient.

Operating room in Paris equipped with the da Vinci robotic surgical system.
© Courtesy of Intuitive Surgical

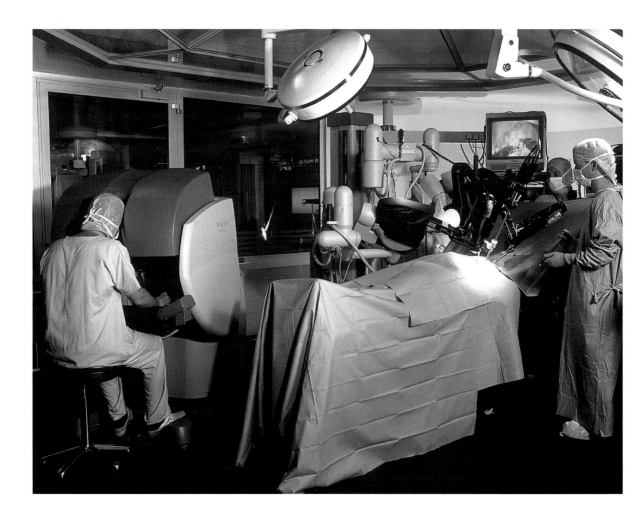

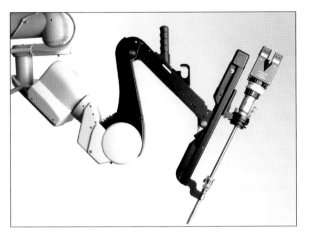

*Robotic arm
for minimally
invasive surgery.*
© Courtesy of
Intuitive Surgical

Minimally invasive operations

The advent of robots was prefaced by a prior revolution, known as minimally invasive surgery, (MIS), which uses a method called endoscopy for examining the inside of the body. MIS is performed through three small incisions made in the body. Into one of them is lowered an endoscope, a small, thin optical-fiber instrument connected to a tiny camera that projects onto a video monitor images of the organs to be operated on. Through the other two incisions the surgeon manipulates small, thin surgical instruments.

Of the three five-millimeter incisions, one is used to insert the optical-fiber endoscope equipped with its tiny camera. Into the other two incisions go the robotic articulated arms that hold the surgical instruments. Seated at the console, the surgeon has a close-up 3D view of the operating field. He or she moves the controls and the system translates them into commands that the robot executes by making the required surgical movements inside the patient.

MIS has spread fast and is now practiced in a wide range of interventions. It offers numerous advantages in addition to not requiring deep cuts that leave unsightly scars. Healthy bodily tissue is left undisturbed, hospital stays are shorter and patients convalescence faster. And MIS has paved the way for robotic surgical systems like ZEUS and da Vinci, which strive to make operations safer, more comfortable and precise.

Though robotic surgical systems might work on a basic master-and-slave principle, they have nevertheless brought great progress. Surgeons are more comfortable, less tired and surer of hand. Images from the endoscopic camera can, if needed, be enlarged 100 times or more, which has enabled surgical operations that were not previously feasible. Any hand tremors are filtered out. "As instruments get longer, the surgeon's natural tremor is augmented," says Canadian cardiovascular surgeon Dr. W. Douglas Boyd. "A good example would be trying to write your name with a pencil by holding the eraser...the accuracy and the ability to write your name holding this long instrument would be substantially reduced." He believes that it was inevitable that robots would intervene in heart operations: "To perform bypass surgery with conventional endoscopic instruments in a closed chest [was] beyond the realm of human manual dexterity."

The first medical robots

Yik San Kwoh, medical research and development director of Long Beach Memorial Hospital, pioneered robotic surgery. In 1985 he developed the first robot-assisted surgical interface after three years of tireless programming and reprogramming on his Apple and IBM computers. Kwoh used a PUMA-type industrial robot (see Chapter 5) to manipulate a surgical instrument in conjunction with a scanner and a sensor. Three years later, a surgeon named Brian Davies used the same machine for a transurethral resection and the system developed into one known as Probot, dedicated to urethral-tract operations.

The first robot to assert itself in the medical world was Robodoc. In 1990 it was used for a hip- replacement intervention on a dog. Robodoc was developed by the orthopedic surgeon William Bargar of Sutter General Hospital in Sacramento and researcher Howard Paul, with help from the University of California. Robodoc is two meters (six and a half feet) high and equipped with an arm that holds a tiny drill. Once the doctor has put all the sensors in place, the robot takes only a few minutes to drill into the femur a hole

that will be used to anchor the new hip joint. It works alone and with absolute precision.

In 1992 Robodoc was successfully used in a hip-replacement operation on a human and became the first robot to be approved by the Food and Drug Administration (FDA). Howard Paul formed the company Integrated Surgical Systems to commercialize the invention and by 1997 he had signed a distribution deal with Japan. Robodoc was also commercialized in a number of European countries.

Despite its promising beginnings, Robodoc never became a robotic surgeon on a large scale. "The operation was perfect, but it involved two stages. One to affix the sensors and the next to operate the robot. It was costly and the sensors increased the risk of infection, which halted operations," says Luc Soler, in charge of R&D at IRCAD, the French Institute for Research into Cancer of the Digestive System.

Nevertheless the system warranted attention because of its merits. "Robots are capable of precision to the nearest tenth of one millimeter instead of the two millimeters in manual operations," says François Aubart of Eaubonne-Montmorency Hospital near Paris. His assistant, Caspar, is similar to Robodoc. Caspar was

designed and developed by the German company Maquet and also works on the two-phase procedure of preoperative preparation followed by surgery.

In the five years since Caspar performed its first prosthetic implant in May 1998, Aubart has conducted 220 similar operations. The benefits are self-evident, he says, for there are no complications, which is enough to make the system economically viable.

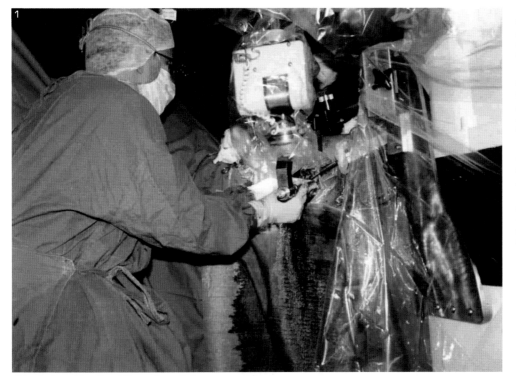

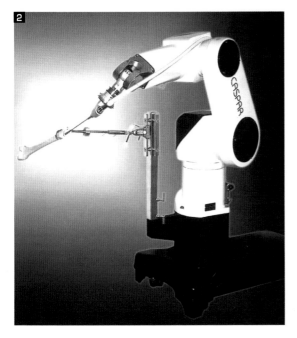

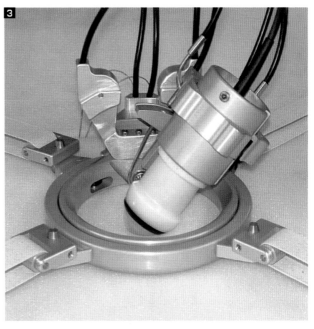

Telesurgery arrives

In the early 1990s NASA began research into telesurgery. The Stanford Research Institute (SRI) swiftly followed suit, developing a telemanipulator. The US army expressed its interest in the work, reasoning that surgeons far from the battlefield might one day be able to operate on wounded troops in a robotically equipped shelter. For the rest of the decade, two California-based companies, Intuitive Surgical and Computer Motion, competed to produce robotic surgical systems.

The Wang brothers founded Computer Motion in the late 1980s to develop and market their voice-actuated Automated Endoscopic System for Optimal Positioning (AESOP) Robotic System. In 1993 it was approved by the FDA. Computer Motion now got busy on a comprehensive robotic telesurgeon, unveiling the great ZEUS in 1998. Intuitive Surgical came into being in 1995, founded by Frederic Moll, Robert Younge, and John Freund. They drew on SRI's work into military robotic surgery and developed a system they called da Vinci.

Both ZEUS and da Vinci are robotic systems that assist in minimally invasive surgery and both have a computer workstation, video display, and hand controls for table-mounted surgical instruments. ZEUS is more compact, while da Vinci is easier to use. "It has the intuitive feel for a surgeon of hands-on, open operations," says Micael Michelin of France's Montpellier-based IT and robotics research unit, LIRMM.

The two companies now cornered a big market with their systems. Although both were American, they developed very rapidly in Europe. Back in the US, the FDA long withheld full clearance. At France's Strasbourg-based IRCAD, Professor Jacques Marescaux acquired three Computer Motion robotic systems in 1996. Two years later he had mastered them and began to carry out test surgical operations on patients.

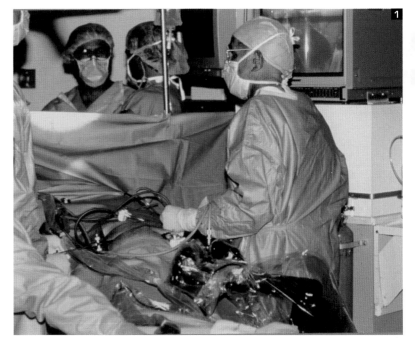

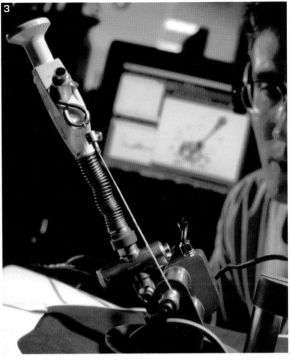

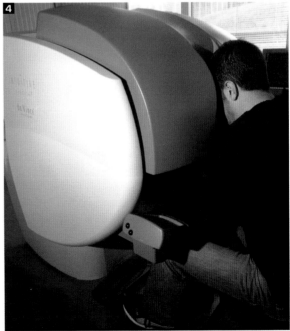

1- 2- AESOP, the voice-controlled endoscopic robot for minimally invasive surgery, developed by Computer Motion with the help of NASA.
© Courtesy of NASA/JPL

3 -The lightweight endoscopic robot REL assists surgeons by positioning the endoscopic camera as it is instructed to.
© Courtesy of CNRS photo library

4- Control console for the da Vinci robotic system.
© Courtesy of INRIA - Photo G. Favier

Robotic surgical systems moved into routine operations in 1998 when Belgian surgeon Dr. Jack Himpens used da Vinci to remove a gall bladder. One year later Dr. Tommaso Falcone and his team were assisted by ZEUS to perform an endoscopic fallopian tube reconnection procedure at Cleveland Clinic. The first minimally invasive coronary bypasses were also carried out in 1999 by Dr. W. Douglas Boyd in Canada and by Drs. Didier Loulmet and Alain Carpentier using da Vinci. With an endowment from French business magnate François Pinault, Henri Mondor Hospital outside Paris acquired a da Vinci system in 2000. It went into action the following year, providing Professor Clément-Claude Abbou with assistance to remove a prostate gland.

One by one, the string of successful surgical outcomes lent credibility to da Vinci and ZEUS, and although they cost over one million dollars, numerous European hospitals acquired them. And once the FDA had given its stamp of approval in 1999, American hospitals followed suit.

Both the smart surgical systems had proved themselves. The unerringly precise, scaled-down motion of robotic arms was steadier and surer. Furthermore, the systems brought savings that ran into thousands of dollars for every patient because hospital stays were shorter and there were fewer post-op complications. Robosurgery soon widened its range to knee, spine and... legal procedures.

In 2000 Intuitive, Computer Motion and IBM became embroiled in a tangle of wrangles. At the same time IBM and Intuitive Surgical initiated proceedings against Computer Motion for infringement of IBM's voice-control patent, licensed exclusively to Intuitive. A court eventually ruled in favor of Computer Motion, which, meantime, had filed a suit against Intuitive for infringing multiple patents, including its articulated telemanipulated arm. Intuitive won the case, but Computer Motion then won another, and so on. The two companies were locked in such lengthy litigation that *Business Week* talked about the "the battle of the robosurgeons".

Then in March, just after Intuitive had been ordered to pay a large fine, Computer Motion's president and COO, Joe DeVivo, took everyone by surprise when he offered to merge with Intuitive Surgical, which would be the majority partner. By June 2003 there was only one major player on the robosurgery market. An Intuitive Surgical survey in 2004 revealed that it had 210 platforms installed worldwide at an average cost of $1.3 million each. Da Vinci's robot even put in a guest appearance in the James Bond movie *Die Another Day*, in which British scientists use it to scan 007 to check his identity.

1- An important component of the da Vinci Surgical System is the patient-side cart. Whereas early surgical robots mimicked a surgeon's hands and had only one or two side-mounted arms, later systems usually have four, which execute the surgeon's commands and afford the robot leverage.

2- This robot is called BRIGIT, short for Bone Resection Instrument Guidance by Interactive Telemanipulator. It helps to guide orthopedic surgery instruments for milling, drilling and dissecting bone.

© *Courtesy of the CNRS photo library.*
Photo: Emmanuel Perrin

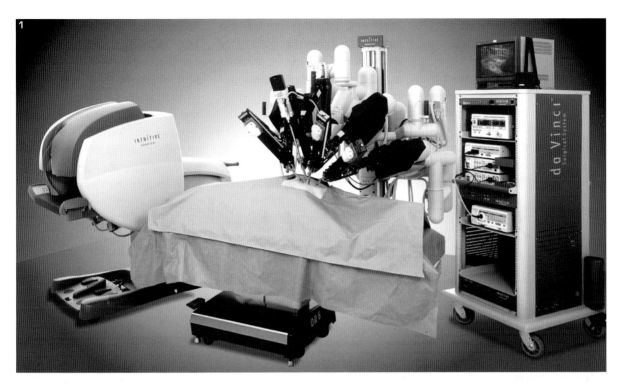

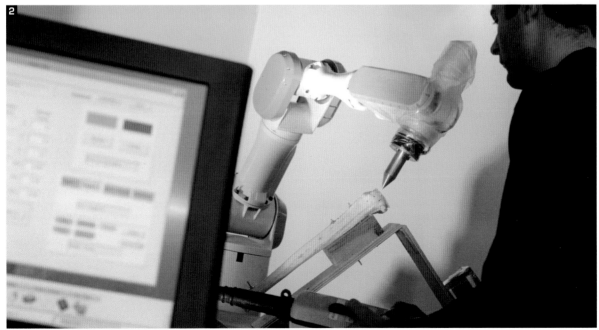

Long-distance surgery

In minimally invasive surgery the surgeon's console is at a remove from the operating table. With the development of telecommunications it could be remote—outside the operating room, in another city, country and even continent. Like the first transatlantic air crossing, the first transatlantic surgical operation was to usher in a new era.

In 1995 in a presentation of the European Eureka Master Project, Professor Jacques Marescaux (*see interview on page 386*) of Strasbourg-based IRCAD argued that telesurgery was feasible. To set up the operation he and his team from IRCAD had to contend with a host of technical and bureaucratic obstacles, not to mention a consensus of scoffing. Two months before the operation's scheduled date in September 2001, the scientific director of the US defense R&D agency, DARPA, stated that it would be

Simulating a liver operation

A scanner can be used to display 3D reconstructions of a liver so that surgeons can view it before operating. A future improvement to such a system will be to represent bleeding realistically. The ultimate step could be semiautomated or even fully automated telesurgery.

© Courtesy of INRIA - Photo: A. Eidelman

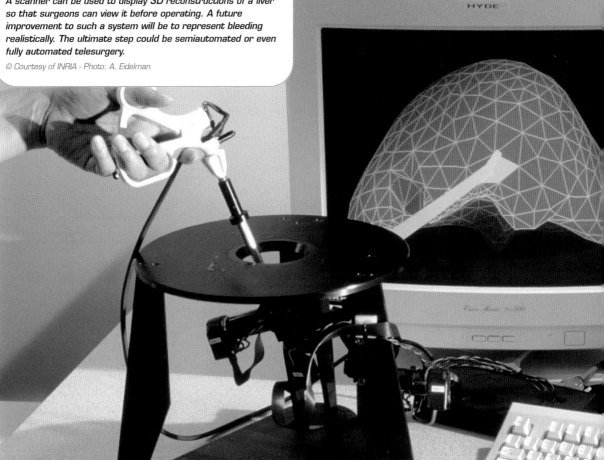

another five years before telesurgery became a reality and would not, in any case, be feasible for distances in excess of 200 miles.

On September 7, 2001 Professor Marescaux proved the skeptics wrong when, operating from a ZEUS control console in New York, he controlled robotic surgi-

cal instruments to remove the gall bladder of a female patient in the French city of Strasbourg.

The world's first complete telesurgery procedure was successfully performed to worldwide acclaim. It was dubbed the "Lindbergh Operation" and was the cover story in *Nature* and countless other scientific and medical reviews.

In the wake of Marescaux's watershed achievement, Canada's Hamilton University has conducted some thirty telesurgery operations. "That was good for us," says Professor Marescaux, "because Canada is a typical example of a country with few expert surgeons. The operations have helped validate our system and shown that the technology is well and truly applicable."

Telesurgey radically changed mindsets. For some patients it meant that they could choose the surgeon they wanted wherever he or she lived and without having to spend time and money abroad. For surgeons, it was their introduction to telecollaboration and the sharing of knowledge with their peers. In a sense the robot had brought them out of their ivory towers.

Prior to surgery there is diagnosis. Telediagnosis could be useful for people living in isolated communities with no access to specialist medical knowledge.

Otelo, a telecommunications-based portable ultrasound probe-holding robotic system, has been developed to that end by the vision and robotics laboratory LRV, based in the central French city of Tours, and that city's hospital. A robot on the remote site reproduces the movements of the specialist manipulating the controls to perform an ultrasound scan. At the control center the specialist can view ultrasound images. Tested in harsh conditions between Tours, Katmandu and Cyprus, Otelo is due to blast off for use in the International Space Station in 2005. The company Sinters is to market the system.

The Canadian Space Agency has followed with interest developments in end-to-end telemedicine. In October 2004, it participated in an experiment in which its control center telecommunicated with a submarine at a depth of 19 meters (62 feet) off Key Largo, Florida. "The extreme nature of conditions on a long-endurance underwater mission is similar to those found in space," said one project leader.

Ultimately the ambition is to be able to carry out telesurgery with no doctor on the remote site and to enable an astronaut with no medical training to provide emergency care with the assistance of

Precision surgery

Precision surgery is where robots come into their own because they can handle instruments of Lilliputian proportions, which our (relatively) Brobdignagian mitts are too clumsy to hold.

A joint venture between a team from NASA's Jet Propulsion Laboratory (JPL), headed by Paul Schenckerd, and the company MicroDexterity Systems produced a tiny remote-controlled robot. Called MicroDexterity it has intervened in microsurgery operations that demanded such minute finesse that they had previously been impossible.

The Lapeyronie Hospital in the city of Montpellier in southwest France often has to practice emergency skin grafts on people with third-degree burns that can cover up to 50 square centimeters. The donor skin used is surgically removed from elsewhere on the burned person's body by a dermatome, a tool with a sharp razor blade. Applied with constant pressure at constant speed it shears off the skin. There must not be one false move. That is where Dermarob comes in. Designed by the Montpellier-based robotics and microelectronics research laboratory LIRMM Dermarob is a robotic

arm that in one fell swoop shears off a designated area of donor skin of the right thickness to the nearest two-tenths of a millimeter.

The LAB is an automation research unit in the eastern French city of Besançon and the birthplace of a microrobot that uses a magnet to move medical instruments measured in tenths of a micrometer (thousandth of one millimeter). It is of invaluable aid to doctors who are thus able to manipulate oocytes without damaging them and carry out safer in vitro fertilization.

The LAB has also designed robotic microtweezers that work in conjunction with an electronic scanning microscope and can manipulate objects the size of a grain of salt.

One of the most impressive miniaturized systems recently developed is a minute, battery-less endoscope, Norika3. It is the work of the Japanese company RF SYSTEM Lab and consists of a camera in a capsule

that is 2.3 centimeters (one inche) long. When swallowed the capsule releases the camera, which navigates through the digestive system transmitting pictures as it goes.

Norika3 foreshadows the coming of nanorobots which, at one-millionth of a millimeter, are molecule size. And they will be everywhere, believe some. The World Future Society's publication *The Futurist* has described toothpaste in the 2020 as packed with nanorobots that will chip away at dental plaque like the seven dwarfs in their mountain caves.

Page right:
SCALPP is an automated system for cutting away donor skin for grafts and orthopedic surgery. It moves its dermatome across a delineated patch of skin at a constant pressure and speed so that the piece it shears off is the right shape and of constant thickness.
© *Courtesy of the CNRS photo library - Photo: Leif Carlson*

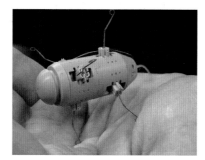

This gastrointestinal analysis capsule was developed by the Italian-based research center CRIM as part of a project called Emiloc, short for Endoscopic Microcapsule Locomotion and Control. The capsule is 20 millimeters long and 10 in diameter. It will house a camera in its middle section.
© *Courtesy of Lucia Beccai*

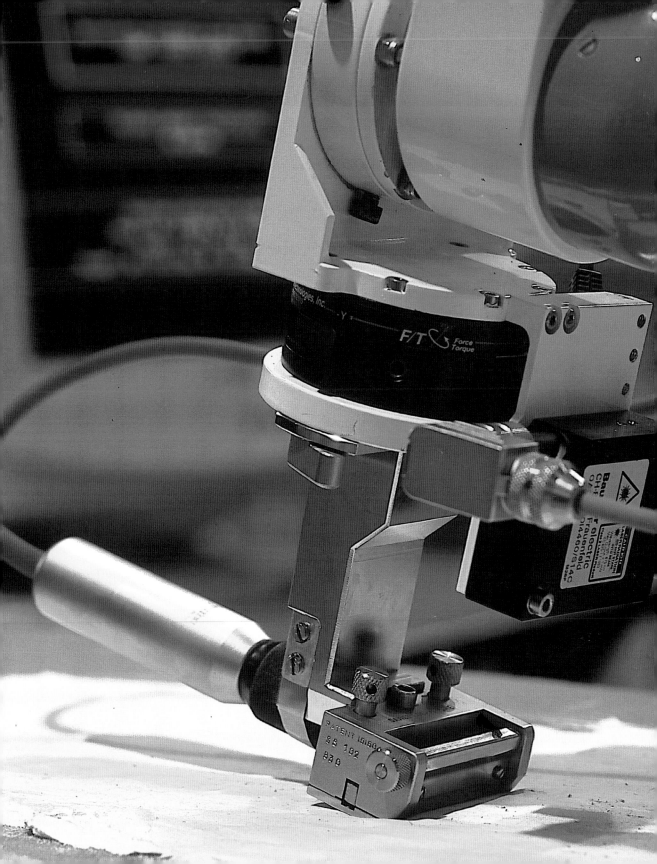

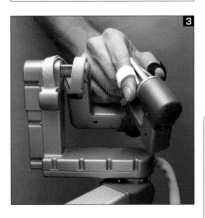

Microsurgery

1- 2- MIPS: parallel robot, 7 millimeters in diameter.
© Courtesy of INRIA - Photo Sébastien Paris

3- Robot surgical system control console levers.
© Courtesy of intuitive Surgical

4 - Robotic eye surgery system. The robot's arm assists surgeons in particularly delicate procedures.
© Courtesy of JPL/NASA

5- Robotic microsurgery.
© Courtesy of JPL/NASA

6- Endoscope.
© Courtesy of Intuitive Surgical

7- Black Diamond Instrument.
© Courtesy of Intuitive Surgical

8- Endowrist® surgical instruments are designed with seven degrees of motion that mimic the dexterity of the human hand and wrist.
© Courtesy of Intuitive Surgical

9- Suture performed by robotic surgical instruments.
© Courtesy of Intuitive Surgical

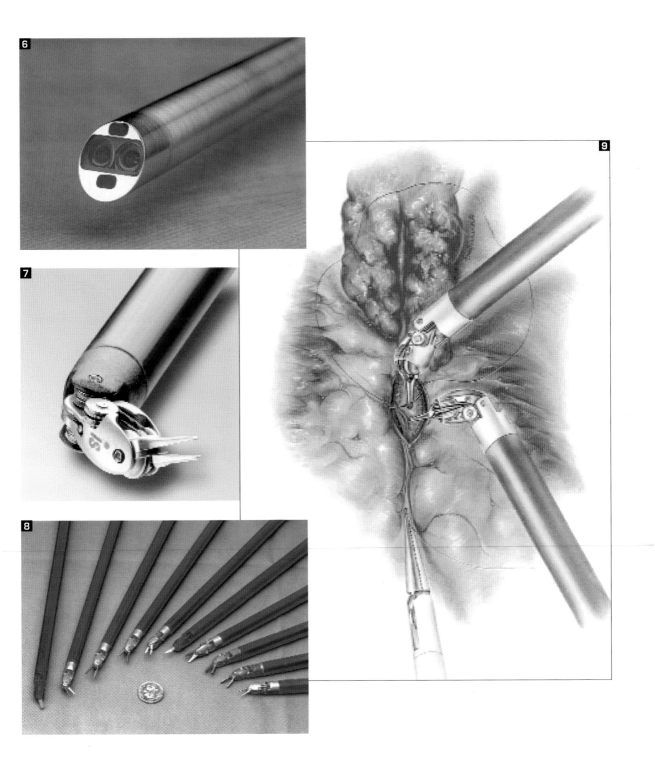

Robotic medicine here to stay

The benefits of robotically assisted minimally invasive surgery are undeniable, say practitioners and proponents. "The winners in all this are the people who are able to get up, move themselves off the operating table, be up walking the night of surgery and be home with their families within days with very minimal pain and suffering," states Dr. W. Douglas Boyd. He adds that after robotic bypass surgery patients spend a couple of days in hospital rather than a week or more, and that recovery times are measured in weeks not months.

In a study published in the *Journal of Urology*, Rossweler shows that laparoscopic, or keyhole, surgery in the abdomen results in complication rates of only 6 percent compared to 19 percent for conventional surgery, while only 10 percent of patients require blood transfusion compared to 56 percent. Furthermore, they can resume normal activity after 27 rather than 52 days. The boons for surgeons are much improved vision of the surgical area and additional degrees of freedom for instrument tips.

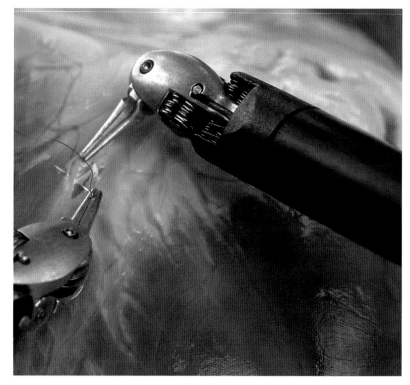

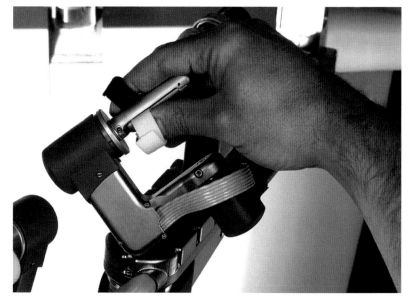

Robotic surgical systems are, however, far from perfect. Luc Soler, who heads R&D at France's cancer research institute, IRCAD, believes that they will eventually disappear. "They are first-generation devices that haven't changed surgical instruments. They just hold them in position from the outside. If further progress is to be made, tools have to change."

One significant drawback is that surgeons have to work in reverse-motion. So to execute a left-to-right surgical incision, for example, he or she must move the controls from right to left. The most widely used surgical systems, such as ZEUS, rob surgeons of the hands-on, open-surgery feel. Automated systems in fields like gaming and industry use force and tactile feedback to reproduce the sense of touch. The failing has been addressed by the MIT-invented haptic robotic arm Phantom, which is marketed by SensAble Technologies, Inc. The arm imparts a sensation to the user's fingers that is similar to touch. German endoscope manufacturer Karl Storz makes force-feedback surgical instruments.

Some doctors believe that surgical systems like da Vinci are not designed for beating-heart bypass operations. One doctor with such misgivings is Professor Daniel Loisance who has used da Vinci for four years at the Henri Mondor Hospital. He believes that the most delicate, demanding operations are beyond the capabilities of robots. To fill that gap, Professor Michel de Mathelin's team at the Louis Pasteur University in Strasbourg are developing a robot that can filter out motion so as to operate in a moving surgical area like a beating heart.

Despite perceived limitations in robotic medical systems, research worldwide has never slackened its pace in a bid to take them to the next stage in their evolution.

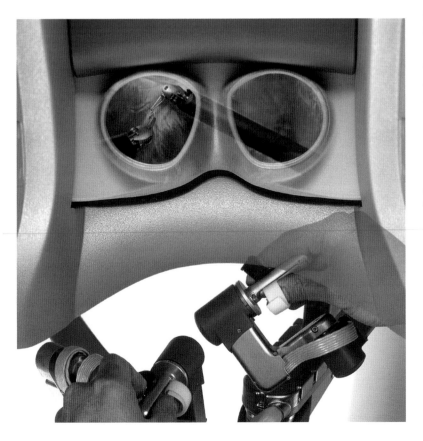

EndoWrist instruments faithfully reproduce a surgeon's hand movements.
© Courtesy of Intuitive Surgical

Back in September 2000, France's National Network of Health Technologies (RNTS) gave its go-ahead to a project called Endoxyrob, undertaken by the French company Sinters. When commercialized, the system will be more compact than its predecessors, manipulate miniaturized instruments, offer enhanced ease of use with finessed force feedback...and all for an extremely competitive price. Hitachi has adopted a radically new tack with a robotic system that further minimizes minimally invasive surgery to a single incision into which a micromanipulator handling three instruments can be inserted.

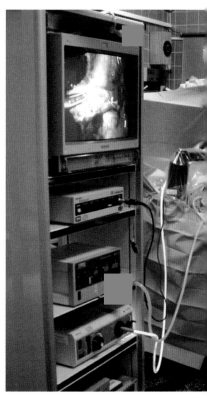

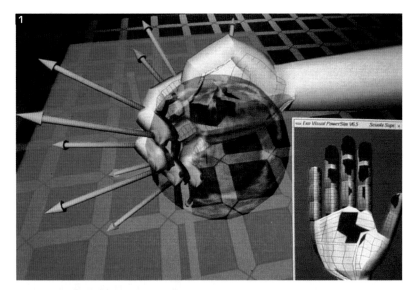

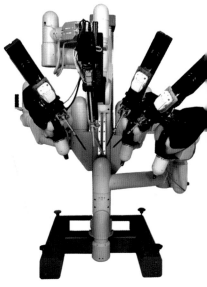

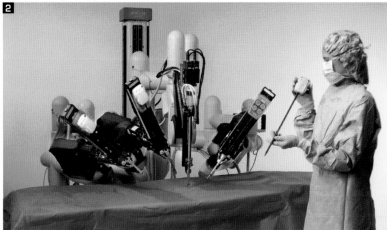

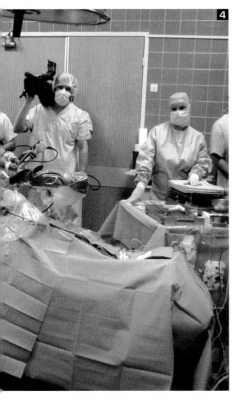

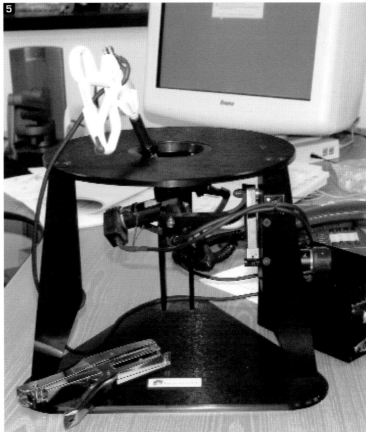

1- Study into force feedback.
© Courtesy of INRIA

2- 3- The patient-side cart for robotic arms holds minimally invasive surgical instruments.
© Courtesy of Intuitive Surgical

4- Robot-assisted surgical procedure.
© Courtesy of IRCAD

5- Development and modeling research work on an endoscopic instrument at IRCAD.
© Courtesy of IRCAD

6- 7- Da Vinci control console. The InSite vision system provides surgeons with minimally invasive 3D views of procedures.
© Courtesy of Intuitive Surgical

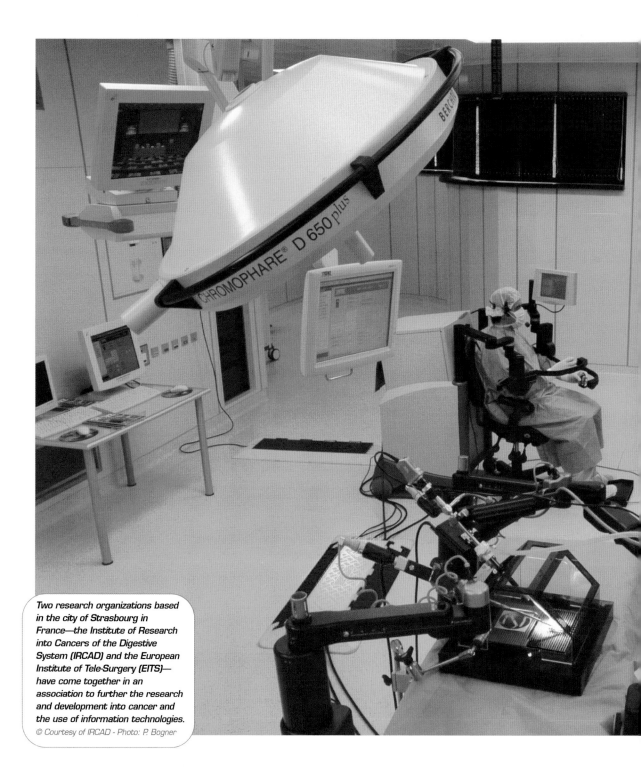

Two research organizations based in the city of Strasbourg in France—the Institute of Research into Cancers of the Digestive System (IRCAD) and the European Institute of Tele-Surgery (EITS)—have come together in an association to further the research and development into cancer and the use of information technologies.

© Courtesy of IRCAD - Photo: P. Bogner

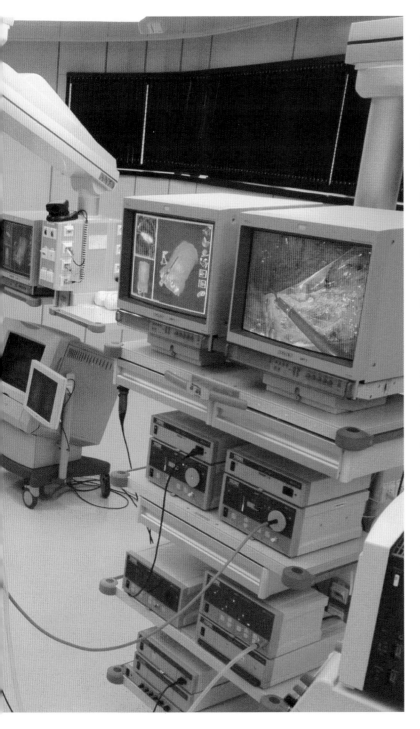

IRCAD focuses on providing a comprehensive package of computer—and robotics-based surgical solutions within the cancer research project Odysseus that the European consortium Eureka has endowed with $2.6 million. IRCAD's pioneering work is widely acknowledged: in 1999 and 2000 it received the Smithsonian Institute's Computer World Award and in 2003 the UN's World Summit Award for its project Argonaute 3D. Drawing on its expertise, it seeks to use robotics technology in a range of surgical assistance solutions:

1- **Digital 3D modeling of patients from medical imaging**
2- **Diagnostic and preoperative planning software**
3- **Radiological and surgical simulators**
4- **Augmented reality systems,**
5- **Robotic systems for automated surgical motion**
6- **Remote cooperative work systems.**

IRCAD's work presages profound changes in surgical practice. Similar thrusts are driving research efforts worldwide, some of which overlap IRCAD's focus.

Digital 3D modeling

Research into filmless medical imaging seeks to make specialist diagnoses faster and more efficient by digitally "cloning" images of patients' insides and their trouble spots.

"Radiologists currently take about 15 minutes to diagnose a patient. If a computer speeds up the actual diagnosis time, then surgeons are freed up to think for most of those 15 minutes," says Luc Soler. IRCAD has designed and developed a 3D patient modeling system that runs on a PC-based Web browser and can turn the digitized model every which way. The color-coded organs make it easy for patients to recognize them. "Patients can look at 3D images, ask the surgeon questions, and identify with them. So the surgeon can say, look, that little green ball is your cancer and we're going to cut it out," elaborates Soler. The surgeon can also take a patient on a conducted virtual tour of his or her body's passages and tubes, such as the colon.

Other researchers are working in a similar direction. At Leyden University's Medical Research Center in the Netherlands a mobile robot produces high-precision delineation of the heart's wall and other designated parts. Magnetic resonance images make up the

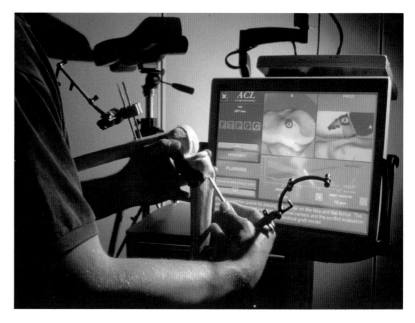

virtual world in which the robot navigates to explore and map the heart. Traditionally, doctors have to view some 200 images of a heart, then draw its contours by hand. The thinnest parts are those in which they are interested because they are the ones starved of oxygen. The Leyden robot saves time by automating the painstaking jobs and outlining the heart wall to point up the thin oxygen-starved parts.

Diagnostic and preoperative planning software

Surgical strategy requires planning ahead. To arrive at a diagnosis, a surgeon can view a 3D image of a patient's organs digitally reconstituted from an MRI scan. He or she highlights, colors, brightens and erases parts in order to home in on, say, the tumor he or she is seeking. Conventionally the process takes three months of chemotherapy. To obtain a second opinion and determine the best surgical strategy, the surgeon can use IRCAD's Argonaute 3D platform. Other physicians, surgeons and specialists can access it as long they have a broadband Internet connection. The attendees at the virtual planning meeting access the networked platform where they can see and intervene on the same volumetric view. They give and receive their opinions in real time.

Page left: Digitizing an imaged knee ahead of preoperative planning. The system pictured was jointly developed in France by the TIMC-IMAG Laboratory, the Grenoble Mutualist Clinic, Grenoble University Hospital, and the company Praxim.
© Courtesy of CNRS photo library - Photo: Emmanuel Perrin

Left: A broadband connection allows remote teams to view an operating field and jointly work out a strategy for surgical procedure.
© Courtesy of IRCAD

Radiological and surgical simulators

A particular benefit of surgical systems like da Vinci or ZEUS is that they can be used to simulate procedures in much the same way as flight simulators. A volumetric model of a body part replaces the image of the real thing, while surgery simulation software supplies visual feedback.

The Norwegian company Sim-Surgery, has developed an educational software platform that has been developed and tested on the ZEUS surgical system. With simulation software a trainee can control robotic instruments from the master console, while visual feedback of, for example, a suture, shows the needle at work and the response of tissue. Work to introduce force feedback that would simulate touch and organ response is under development.

Augmented reality

In Japan, the US and Europe research is ongoing into head-mounted and other interfaces that superimpose graphics, audio and other sentient and informative enhancements over a real-world environment in real-time. Elekta Instruments' robot Surgiscope draws on augmented-reality technology to guide surgical instruments towards lesions previously considered too small to operate on. "Developments in medical imaging have made it possible to visualize body parts in different ways," says Micael Michelin of LIRMM. They can even reveal hitherto unsuspected pathologies. He adds that by using localization and robotic guidance systems on such images, a surgeon can manipulate surgical instruments and watch them at work on anatomical parts.

IRCARD has developed new software that aims to superimpose a model patient over a surgeon's real-world view, while Eve Coste Manière of the French artificial intelligence research institute INRIA, is also developing augmented-reality software for medical applications.

Ultimately a robot may be able to use augmented-reality data that has been transmitted to it to perform a surgical operation itself. It will first have to be positioned correctly and the surgeon will determine the surgical instruments' entry point and trajectory plan and will supervise the operation.

Long-distance surgery

It is in the sphere of telesurgery that robotic surgical systems like ZEUS and da Vinci really come into their own. They enhance and reproduce surgeons' movements either in real time or with a time lag, position instruments and perform sutures. "Eventually a surgeon might be able to say, 'I want to take out that tumor' and the robot will follow the right trajectory, factoring in breathing, and take it out," says Luc Soler. "Robots might even destroy some tumors through radiofrequency ablation, which burns the tumor from the inside. They would insert the rod that generates heat from its tip."

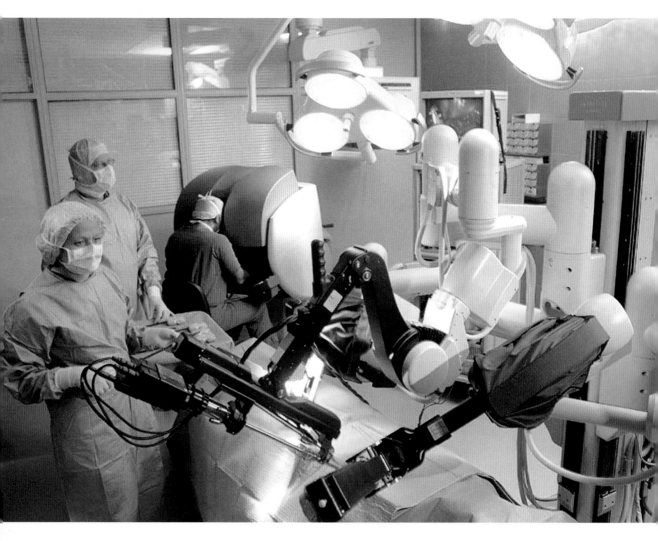

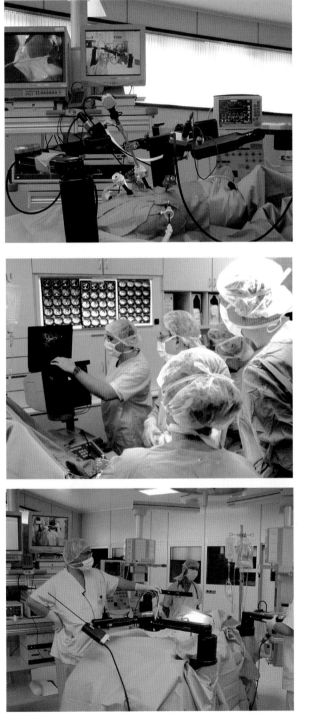

Robots and telemanipulation have been a watershed in medicine, with more and more surgeons seeking out the opinion of a distant specialist during an operation. As W. Douglas Boyd puts it: "For difficult cases we have the capabilities now of bringing the world's experts into any surgeon's operating room."

Telemanipulation is still in its infancy, however, and Georgi Graschev of the University of Medicine in Berlin looks even further ahead: "If manipulation is remote controlled, it might even be possible to let an outside party perform certain precise movements."

The robot is in the process of asserting itself as a high-precision aide for humans, who nonetheless retain their place as master craftsmen. It is a relationship that will be the pattern of collaboration for a long time to come, because, as Luc Soler points out, "In medical matters you can't ask a computer or a robot to have a say in a patient's life."

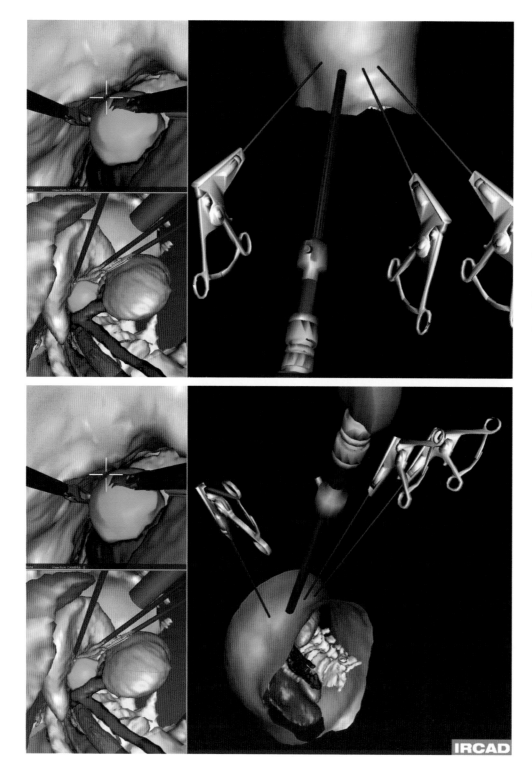

The 3D reproduction of a patient's anatomy used in a surgical operation is derived from two medically imaged sections. The organs are highlighted in different colors for enhanced visibility.

By immersing themselves and navigating in a surgical virtual reality environment users can see how organs react and plan surgical procedure.

© Courtesy of IRCAD

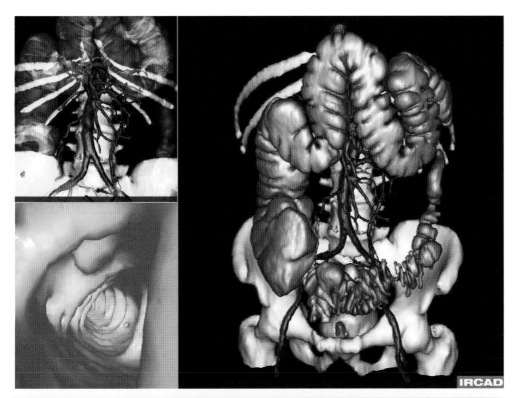

IRCAD

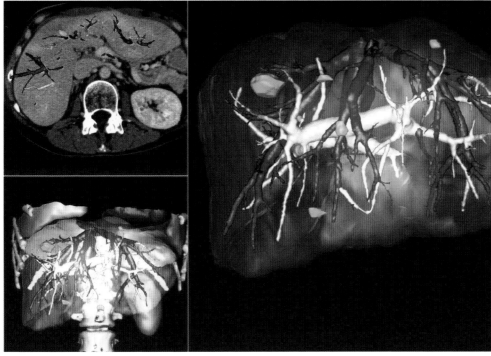

Virtual reality in surgery involves four phases: immersion, navigation, interaction and planning.

1- Immersion is the sensation of venturing into the human body.

2- Navigation involves moving around within the 3D image.

3- Interacting consists of manipulating, tweaking and transforming the 3D image with real-time visual feedback. Pathology or anatomical variations can be created.

4- Planning refers to choosing the best instrument trajectory for optimal precision.

© Courtesy of IRCAD

Medical care robots

As more Japanese are living longer and longer they are generating needs that are being addressed through the development of robotics applications specifically for the elderly. Many of the care robots look toylike enough to be meant for children, but they are intended for diverse forms of looking after seniors in hospital, retirement homes and at home.

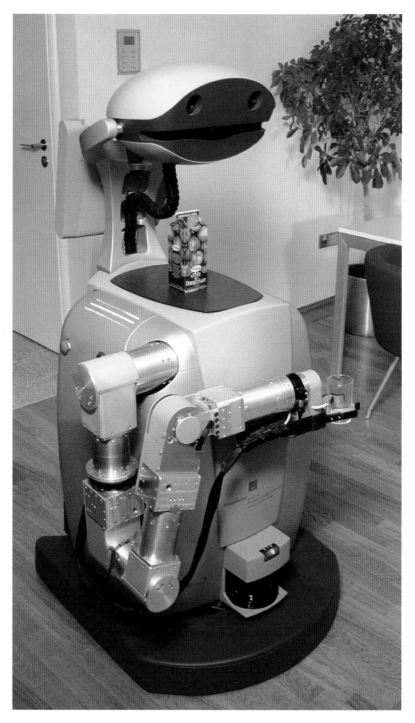

Care-O-Bot II.
In the near future robot helpers will be needed for the elderly and people with reduced mobility. They will allow them to continue to live in their family environments while limiting and reducing costs in retirement homes, which are likely to rise with the size of the aging population.
© Courtesy of Fraunhofer IPA

Home-help robots

Many of the personal robots described in Chapter 4 would also have their place in this one. They include pets and bots designed to hold conversations with elderly people; examples are Mitsubishi's Wakamaru and Toyota's Partners, one of which has been specially made to provide the elderly with care, companionship, and even to carry them.

Some personal robots are specifically designed to provide home healthcare. One is Sanyo Electric's nurse bot, Hopis, which can take the pulse and temperature, measure blood sugar and eye conditions, and ask patients questions, all of which it transmits to a health center.

Automated assistance for the elderly is not an issue in Japan alone. In the US, Carnegie Mellon University is developing a robot assistant called Pearl, designed to keep the elderly company, remind them to take their medicine and, if necessary, to call for help.

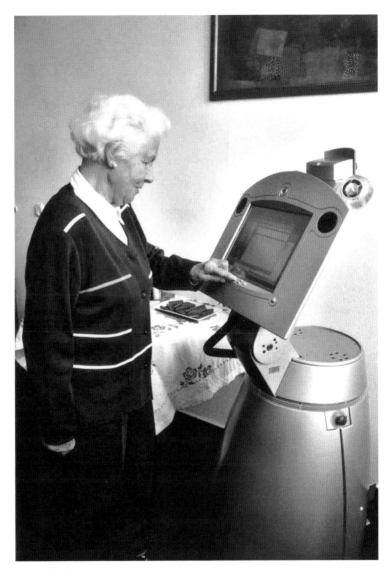

Care-O-Bot I.
© Courtesy of
Fraunhofer IPA

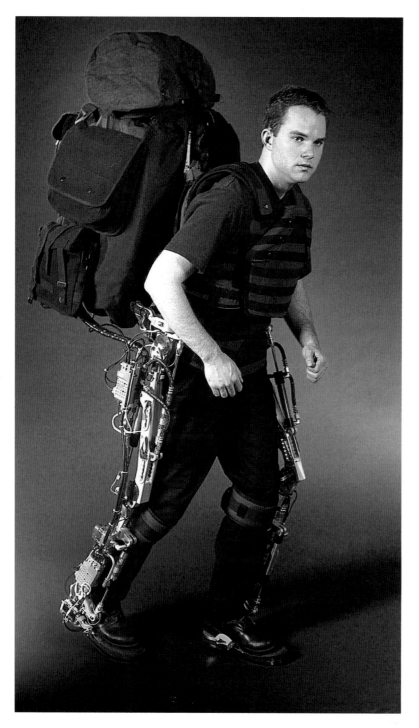

Leg-assisting robots

A promising robotics application is orthotics, robotic legs that help people with weak or impaired legs to walk. In the US alone 1.7 million people are thought to have some form of leg disability and some 200,000 are confined to wheelchairs with no use of their lower limbs due to spinal injury. Yobotics develops orthotic devices to help those with muscular weakness to walk. One currently under development is the wearable, powered device RoboWalker.

In France the Laboratoire Robotique de Paris has developed Monimad, an intelligent wheeled "deambulator", whose primary aim is to help those who have difficulty walking to regain confidence in themselves. It begins by helping them get to their feet from a sitting position by lowering its handles to provide leverage. It is keenly attentive to posture, self-adjusting to the walker's stance. Should he or she tip to the right, Monimad exerts enough pressure to the left to prevent a fall. A further innovation under development by the University of Tsukuba and Mitsui and Co. in Japan, is the robot suit, or "powered suit", which supplements the strength of the aged and disabled, helping them to walk, climb stairs and rest standing up.

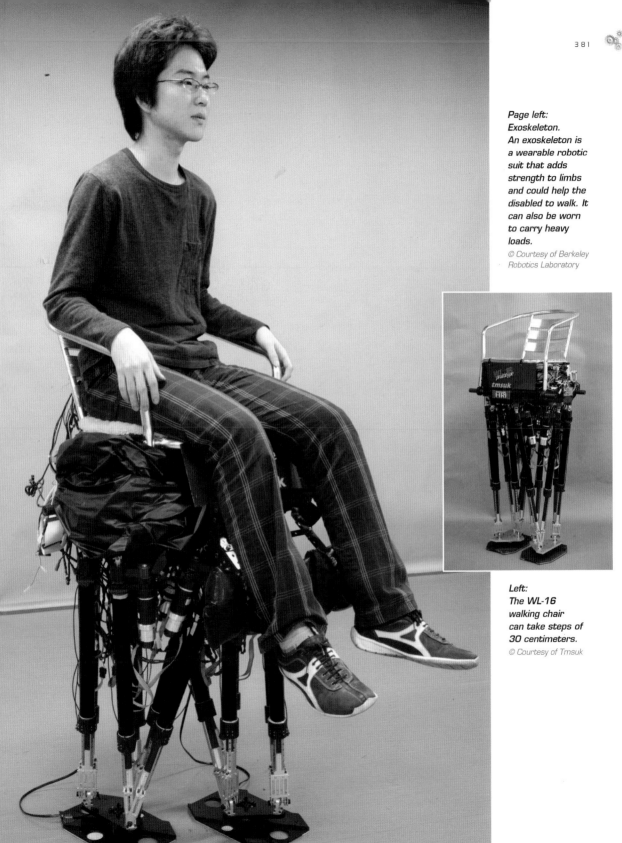

Page left:
Exoskeleton.
An exoskeleton is a wearable robotic suit that adds strength to limbs and could help the disabled to walk. It can also be worn to carry heavy loads.
© Courtesy of Berkeley Robotics Laboratory

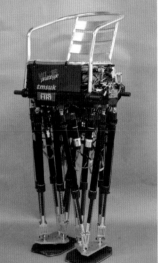

Left:
The WL-16 walking chair can take steps of 30 centimeters.
© Courtesy of Tmsuk

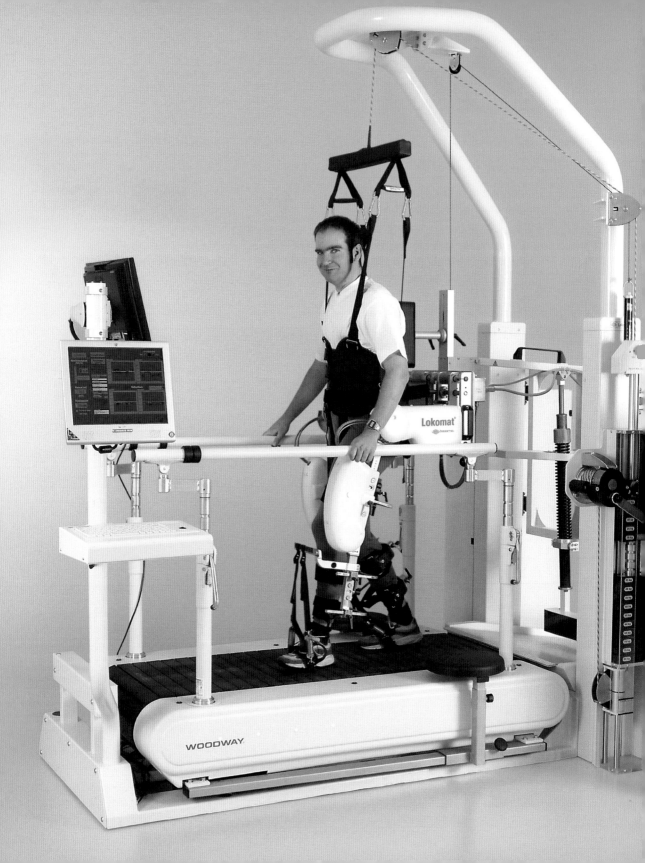

The Robot Scientist Project

The project conducts research into computer and biological sciences by developing computing and artificial-intelligence techniques for application to biology.
The robot scientist can interpret and design biochemistry experiments without any human help and experiments to test them.

© Courtesy of Ken Whelan - University of Wales, Aberystwyth

If robots could add a muscle stimulation function to their capabilities, they could help paraplegics to walk. The Demar project at LIRMM research facility in Montpellier aims to use stimulators that send electric charges into leg muscles to arouse a response. The stimulators are connected to a unit that measures the voltage needed to elicit a response from muscles.

Left:
The Lokomat, a robotic rehabilitation system.
© Courtesy of Hocomat

Autonomous robots in hospitals

Previously we discussed how robots assist surgeons in highly delicate operations and how they might even undertake procedures themselves. However, it might be much longer before they develop the bedside manner of devoted nursedom, where interpersonal skills count for so much. For the time being they are confined to ancillary tasks like couriering medical supplies and drugs and fetching dirty linen and waste. Their work frees up

nurses, but does nothing to off-set a nursing shortage. It is thought that if robots ever do replace nurses, it will be to supplement growing shortages.

The Pyxis Helpmate brings patients their meals and medication, gliding from ward to ward and floor to floor. The Care-O-Robot is an autonomous vehicle, developed by the Fraunhofer Institute in Stuttgart, which can also take meals to patients and dirty linen to the laundry, riding elevators that it operates by means of an infrared sensor. And it can scan patients' electronic bracelets and send the data to the doctor.

The automatic vehicle Robu-CAB, designed by the French company Robosoft, is intended to shuttle people over short distances on large sites like airports and business parks. It would be an ideal people carrier in a large hospital.

The cost of hospital assistance robots is at first prohibitive. Help-mate is priced at $70,000, yet about 100 hospitals have already acquired one and recouped their outlay after only a year.

Healing robots on the horizon

Healthcare robots are not confined purely to assistance and ancillary tasks. They can also have a therapeutic effect. It has emerged that Aibo has brought smiles to the faces of the lonely—old people and the mentally handicapped. So, too, reportedly, has Paro, a robot seal invented by Takanori Shibata. It has only to flap its flippers and blink its eyes to elicit a response from patients and even to relieve depression. Brian Scassellati of Yale Child Study Center's Autist Group is developing an android specifically to interact with autistic children and focus their attention.

Traditionalists and animal lovers argue, of course, that there is no need for robotic companionship, when animals have soothed and amused the lonely, elderly and sick since time immemorial. To which the advocates of Aibo and his ilk retort that they are cleaner, safer (they don't bite or cause allergies) and easier to look after. Europeans would probably side with the pet lovers, because of their deeply ingrained culture of respect for natural life forms, which they consider superior to mechanical beings. The animist tradition of the Japanese however sees a life force in all things, animal, vegetable, min-eral and mechanical. Although robots can and do provide company and comfort, research has not pursued their possible therapeutic value. That remains the province of robots in fictional writings like those of Asimov: only there do they lovingly accompany their human masters to the end of the road.

1- The Pyxis HelpMate SP Robotic Courier System fetches and carries meals for patients, medicine, lab samples, supplies, and medical records. And it can ride elevators.
© Courtesy of Pyxis Products/Cardinal Health

2- The HTS-MiniLab is a fully integrated robotic laboratory.
© Courtesy of SSI Robotics

3- 4- The Freedom EVO is an open robotic platform developed by Tecan that offers automated procedures for genomics, proteomics, drug discovery applications and other analyses, whether in vitro or in vivo.
© Courtesy of Tecan

Interview
Jacques Marescaux

An internationally renowned surgeon and founding president of IRCAD, Professor Jacques Marescaux stepped into the global spotlight when he performed the first ever robot-assisted trans-Atlantic telesurgery operation.

Jacques Marescaux. Photo : FYP

What did it change for you as a practitioner to work on an image on a screen instead of operating on the body of a patient?
To begin with it was very difficult. The surgeon in me had to be humble and learn from scratch. I was used to working hands-on and suddenly I had to manipulate long surgical instruments and obviously I was clumsy at times. I couldn't help the tremor in my hands. Image quality was poor at first and, for a while, we lost the 3D view. And we lost the force feedback, so we couldn't feel the instruments at work. Yes, tough beginnings.

How did endoscopy pave the way for the advent of robotics?
Minimally invasive surgery introduced us to working at a remove from the patient and we realized that our instruments weren't really adapted to the way we worked. We weren't

comfortable with them and we knew we couldn't go on working like that. Some operations could last for four or five hours and we were absolutely exhausted. The idea of robotically assisted surgery was born from there. It gave rise to equipment that enabled the surgeon to sit comfortably and to an interface that analyzed his or her hand movements, pared them down and transmitted them to three robotic arms. Telemanipulation was born. Another argument for robotically assisted procedures was the need for sutures on a microscopic scale. Because of hand tremor no surgeon could perform them.

What were the earliest robotically-assisted machines that you tested?
In 1995 I was at Stanford and I had the opportunity to handle a prototype of the da Vinci surgical

system, which had been developed by the Intuitive Surgicals Company for the US army. I used the system to operate on an animal on the other side of the room. I was so totally immersed in the 3D world I saw before me that at one point I dropped the swab I was holding in my tweezers. It was only when I bent down to pick it up that I realized, of course, it was on the other side of the room. When I went back to France I was still under the influence. At the time the da Vinci prototype hadn't been fully developed, but then, in 1996, I had the good fortune to meet the Wang brothers who had created Computer Motion in San Diego. When I laid eyes on ZEUS, I thought it was phenomenal. It weighed 150 kilograms, whereas da Vinci was one ton. The arms were fixed to the operating table and the camera was voice-controlled. It was like a real-life operation where you give instructions to your assistant. I was bowled over by ZEUS even though da Vinci was easier to manipulate. We liked the Wang brothers, who are tremendous engineers. On their side, they were convinced that IRCAD had potential—we do train 3,000 surgeons a year, after all. We had three ZEUS robots shipped over and the Wang brothers set up a subsidiary. From then on we were involved in developing ZEUS.

Did you test the ZEUS systems in your operating room?

Yes, we tested ZEUS many times, and da Vinci, in fact, which was easier to use. What we wanted to know most of all was how they impacted on the medical team. When a robot comes into the operating room, how do the surgeons, the anesthetists and nurses react? I mean, it was the first time a surgeon had ever brought such a contraption into the operating theater. Yet they took to it immediately. It's an easy matter to set up ZEUS, it only takes ten minutes.

As a surgeon, how do you manage to cope with systems without force feedback?

It's a big debate. We've realized that you can do without force feedback. The quality of image is excellent, it's five times better than the naked eye. When you see an organ changing shape under the pressure of an instrument, it's as if you can feel the force feedback. But it's true, surgical systems must evolve. They're still not properly adapted to our craft. Future systems will deliver force feedback, but the priority is still quality 3D imaging.

When did you first feel that it would be possible to use robotic endoscopic systems for telesurgery procedures?

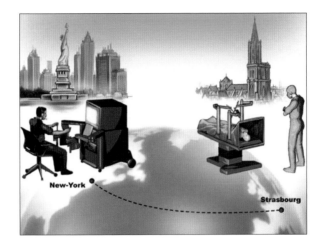

Diagram of Operation Lindbergh.
© Courtesy of Jacques Marescaux

It was in 1993. We'd started working on a project called MASTER[1]. It was part of the big European research consortium Eureka. It was the biggest project they had ever funded and it was finally accepted in 1994. To tell the truth, there were no robots in the original project. It focused on minimally invasive surgery and virtual reality. But, quite by chance, when I went to see someone at ANVAR[2] I got lucky, because the person I saw said, "We want to fund innovation; it would be good if there were remote robotics in the project." To begin with, I thought nobody would take us seriously, but Alcatel was interested, in fact. Then the three ministries concerned gave their approval. They were Industry, Research and Health. Then the Health Ministry wanted to shelve the project, but there was a second assessment and it concluded that it would be valuable to go ahead and attempt telesurgery. And we were already talking to engineers at France Telecom and Alcatel, which had cleared any technical problems. And at Strasbourg we were, after all, one of the earliest pioneers of minimally invasive surgery. The concept for Operation Lindbergh took shape at a time when there were no robots or remote surgery.

What were the technical problems you had to overcome to set up Operation Lindbergh?

They were dreadful and, in fact, DARPA's scientific director said just two months before the operation which was set for September 7, 2001, that it was impossible. They said there was no way to make data transmission any faster than 600 milliseconds via satellite. It was a real sticking point for everyone. What we wanted was to use ATM[3] by cable.

At a meeting at France Telecom's research center we found ourselves with three engineers. They just smiled and said, "Where's the problem, we can get data transmission speed faster than 200 milliseconds." We ran tests and they were conclusive. Outgoing images just had to be compressed and incoming ones decompressed and we solved that problem with software we bought off the shelf. The first teleoperation we tried was in 2000 between Paris and Strasbourg. The patient was a pig and we removed its gall bladder by telesurgery. The procedure lasted four hours and half. I was totally exhausted by the time it was over. Data transmission speed varied, sometimes reaching 400 milliseconds. It was maddening. You'd start inserting your scalpel, then you'd stop,

but in real life the scalpel was still going in. It was a disaster. I was so depressed. What's more, President Chirac kept asking how far we'd got. But in the end, after lots of brainstorming sessions, we were ready for the real-life test between New York and Strasbourg on another pig. We did it six times with no problems and with data flowing at 200 milliseconds. The next step was to teleoperate on a human being.

Didn't you feel discouraged at any time?

I had to be stubborn, but I was kept going by my teams and the faxes I received from President Chirac. I felt that I had to do it. Obviously, there were all sorts of problems to solve. The background research was laborious and reams of red tape had to be sorted out, too—that's a two-year full-time job in its own right. Then, two days before the operation, the FDA said no, it couldn't go ahead. But they had already approved the robot themselves and the patient had given her written consent. I think they were annoyed that a French surgeon was going to do this world first. It was frightening. We had to phone the embassy to get them to intervene. We even threatened the Americans with performing the operation from the French Embassy in New York,

which is French territory. That was when everything sorted itself out.

Could you tell us about Operation Lindbergh?

It felt like an ordinary operation, which gall bladder removal is. It lasted 45 minutes. We were well prepared, though. A local team in New York was ready to step in and take over at the slightest telecommunications glitch. But France Telecom had planned a second emergency line just in case. If the operation was to go smoothly, we needed a response time of 200 milliseconds and we got 135.

How did 9/11, just a few days later, affect media coverage?

The operation took place on September 7, 2001, and the news conference was planned for September 11. There was to be a live link between New York, Paris and Strasbourg and the astronaut Neil Armstrong was to be there. It was nine o'clock in the morning when we heard about the Twin Towers. So we postponed the press conference for four days. We still managed to get coverage from 40 television channels and there were hundreds of press pieces. The magazine *Nature* gave our account of the teleoperation pride of place. It was the first time in their history they had ever agreed to run that kind of article.

What was the message you wanted to get across after your remarkable performance?

That individualism had had its day and the use of video technology had helped to finish it off. It showed that surgery was a skill like any other. According to a report submitted to Bill Clinton in 1999, medical and surgical malpractice was costing the US billions of dollars and 100,000 lives a year. More than half of those lives could have been saved if there had been a second surgeon. It's awful to think that the outcome of a rectal cancer operation depends entirely on the surgeon. Worldwide 30 percent of patients relapse. It's unacceptable. With the advent of current systems, surgery can be an exercise in telecollaboration. A coronary microsuture can be between 80 percent and 100 percent of the quality required. If it's only 80 percent the patient has another heart attack. The day will come when a surgeon will seek out a second specialist opinion and will teleoperate on a virtual-reality clone before performing surgery on a real person. Telemedicine has ushered in the sharing of knowledge, intelligence and skill.

What sort of operation could a robot perform by itself?

Simulation involves operating on the virtual clone of a patient. If the surgeon makes a mistake, he or she can start over and over again until the procedure is perfect. Once the perfect operation has been determined, it can be stored in the computer. However, I think it might be more promising to superimpose the real patient over his or her virtual self.

How do patients perceive the arrival of robots in medicine?

We explain to them that robots help to "augment intelligence" so that they understand that robots add that extra something. They have been very receptive and no patient has ever refused robotically assisted surgery. That might not be the case if an operation were one day to be fully automated.

French president Jacques Chirac visits IRCAD.
© Courtesy of
Jacques Marescaux

---

(1) MASTER: Minimal Access Surgery by Telecom and Robotics.
(2) ANVAR: Agence nationale de valorisation de la recherche, a French government agency for the promotion of research.
(3) ATM: Asynchronous Transmission Mode, broadband technology for voice and data transmission.

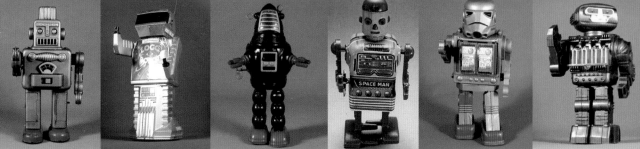

1950   1960   1960   1961   1970   1975

1975   1978   1978   1978   1980

1980   1985   1985   1985

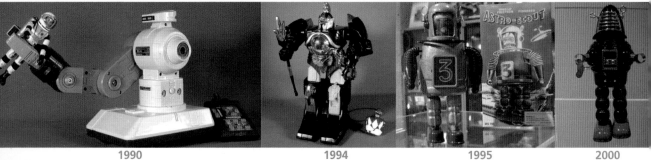

1990   1994   1995   2000

2001

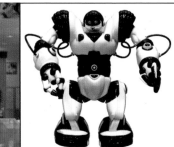

2003

2004

50 years of robot toys

For us, these robots with their simple mechanisms cannot be anything other than toys, but they delighted generations of young science buffs for whom robots were only a dream.

1950–1994:
Jean-Pierre Hartmann's collection of robot toys
1995–2004:
Robot toys, Courtesy of Robopolis.com

PLAYFUL ROBOTS

Robots take us by surprise, where we're least expecting them. They slip into our houses in the most natural way in the world at Christmas and on birthdays. The invasion is so stealthy that many adults, going quietly about their daily business, notice nothing.

Open your eyes: what do the impish Poo-Chi, the waddling Robosapiens and the cute kittens such as NeCoRo look like? Beneath the veneer of the "smart toy" these are actually little robots.

While they're fraternizing with these self-evolving toys that obey their every command, an entire upcoming generation is getting used to having around objects capable of reacting to external events. It's a fair bet that these kids will be big fans of robotics by the time they reach their teens.

In fact, their big brothers and sisters are already showing the way, as the numbers flocking to high school robotics clubs shows. This precocious talent is being encouraged by the multitude of programmable robot kits coming on the market.

The best and brightest among them take part in high-profile competitions such as RoboCup, in which robot soccer teams play against each other. By the time they reach twenty, they'll find it the most natural thing in the world to tell the android to go and wash the car, weed the lawn, or feed the hamster.

This enthusiasm is strikingly similar to the early days of micro-computing, which saw the birth of Apple and Microsoft.

Automata

Interview
Frédéric Marchand

Frédéric Marchand collects and sells old toys, and has written several books on the subject, including one on the history of Martin Toys, the first automata made as toys.

When were the first toy automata made?

The end of the 19th century. The father of the "two-bit toy" and the wind-up automat was Fernand Martin, a Frenchman, who built a factory, The Martin Works, and produced his first models in 1870. He revolutionized the business by being the first to scale down automata. Previously only rich families could afford the leading brands. Martin's toys were sold by street merchants, traveling salesmen, for the equivalent of half a day's wage for a working man. His automata represented a number of trades—carpenters, smithies, and less-skilled trades too. They were made of hand-painted metal with clothes made from real cloth and used clockwork mechanisms.

Were children interested in these automata showing adults at work?

Yes, because they saw something of their parents in them, and they were a big success straight away, since they were cheap and original. Later Martin was copied by German firms, notably Lehmann's, who started producing more robust models with enamel paint. And since they set up sales franchises all over the world, they soon became a serious rival. All the automata makers then started making toys as well. Nonetheless, Martin's toys were more original and were nothing like their rivals, being inspired by Parisian life, e.g., street violinists, dancing bears, drunks, pig sellers, market porters and so on.

How long did Fernand Martin go on doing this?

Until 1912, when he sold the company to the son of a banker, Flerscheim, who died unfortunately in 1916. Between times, the Martin Works had grown into a major industry, making 80,000 pieces a year, all assembled and finished by hand. They even had an outlet in the US. As for Fernand Martin, he was the president of the toymakers' trade association, and in 1900 was also behind the Lepine competition for new inventions that continues to this day, saying: "There are people with great ideas, but we industrialists don't always know where they are". Later, Martin's was bought by the Bonnet brothers, toy wholesalers, and continued until 1960 as Martin Bonnet.

How did we get from automata to models resembling robots?

One of the first known robots is a tin German model from 1936 with a clockwork mechanism. Then after 1945 the Japanese, with their passion for science fiction, started producing toy robots using a battery-driven electric motor. There was no need to wind them up anymore. By the end of the 1970s, the Japanese had cornered over 90 percent of the market.

This page:
the gay violinist,
Jouet Martin, 1902.

Following pages:
1- The Penny Toy,
Christmas tree toy,
Philip Meyer, 1900
2- Toy airplane,
The Spirit of St Louis,
Schuco (Germany) 1928
3- Elephant, Blumer
Schuller (Germany) 1935
4- Autopede, Victor
Bonnet, Martin's
successor, 1920
5- Clown, Koeler
(Germany) 1950
6- Motorcycle automat,
Keleruan, 1946
7- Gunterman's Billiards,
(Germany) 1920,
with perpetual-motion
mechanism by
Georges Lévy

© Frédéric Marchand
Photos : Christophe Recoura.

5

6

7

Interactive toys

Many children have abandoned the traditional electric train in favor of a Gameboy or console. The toy business underwent a revolution with the arrival of digital toys springing from the meeting of two disciplines: artificial-life software and robotics. During the 1990s, video-game programmers notched up striking achievements in terms of artificial intelligence and virtual reality, developing hyperrealistic simulations in domains as varied as urban planning (Sim City), military strategy (Age of Empires) or even the growth and evolution of living beings (Creatures).

AI was there, just waiting to be integrated into tactile objects like robots. As things worked out, the toy industry served as a full-scale testing ground for this new market. It soon became obvious that it was huge.

The Tamagotchi reveals a market

The hyperminimalist Tamagotchi seems prehistoric today given how limited it was. However, this trial run had at least one merit: it showed how the public could be attracted to virtual beings. The little plastic egg was an astonishing success, despite its rudimentary style and poor esthetics. In 1997, Aki Maita, the Tamagotchi's creator, explained his thinking: "I got the idea of a virtual pet animal because Japanese apartments are often too small to have a real pet".

The Tamagotchi craze went far beyond Japan. In cities such as San Francisco, queues formed outside toyshops on the day of its launch. The most astonishing thing is that these creatures were not seen as simple toys. Japanese executives were known to suspend a meeting to "go feed the little'un". However, it soon became clear that such a gadget was too limited to sustain interest for long.

Sim City 4
In 1987, Will Wright used works on urban management, town planning, and the mathematical modeling of such phenomena to create the Sim City game.
© Courtesy of Electronic Arts

Furby, the first toy to integrate software

In 1998, Tiger Electronics' Furby was the first toy to show a semblance of intelligence. This 20-centimeter-tall (8 inches) creature has synthetic fur and tender little eyes with lashes that move when it speaks. A Furby expresses itself, complains if it is neglected, learns certain things, gives you a kiss, and sings and dances. This toy conquered the world—40 million were sold! It appealed to four year-olds and forty-somethings, and marks the first successful attempt to create an artificial pet.

Even though it is as static as a Tamagotchi, Furby stands out by the level of language: "Dah doo-ay way!" "Boh-bay," "Kay may-may u-nye" meaning "I'm having lots of fun," "I'm worried," "I love you." If the animal is hungry, it says so in its language and you have to slide a finger into its mouth to feed it. Another distinguishing trait is that when a Furby meets another Furby, they communicate!

Sony dreamed it and did it

1999 is the official date of birth of the leisure robot following the sale of Sony's famous dog Aibo, which showed the public's ravenous appetite for virtual pets, with 135,000 orders in a week for the $2,600 creature (see chapter 4). Toshitada Doi, Aibo's creator, wasn't trying to create a toy. But the manufacturers stressed that Aibo is a playful dog and targeted the children's market with low-price models adapted for Christmas.

The race was on to produce affordable robots for young consumers. The most common characteristics of this new type of toy are:

- **Looks like an animal or small human**
- **Can guide itself around obstacles**
- **Voice detection and the ability to move toward a voice**
- **Recognizes its own name: the animal reacts when called**
- **Expresses feelings: joy, sadness, boredom, etc.**
- **Communicates with another robot of the same family**
- **Ability to evolve over time, notably depending on the degree of patting or attention received.**

These characteristics are shared by dogs such as i-Cybie or Dog.com and similar toys.

Furby. With his tender eyes and tuft of hair, Furby was the first smart toy to appeal to millions of children.
© Courtesy of Hasbro

1

Of toys and robots...

Unlike traditional toys, robots interact with the child, and their attitude depends on the amount of attention they get.

1. *Poo-Chi, Hasbro*
2. *Dog.com, Tomy*
3. *Micro Babies, Tomy*
4. *Popee, Hasbro*
5. *Lovies, Tomy*
6. *Biobug, Hasbro*
7. *Micro Pets, Tomy*

7

6

i-Cybie, a real robot for children

In spring 2000 a new phenomenon swept trade shows in Europe and the US. Known as i-Cybie, it was made in Hong Kong by Silverlit. The US company Tiger Electronics won the production license and presented the animal in public at the International Toy Fair. The intelligent dog got good press, since it had everything: personality, a need to be petted, the ability to go towards its master, etc. i-Cybie even fell asleep at sunset and awoke at sunup. There was no doubt about it: the day of the robot toy seemed to have dawned thanks to this low-priced mass-market model.

But then Tiger's executives decided i-Cybie needed a facelift before being offered to American kids, and the final product wasn't launched until August 2001. It now has an angular face that gives it that "nice doggy" look. It weighs 1.3 kg and is 35 cm tall. In France, i-Cybie was a runaway success, and the importer was sold out before Christmas.

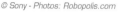

Dogfaced Aibo

Although designed for adults, Aibo indirectly encouraged the emergence of the robot toy market.
Opposite: Version ERS-312 Macaron, Sony
Below: Version ERS-7.
Bottom: ERS 220
© Sony - Photos: Robopolis.com

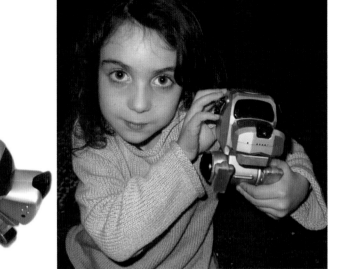

A robot zoo

In the meantime, a little zoo, featuring dogs mostly, had come to life. The Japanese company Tomy started the ball rolling with its Lovies, talking owls that could also produce offspring. Its success was there for all to see: the bestseller in France at Christmas 2000 and number one in the US for six months. Tomy followed this up with Dog.com, a Dachshund that needs taking care of and that becomes, depending on the circumstances, a joker, lazy, greedy, shy, etc. This graceful mutt has a three-stage development. At the infant stage it produces a few sounds. During adolescence, it mixes barking with some human expressions. Then (surprise, surprise) the adult Dog.com communicates only through speech, alternating various personalities. In his joker persona, for example, he threatens to report his owner to the Humane Society.

In the US, Lansey's Web Web outstripped the competition and even made the cover of *Time* magazine with 2.2 million sales. The manufacturer then introduced the dog's cousins, notably a cat and a parrot, Coco-Web, who, as you might expect, repeats whatever it hears.

Web Web's main rival is Poo-Chi, an adorable dog sold for the princely sum of $39. The toy is hardly out of the box before the kids are fawning over it, with its big heart-shaped eyes when it's happy, singing one of its songs when it's celebrating, and most of all, it liking to be looked after. Poo-Chi was launched in April 2000 by Sega Toys, and sales had already reached 10 million by November. In December, a spokesperson for Sears said that Poo-Chi was their bestseller, with demand two and half times greater than forecast.

i-Cybie.
Courtesy of Hasbro

Dogs were the frontrunners during the early stages of market exploration, with, apart from i-Cybie, Web Web and Poo-Chi, models such as Nikko's Cyberdog capable of barking at intruders when its owner was sleeping (the rest of the time indulging in the usual canine pastimes of barking and looking for food). The family of electronic animals gradually expanded, though. Cats followed the dogs, with models such as Bandai's BN-1 that recognized its owner and reacted to affection, but could also play with other felines of the same species. After that came panthers, ladybirds, fish, parrots, dinosaurs and so on, and even a tango-dancing lizard from Tomy.

The Bio-Bug insect is rather worrying: as soon as it is switched on, it starts exploring its environment, to the accompaniment of piercing cries. Fed and trained, it evolves from a newborn to an adult, changing color at each stage. The Bio-Bug is born to fight its own kinsmen (Predator, Crusher, Acceleraider) changing the living room carpet into a remake of the movie *Antz*! The more peaceful little Micro-bug is happy just looking permanently for light.

Mitsubishi Heavy Industries produced an artificial fish based on a species that has been extinct for 500 million years and intends to build robot versions of other extinct fish for museums, aquariums and households.

Each year a new batch of pets appears. In 2003, the Japanese company Omron came up with a cat called NeCoRo, complete with fur, which is fairly realistic except for its groggy stupor that gives the impression it's in a state of shock. The following year, Sega Toys produced its own furry cat called Nearme.

Gigamodus Megatron, Armada Transformer robot, Tiger Electronics
© Courtsy of Hasbro

*Opposite: Anim'
Animals robot cats
from Tiger*
© Courtesy of Hasbro

*Below: NeCoRo
cats, Omron*
© Courtesy of Omron

From the artificial baby
to the android

Science-fiction–type android robots capable of moving forward and turning while emitting a few beeps and flashes are among the commoner artificial creatures, but there are also some babies, at least one of which is quite disturbing.

My Real Baby was a joint development by Hasbro and iRobot, founded by Rodney A. Brooks, director of MIT's artificial intelligence laboratory. My Real Baby integrates the lab's findings on behavioral research, giving a highly realistic doll that reacts to its environment and shows its mood via hundreds of facial expressions.

Robosapiens is more open about its synthetic nature. This 35-centimeter-high (14 inches) astronaut, distributed by Wow Wee in the US, reminds you of the one in *Toy Story*, minus the face. Robosapiens' creator, NASA scientist Mark W. Tilden, also worked for DARPA (Defense Advanced Research Projects Agency). Tilden is behind an electronics technology inspired by biology and claims to have used the principles of advanced robotics to simulate human locomotion and movements. "I worked for 16 years on the physics of robotics," he explains, adding that he developed the basis for "biomorphic robotics" around 1988. "Rather than using a computer for each movement, Robosapiens uses biology-based techniques that allow the robot to move in a more or less natural manner thanks to seven motors". Although Robosapiens is animated by a relatively simple program, it moves fluidly: its adaptive suspension is anchored in its very design at the physical level, inspired by human and animal morphology. But even though Robosapiens is capable of a certain number of autonomous movements, it requires some dexterity with the remote control to make it dance, grasp objects, then lift and throw them, or else to try some karate moves.

Laboratoire d'automatique de Besançon
(LAB, Besançon Automatics Laboratory)

These ladybirds have a number of sensors that enable them to avoid objects, recognize a caress, detect darker places to hide in, but that also allow interactions between a pair of them (the robot child follows the robot mother). The implanted elementary behaviors allow the ladybirds to send different messages according to their state (happy, hungry, lost). This technology is being developed within the framework of a partnership with the SMOBY company that specializes in children's games.

© CNRS Photothèque
Photo: Jérome Chatin

Robosapiens

The quality of Robosapiens' locomotion is due to the mechanical ingeniousness of its suspension rather than its software.

© Courtesy of Mark Tilden

A broadening
of the toy market

The market for digital toys is expanding rapidly. According to the Yano Research Institute, its value increased twentyfold from 1995 to 2002. The target for these robot toys is the 3-to 9-year-old age group, with prices starting around $19.50.

Nonetheless it is amusing to see how adults in the stores quickly get into the spirit of things, squawking at the parrot or meowing to a metal cat, even though they only came to get a toy for the child. Make no mistake, the young parents who come to get Robosapiens for their offspring are often hooked themselves and the children have to fight to get their turn at the remote control!

Could the arrival of robots widen the toy market? Undoubtedly. Sega Toys broke down buying behavior by age for its main robot toys. For Yume Neko the furry cat, although the main audience is children, almost 23 percent of owners were aged over 40. Another market segment that interests manufacturers such as Sega is toys with a possible therapeutic role. The Japanese National Institute for Advanced Industrial Science and Technology AIST (see chapter 3) carried out experiments using a little robot seal called PARO. During the five weeks of testing, patients given the robot were more cheerful and less sad than the others. "Although the therapeutic benefits of pets is well known, it's difficult to bring them into hospitals or retirement homes because of fears of allergies, or scratching or biting, etc. Likewise, it's often difficult for people living alone to look after their pets, which are often not allowed in collective housing. PARO was designed to overcome these difficulties," according to a document from AIST, running the risk of these artificial animals taking over from real ones.

Are the days of kitty, the Doberman, and the hamster numbered because of these hairy robots that don't need litter boxes or walks? Not in the least, according to Jean Arcady Meyer of AnimatLab, who forecasts the peaceful coexistence of old enemies, e.g., "a cat and a robot dog playing together".

Worrying developments

Toys such as Poo-Chi and Miou-Chi herald the arrival of fake pets in the home. As one might expect, manufacturers play the emotion card for all it's worth—the virtual animal can simulate deep despair if it is neglected. By having the toy act in this manner, manufacturers can indirectly force children to devote time and attention to these artificial entities, thereby encouraging dependence.

Should we worry? Opinions are divided. Japanese manufacturers like to argue that using such toys develops a sense of responsibility, since you have to look after the animal (feed it, encourage it, etc.) and that children will grow out of the attachment naturally. And anyway, if the child tends to think the toy is an actual animal, removing the batteries should supply a reality check.

The problem is that some people seem prepared to consider these objects as living beings. This is not a new phenomenon. Warner Brothers received sackloads of mail for Bugs Bunny in the 1950s, some of it from adults. When Bob Clampett, who directed several episodes, showed visitors around the studio, he was often asked by adults and children alike when they would meet Bugs! More recently,

PARO was developed to satisfy the demand for a robot pet that could live with humans. Takanori Shibata, its creator, spent $9 million developing it. PARO's eyes blink and it wriggles, but the body remains stationary.
© Courtesy of AIST

Core Design, manufacturers of the Tomb Raider game, received e-mails asking whether Lara Croft had a boyfriend, and it seems that flesh and blood males confessed to being in love with the virtual heroine. The phenomenon was even worse with the Tamagotchi: virtual graveyards appeared on the Internet, where grief-stricken owners could bitterly lament not looking after the little one as well as they should have.

Asiaweek magazine reported the case of a 74-year-old woman, Tomoko Komiyama, who took part in a year-long experiment on the interaction with a robot koala teddy bear, Wandakun. After staring for so long into Wandakun's big brown eyes, the old lady admitted falling in love with her companion, after so many years of loneliness. She not only chatted to it regularly, but also knitted it jackets to keep its circuits warm in winter. Funnily enough, the doctor consulted about this relationship found it positive, claiming that since Japanese culture frowned on revealing emotions to others, the presence of a robot could help people to express their feelings.

Maybe so, but when robots have reached the stage where it becomes feasible to have a robot buddy, or even a Lara Croft in every home, certain persons may be tempted to consider the machines as living beings and prefer their company to the riskier business of dealing with real people.

It bears thinking about, and could lead to the creation of secondary laws of robotics along these lines: "A robot must always make it clear that it is an artificial creature and must oppose efforts by a human to ignore this".

Teaching robots

Teaching robots are the natural successors to today's simulators and educational playware. The advantages of interactive training have been shown over the past decade. The next stage will be to integrate the software into articulated, mobile objects, allowing a higher degree of realism: to learn a dance step, an android

that can lead its partner would be more effective than looking at an image on a screen.

Simulators and edutoys

The teaching quality of computer-assisted learning has already been proved. The expansion of the educational software toy market shows the benefits. This is the second most important soft-

ware market, and at periods such as the start of the school year, sales even outstrip those of video games. A number of sociological studies show the advantages of this technology: the program will endlessly repeat the same lesson without getting irritated and can adapt to the user's level. According to studies carried out by Dataquest, use of such software allows a better retention rate because the association of sound and vision has a greater impact than simple text.

Video games provide further proof. Flight buffs install real cockpits in their homes and use simulators such as Flight to perfect their piloting skills. Likewise, some Formula 1 drivers admit they can memorize the particularities of a given circuit by practicing on a PlayStation 2. They do insist though that the real sensations of a race cannot be reproduced on a simulator. Clearly, robot-assisted learning has a future, either in the shape of androids playing the role of teachers, or equipment with functions automated for teaching purposes. There are few applications at this time, but some of them show the way ahead.

Programming the movements of a Mindstorm robot
© Courtesy of Lego

From robot teachers to robot art lovers

It should soon be possible to perfect your table football skills using an invention from the University of Freiburg in Germany and expected to be on the market in 2005. A transparent surface allows an onboard computer to follow the ball's trajectory at 50 images per second and to activate the player most likely to score. As with video games, players can select a skill level, and according to project leader Bernhard Nebel, such a machine will be capable of beating the world champion within five years.

At Tokohu University in Japan, Professor Kazuhiro Kosuge's team has developed a dance master, the MS-DanceR (Mobile Smart Dance Robot). Although it can take its partner by the arm, the robot is not a real android, since its base consists of a platform on four casters to enable it to move in all directions, but that doesn't stop it being the king of the waltz: MS-DanceR is programmed to dance to Strauss's *Blue Danube*.

At first sight Pekee's ovoid appearance hides the fact that it is a teaching robot. The 40 centimeters (16 inches) long wheeled machine integrates all the latest technologies: WiFi, infrared distance measuring linked to GPS, gyro, color video camera and so on.

The KiRo robot is designed to teach. It's a real table football coach in the home, adapting its level to that of the player.
© *Courtesy of Thilo Weigel, Freiburg University*

Pekee was born in France at Wany, a company created in 2000 by researchers from the Montpellier Laboratory of Computer Science, Robotics, and Microelectronics (LIRMM).

Erwann Lavarec and his colleagues were seeking to develop a robot to help in teaching robotics at the university. They used an open platform that could evolve: Pekee is transformable, and can adapt to the needs of students or teachers. It is remote controlled via WiFi, and uses infrared position sensors. Although initially designed for teaching, Pekee attracted the interest of industry and research to the extent that Wany is now exploring contacts with Japanese manufacturers of robots and household goods.

Diligent, a robot museum guide, was tested at the French Institute for Spoken Communication (ICP) within the framework of the French robotics project Robea. Diligent does not just lead the visitor to a given painting and provide illuminating comment; if it sees a quizzical look on the art lover's face, it is supposed to provide spontaneous explanations about the area of the canvas looked at.

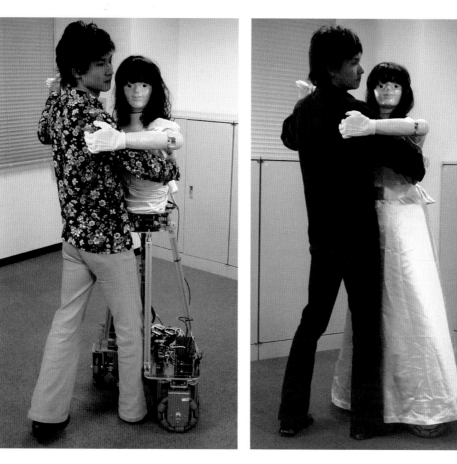

MS-DanceR
If you're too shy to take dancing lessons, learn to waltz in the arms of this robot. There's no danger of stepping on its toes!
© Courtesy of Yasuhisa Hirata

Laboratory for electronics and automatics science and materials (LASMEA)

Pekee, an evolving robot from Wany Robotics, communicates with the user via a radio link. The RNTL WACIF project concerns the development of an indoors robot with autonomous navigation, remote operation, and alarm and safety functions.

© CNRS Photothèque
Photo: Jérome Chatin

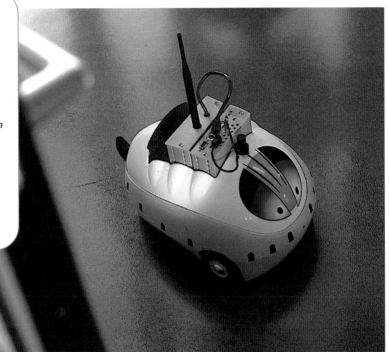

Systems Analysis and Architecture Laboratory (LAAS) Toulouse, France

Diligent, the future personal robot. The goal of research with this robot is navigation in a structured environment: a building with rooms, corridors, hallways, various obstacles, in the presence of and interaction with a human being.

© CNRS Photothèque
Photo: Laurence Medard

Internet-guided robot

It is worth noting that the education sector sometimes uses an unusual teaching application: a robot controlled via the Internet. Yahoo! lists a dozen applications of this type. Khep on the Web guides a little robot equipped with a camera through a labyrinth. Pumapaint allows users at home to make a robot create a painting using four colors. The more impressive Robotic Tele-Excavation, active between 1994 and 1995, guided a robot arm on an archeological site. The first robots that could be remotely controlled include Taylor and Trevelyan's Telerobot at the University of Western Australia: www.mecha.uwa.edu.au/jpt/.

"My students solved huge technical problems," says professor James Trevelyan, "since they had to provide public access from a Web browser to whomever wanted to guide a powerful, complicated robot, while at the same time ensuring long-term operation and protecting the robot against accidental damage". One challenge was to enable several users to cooperate efficiently in controlling the robot. The site averages 2000 visitors a week. "A laboratory with remote access gives students the chance to interact with equipment in a context similar to that in industry," comments Professor Trevelyan.

An ideal teacher in the home

In the longer term, it seems reasonable to predict the advent of real teachers in the home, capable of teaching the desired subject at the flick of a program, with the patience of Job, and even better, able to adapt the lesson by understanding the student's reactions, e.g., a frown would solicit a simpler or more playful approach to the subject in question.

A robot of this type is already operating: General Robotics' RB-5X is a bit like a green R2D2 designed to teach English and math. If General Robotics is to be believed, a six-month trial of RB-5X with 7- to 13-year-olds at Washington Elementary School in Las Cruces, New Mexico, increased understanding of mathematics by 40 percent for boys and 80 percent for girls. If proof were needed of the interest of such an approach, this is surely it.

James Trevelyan's robot arm is one of the first users could guide via the Internet.
© Courtesy of James Trevelyan

Robots in programmable kits

With Mindstorms, Lego made robot construction and programming available to the general public.
© Lego Mindstorms

The future Bill Gates may be living just around the block, an adolescent like any other, except a bit smarter. In his or her room (s)he is building robots, and not just any old robot: automata that can move, capture signals from their environment, and react accordingly. Once the bricks are assembled, certain actions can be programmed on the PC or Mac. Whereas their parents copied long listings of BASIC in the 1980s, today's adolescents invent robots, imagine new applications, and animate their creations, mouse in hand.

Lego Mindstorms

Lego's programmable robots appeared in the US in 1998, and represent the fruits of a decade's joint research with MIT in Boston. The enthusiasm they aroused is shown by the practically instantaneous 10 million connections to their site, with an average visit of 19 minutes. Lego's ambition? Allow children to build robots that do what the children want. The kits sold to the public contain a vast array of Lego bricks, and it's up to the budding geniuses to build the model of their choice. But the main difference comes from modules that can be programmed to render the robot mobile and interactive. Lego's hunch paid off when it released these kits for all and sundry, since very soon astonishing creations began to appear. Although the target audience was the 9-to-16 age group, 40 percent of users are adults incapable of resisting the temptation to make Asimov's dreams a reality.

At the end of 1999, Lego exhibited some spectacular robots built by users for the European launch of Mindstorms, and the effect was stunning. The show started with climbing robots. Two robots wriggle like large insects, elegantly swaying their lower sections to help them climb, and finally manage to scramble up a vertical surface with handholds in it. Once at the top, each one throws down a rope and slides to the ground. Impressive is not the word!

Later a robot that thinks it's von Karajan conducts a band of six musicians in an iconoclastic version of the *Blue Danube Waltz*, complete with unforgettable drum solo. Elsewhere, robot painters dabble on the floor. These robots lack eye appeal, but this is soon supplied by a catwalk of robot supermodels, flouncing around in their gold costumes and big celluloid hair. Pure science fiction!

One year later, Lego Mindstorms had a camera enabling them to have vision. This time, the star of the show was a robot

barman. Show him a blue coupon and he'll pass on your order to one of the bar staff who pours you a beer. The barman takes the glass and serves you with the grace of a karate expert. Show him a red coupon and the same thing happens again, but with a fruit juice.

The Mindstorms site (mindstorms.lego.com) shows off the latest creations from nerds the world over—a car that follows a white line on the ground, prototype hovercraft, hockey players, CD throwers, and even one that

gathers up Lego bricks. One of the most fearsome throws everything off the table once it's activated. Elsewhere, we learn that 12- to 14-year-old girls programmed robots to clean solar panels.

Thirty-year-old J.P. Brown is a typical example of someone who has caught the robotics bug. At night the mild-mannered Chicago archeological museum curator builds Mindstorms robots, one of which tries to solve the infernal Rubik's Cube. One of Brown's main problems in this exercise was finding a system that would allow the robot to hold and rotate the faces of the cube. Among other things, the solution involved coating the cube in grease, adding rubber bands, programming the robot for color recognition, and incorporating a software he found on the Web for solving Rubik's Cube in under twenty twists.

The construction of a Mindstorms robot introduces disciplines as varied as electronics, computing and mechanics.
© Courtesy of Lego

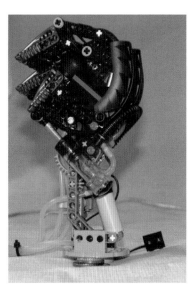

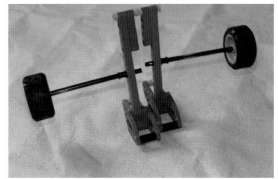

Christophe Masclet "the ingenious"

All over the world, tens of thousands of people like Christophe Masclet rush home from school or work to build the robots and the other artificial creatures they dreamed up.

© Christophe Masclet
Photos: FYP

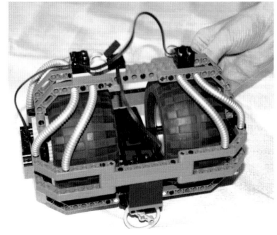

Interview
Christophe Masclet

It takes all kinds! Christophe Masclet likes nothing better than to build surprisingly ingenious robots, requiring the mastery of a number of skills.

What do you like about building robots?
For me, the technical challenge is the main thing. The kit I use, Lego Mindstorms, is easily accessible, but can evolve, and it's allowed me to tackle a huge range of technologies such as electronics for the light sensors, to learn concepts of mechanics, and to improve my programming skills. To put it another way, it's a learning tool that incorporates a number of different know-hows and forces you to get to the bottom of a number of domains.

Do you ever have to create your own pieces?
With Lego Mindstorms, sooner or later you are limited by the mechanics, the motors or the power of certain pieces. There are also things Lego doesn't make and you have to find elsewhere or make yourself. Sometimes I also make pieces Lego supplies because it's cheaper.

Christophe Masclet.
Photo : FYP

You haven't kept most of your creations. Why is that?
Once I've built a robot, I lose interest. Of course, you feel something when you dismantle one, but I don't spend days looking at them either, it's building them I'm interested in. I spend a lot of time on the Web checking out what's happening. Each time I see an innovation, I wonder if it's possible to reproduce it using Lego.

What kinds of robot have you built?
The first robot I built with Mindstorms was a little four-legged walker. After that I built a radio-controlled pitcher, an adaptive auto hand, a 50-cm-high (20 inches) humanoid bust, etc.

Do you ever exhibit your creations?
Yes, but in this context, you discover something else: the general public doesn't care about the technology; people want to see something that moves like in the movies. If you show a brilliant concept in an exhibition, but without some fancy presentation, nobody pays any attention. So to arouse public interest, robots are better off being humanoids.

How do you see recreational robots evolving?
They'll become ever smarter thanks to the programming. But you should never forget that robots are not human beings; they don't have a soul. You can simulate some aspects of emotion, but only up to a certain point. Otherwise, personally, I'd like to see some kind of standard emerging, like Linux in computing, so that elements from different manufacturers would be mutually compatible.

Would you like to make your living building robots?
If it could become a full-time job, I'd be happy, it would be a dream come true. There are surely things to be done, e.g., making robots for theme parks.

Which creation are you proudest of?

The radio-controlled hand. I injured my hand—a tendon got cut and I was off work for a month for reeducation. At that time I wanted to create a large robot, and first of all a hand. Because of my situation, I was able to pass a whole month searching Web sites dealing with biology and robotics to see what had been done up till then. I then built my own model. I wanted it to look like a real hand, to be articulated. I also wanted it to look good from the outside, not just mechanically. For me, this hand proves that Lego can be used to build something that goes beyond a toy.

Its head turns, its arms move, there's a camera hidden in its torso and its eyes blink. This highly original creation of Christophe Masclet's took two months of hard work.

© Christophe Masclet
Photos: FYP

Development kits

Lego Mindstorms showed the market potential that teenage and adult programmers represented. Other companies, including Acroname and Fisher Technologies, followed suit, but not necessarily proposing robots built from Legolike bricks. Charmed Labs' highly original kit combines Lego, a Nintendo Game Boy Advance and Bluetooth. Despite its playful connotations, this kit is based on professional programming languages like C or C++.

Budding programmers will prefer a robot such as Hemisson from the Swiss company K-Team, a kind of rolling platform that can be connected to a PC by a cable or via wireless. Hemisson's programmable actions can thus be easily downloaded. K-Team's Khepera II integrates a 32-bit processor and is designed for a more demanding audience, and is actually used in mobile robotics, vision and AI teaching and research. One of its strengths comes from the contributions of a vast user network, including 500 universities worldwide.

Some teams have developed algorithms (programming methods) for image processing that allow Khepera II to recognize objects and where it is. Another algorithm is used for obstacle avoidance. A teaching robot from K-Team, Koala, has six wheels and a series of infrared sensors. It is compatible with Khepera II, but is six times more autonomous.

Khepera II is a miniature mobile robot used for collective robotics experiments.
©K-Team

KoreBot 2003. The KoreBot range is a collection of basic building blocks enabling robots to be created easily.
© K-Team

Hemisson

The Hemisson miniature educational robot is a fun tool for learning programming and the basics of robotics.

© K-Team

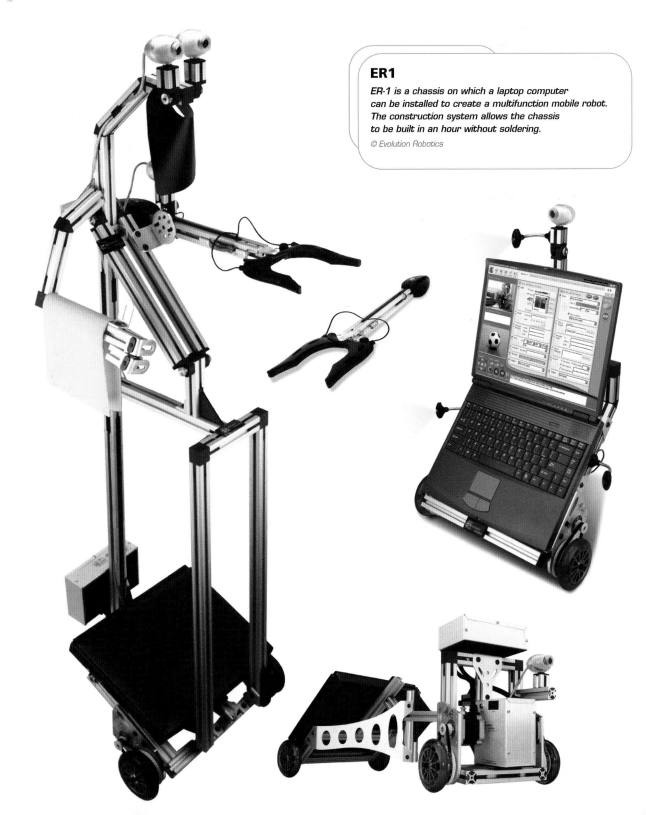

ER1

ER-1 is a chassis on which a laptop computer can be installed to create a multifunction mobile robot. The construction system allows the chassis to be built in an hour without soldering.

© Evolution Robotics

Xport

Charmed Labs have developed a kit to create a robot with a Bluetooth module, Lego pieces, and a Nintendo Game Boy Advance. It was designed for robotics research as well as play.

© Charmed Labs LLC

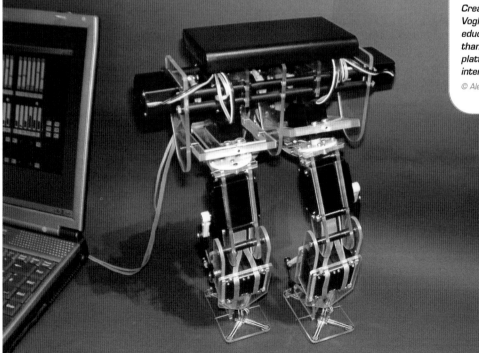

BiPed Robot V-3

Created by Austrian Alexander Vogler, the V-3 is designed for education and research. More than just a toy, the V-3 is an open platform accessible to those interested in robotics research.

© Alexander Vogler

Robot competitions

Every computing school in the world probably now has its robotics club. Universities are not being outdistanced either, with the creation of special sections for the discipline in a manner reminiscent of the heyday of image processing a decade ago. In towns and villages, it's not unusual to see a robotics club appearing from nowhere, driven by the desire to meet other fans. In France alone, the Vie artificielle (Artificial Life) Web site counted forty or so associations in early 2004.

Many competitions hope to stimulate the inventive urge of the upcoming generation. There's nothing new in this: in the early 1960s when robotics was in its infancy, an MIT professor spotted the fun that could be had. Woody Flowers had the bright idea of starting a robot-building competition, explaining that, "I was looking for a way to make my students put their learning into practice."

CIIPS Glory.
Using Professor Thomas Braünl's EyeBot robots, students from the University of Western Australia created a team named CIIPS Glory in honor of their favorite football team, Perth Glory.
© Thomas Braünl, Mobile Robot lab – CIIPS, University of Western Australia.

Soccer
competition:
Humanoid League
during the 2004
RoboCup in Lisbon.
© RoboCup Federation
Photo Courtesy of
Hiroaki Kitano

Soccer-playing robots

The most famous competition of them all is the RoboCup, the brainchild of researcher Hiroaki Kitano *(see interview pages 154–155)*. Kitano first proposed the idea of a yearly competition at a conference on artificial intelligence. The idea came to him in 1993, as a way to stimulate research.

In the early days, the RoboCup concentrated on soccer, for several reasons. First, soccer is popular the world over. But most of all, Kitano wanted to stimulate the development of collective strategies. As in the real world where Zidane or Beckham are up against other players, each robot

has to observe the behavior of its teammates and opponents and take advantage of this to score a goal.

"The RoboCup was created to promote research in smart robotics and AI," according to a spokesperson for the organization. "The aim wasn't to have fun by creating soccer-playing robots; it's a serious scientific project."

The first RoboCup was held in Nagoya in 1997, and the robots were a sorry sight: the most they could do was move a little and spot the ball. The following competitions were like a yardstick of technological progress.

By 2001, the artificial players were moving at a reasonable speed, knew exactly where the ball was, and employed team strategies.

Although it started out as a modest little competition for specialists, the RoboCup has become a major event, appealing to an ever-wider audience. During the RoboCup Japan Open in Osaka in 2004, 135,000 spectators turned out to see the technical wizardry of the miniature Maradonnas with their own eyes.

Soccer remains the main attraction and is now divided into three types of match:

- Small robots—less than 18 cm (7 inches) diameter
- Medium-sized robots— 50 cm (20 inches) diameter maximum
- Four-legged Sony robots, first seen in the competition in 1998, before their official sales launch.

Participants in this last competition must obey strict rules: they are not allowed to modify any mechanism or printed circuit. They have to manage by programming Aibo's animation software.

In 2002, a Humanoid league was added, again on Kitano's initiative. "The humanoid has to have a high degree of freedom if it's to carry out various tasks," he explains. "In the present case, it has to seek out a ball and two goals, and move to the desired spot while avoiding obstacles."

Developed from Robosapiens, the NimbRo RS robot took part in the 2004 RoboCup

© University of Freiburg

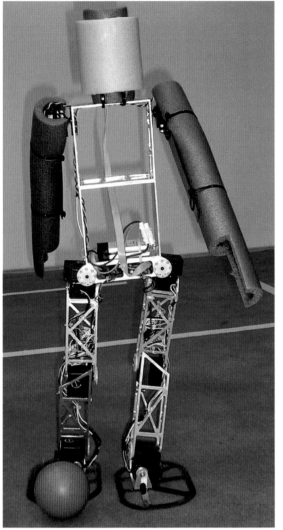

Servoid.
This 74-cm-tall
(29 inches) robot
can move
at 16.5 cm/sec
(6.5 inches).
© University of
Freiburg

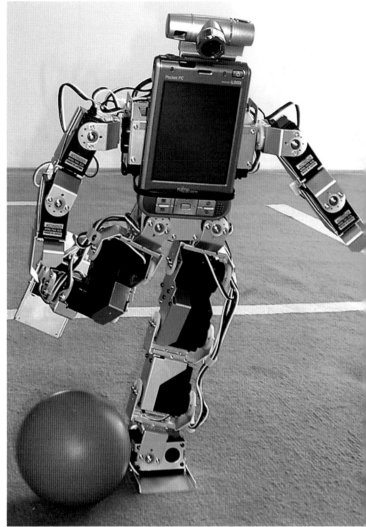

This robot was created
by students from the
University of Freiburg
using the construction
kit supplied by Kondo.
They added a pocket PC
and a camera to make a little
autonomous android.
© University of Freiburg

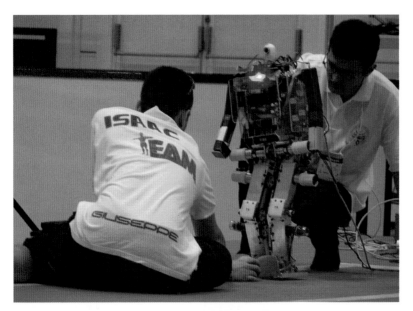

Humanoid League

Soccer playing by robots means not only mastering balance, but also the capacity to manage collective strategies.

© RoboCup Federation
Photo courtesy of Hiroaki Kitano

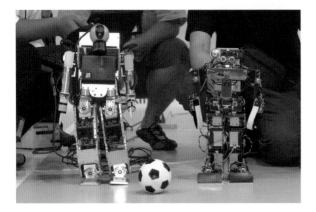

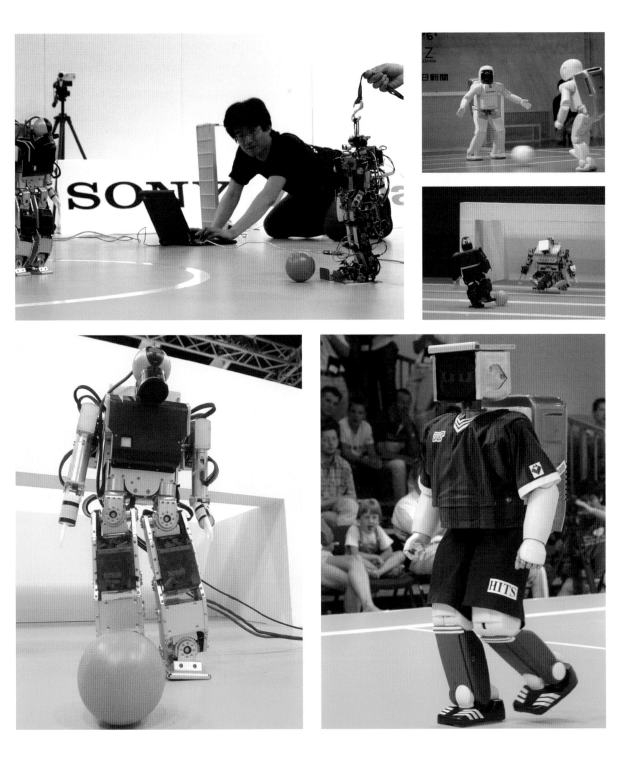

RoboCup

Sony Aibos first appeared in the RoboCup in 1998. The exercise is more exhausting for the programmers than for the artificial quadrupeds!

© RoboCup Federation
Photo courtesy of Hiroaki Kitano

Like the Euro 2004 taking place at the same time, the RoboCup organized in Lisbon in 2004 gave pride of place to the challengers. Three-hundred-forty-six teams of robotics experts met over five days, carrying the colors of thirty-seven different nations. Two of the soccer competitions were won by teams from Iran, while Russia dominated the third. The clash of Sony quadrupeds was won by a team from Bremen University Berlin, who defeated their Australian opponents. But as expected, Japan won the android trophy.

Competitors play to win of course, but they also come to swap information with other teams, a practice encouraged by the RoboCup organizers. As an in-house document puts it: "Improving the capacity of the robots to detect, process and foresee ball movements on the pitch is in the interest of all the teams. This is why the concept of secrets, inherent to Formula One for example, is unknown to RoboCuppers".

The importance the Japanese government attaches to the RoboCup is shown by the fact that the prime minister came along in 1999. In Australia in March 2001 during the RoboCup Junior, the prime minister came again. The ninth RoboCup will be held in Osaka in 2005.

Fifty years to beat a human team

As if to indicate that today's research in robotics is founded on a long-term vision, a spokesperson for Sony stated: "We hope that a team of humanoid players will be capable of beating a human team within the next fifty years." The challenge is not unlike that IBM engineers set themselves in trying to build a computer capable of beating chess grandmaster Gary Kasparov. The organizers of the RoboCup use the following parallel to explain that such a feat can be envisaged: "The Wright Brothers' first flew in 1903. By 1947 the British company De Havilland was selling airliners. Then in 1969 *Apollo 11* landed on the moon. Likewise the first electronic computer, ENIAC, was developed in 1947 and the computer named Big Blue defeated the world chess champion in 1997. Watson and Crick identified the structure of DNA in 1953 and the human genome had been sequenced by 2001. Fifty years seems long enough for a team of humanoids to beat the world soccer champions". Honda, which pioneered the development of walking androids, was the first to set such a target, and even set a closer milestone: by 2030 the company hopes to build a team of robots that could defeat a human team during the World Cup.

Other sports, other competitions

Soccer isn't the only sport that interests competition organizers. Far from it. Rugby, more complex because of its scrums, also has its fans and the game had pride of place at the 11th Robotics Trophies held at the Cité des sciences et de l'industrie (Paris Science Museum) in February and March 2004. The robots had 90 seconds to score as many tries as possible (by placing the ball in the blue zones) or drop goals (sending the ball between two posts.)

An inventory of all the competitions held worldwide would produce a list with neither head nor tail. Most of them spring from American or Japanese initiatives, but they are also to be found in Argentina (soccer), Canada (Eastern Canadian Robot Games), Spain (Hispabot) and Australia (RoboCup Junior).

Some competitions go beyond a simple match between artificial creatures. RoboCup Rescue for example, compares the capacity of robot teams in rescue operations and highlights cooperation between smart machines and humans. As Takahiro Wada of Kagawa University who oversees one of these competitions says: "RoboCup Rescue doesn't just focus on technological skill, it also tries to educate the population about better harmony between robots and humans. Moreover, this competition stresses the ideas of humaneness and solidarity needed in all rescue activities."

Other original ideas include robots manipulating a game of Tetris (AMD Jerry Sanders Creative Design Contest); minimice racing through a maze (APEC); robot vacuum cleaners doing the dusting (Atlanta Robot Rally); a competition for autonomous submarines (International Underwater Robotics Competition); window cleaning (Cleaning Robot Contest); and climbing a cliff (Cliffhanger). The ABU Robocon challenge asks robots to build a bridge then cross it.

And how about wrestling? It's on the bill of a few meetings, given the fighting skills of androids such as Fujitsu's HOAP-2 (an expert in sumo and taijiquan, a Chinese martial art) or MORPH3, from one of the workshops supervised by Kitano. An international competition known as RoboOne, dating from 1999, features bouts between robot boxers built everywhere from university laboratories to fans' bedrooms or garages. The 2004 edition in Japan featured two competitions for remote-controlled robots. The first had eight fighters, with the last 'droid standing declared the winner. The second competition was one on one, with the winner keeping the loser down for a count of ten.

RoboCup Rescue Simulation Project

The robots have to operate in the zone of a simulated natural disaster or industrial accident.

© RoboCup Federation.
Photo: courtesy of Hiroaki Kitano

The Eurobot competition for students aged 18 to 30 brings together the autonomous robots of eight European teams. Eurobot has its origins in the *Coupe de France* (French Cup) a television program broadcast by the French station M6 in 1994. *Coupe de France* was created by Planète Sciences, an association to advance scientific knowledge, along with VM Production, a TV production company who already produced another scientific program, *E=M6*.

The idea for a Europe-wide competition emerged during a competition in Switzerland in 1998, with England and Italy joining France and Switzerland, closely followed by Spain, Serbia-Montenegro, Germany, Austria and the Czech Republic. The only constraint imposed on the participants is that each nation is limited to three teams, which means there are several national-level qualifying rounds. Eurobot took place in La Ferté-Bernard, France, until 2004, but from 2005 on, it will be held elsewhere, starting with Switzerland.

Belgian Robotics Cup

The Cup takes place at the Framaeries Science Adventure Park (PASS). PASS was created on the site of the old Crachet coal mine, rehabilitated by architect Jean Nouvel. It proposes a number of interactive exhibitions that allow visitors to understand science and technology while having fun.

© Courtesy of PASS - Photos Serge Rovenne

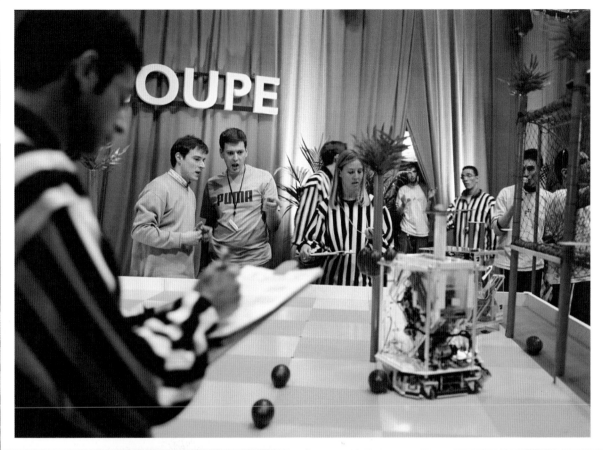

Eurobot

In May 2004 the nonprofit organization Planète Sciences, production company VM Group, the municipality of Ferté-Bernard, and their sponsors created the robotics association Eurobot, with the aim of promoting and organizing robotics activities for young people. Its flagship event is the Eurobot robotics tournament, with its national qualifying rounds and grand international final.

1 - 3 : Courtesy of VM Group.
Photo: Benjamin Turquier
2 - 4 : Courtesy of VM Group.
Photo: Franck Badaire
5 - 6 : Courtesy of ARTEC

Robot Olympics

It was a short step from individual events to Olympic Games for robots. The International Robot Olympiad Committee is a mainly Asian organization. It enjoys wide support from industry and research and its goal is to organize events that will attract a wide audience to promote creation in robotics. The 2004 games were held in San Francisco on March 21st and 22nd. About 400 robots and 173 teams took part in the competitions. The USA won 24 gold medals, Japan 5, while Europe went home empty-handed, except for a silver medal for Germany.

The importance of competitions

Competitions reveal a wealth of hidden talent and creativity, and industrialists are well aware of their utility for research. "A large number of important inventions were first seen during competitions" confirms Jake Mendelssohn of Hartford's Trinity College, which organizes its own competition. Mendelssohn himself opted for this approach for a project he's been working on for a decade: a firefighting robot for the home, able to detect fires and put them out with an onboard extinguisher. "Sooner or later, robot firefighters will be as common as smoke alarms in our

The first edition of the DARPA $1 million challenge did not produce a winner. The aim was to build an autonomous robot capable of covering 24 km (15 miles) in the middle of the desert in under 10 hours. The itinerary was kept secret until three hours before the start of the race. In South Barstow, California, a CD-ROM with 2000 latitude and longitude coordinates allowing the robots to position themselves via GPS was given to the fifteen teams entering robots. Carnegie Mellon's Red Team went the farthest, at 11.90 km (over 7 miles).

© Courtesy of DARPA Challenge

homes," believes Mendelssohn. During the annual competition, the robots have to find their way through a maze to put out a smoldering fire as quickly as possible. Each time the prototype becomes more efficient.

In a competition organized by Intel in Northern Ireland in 2002, Brendan Quinn and Edna Young, then aged 19, developed a self-propelled robot for the

electricity company that whizzed along power lines to stop pigeons landing on them and damaging them with their corrosive droppings.

In order to stimulate inventors, some competitions offer substantial cash prizes. In March 2004, the US defense research agency DARPA initiated a robot race across the desert from Los Angeles to Las Vegas, with a

International Autonomous Underwater Vehicle Competition.

The aim of this competition is to encourage the development of AUV by stimulating the creativity of a new generation of engineers.

© Space and Naval Warfare Systems Center

$1 million prize for the creator of the first robot to complete the race in under 10 hours.

NASA's Jet Propulsion Laboratory is interested in developing muscle functions in robots, and in 2005 is sponsoring the Arm-wrestling Grand Challenge that will pit robots against wrestling champs.

Robot battles

A new kind of spectacle recalling the dramatic intensity of the Roman amphitheatre can be seen in more modern arenas, except that now the gladiatorial killing machines really are machines!

Moving on four wheels, Rosie tries to smash her opponents to smithereens with a kind of hammer. But what's a girl to do about Hypno-Disc, whose enormous circular saw cuts up his opponents in no time? Or that darned Gold-Digger, whirling around with his spade to send opponents flying? Or HammerTime who spikes all who dare approach?

Robot combat says a lot about the survival strategies living beings have developed over the course of evolution. Is it better to be big, heavy and terrifying or tiny, light and agile? Mouser belongs to the latter category, and doesn't do too badly: this stealth mouse slips under the other robots, then upends them, often making them inoperable. Gemini craftily splits itself in two, leaving its adversaries to fight two enemies capable of cooperating.

But to win, you have to know how to defend yourself as well as attack. Rippa Raptor for example has two small saws in front, but has protective spikes on its head and mesh to protect its wheels.

The initiative for combat robots came from Marc Thorpe, who tried to build a radio-controlled vacuum cleaner in 1992. The collateral damage to his kitchen gave him the idea for Robot Wars, a competition for programmed automata. A television program shown in several countries followed, inspiring similar shows such as *BattleBots*. The organization of the matches now benefits from a certain know-how, and the arena is protected by a large metal cage in case one of the robots ever decides to have a go at the audience.

Once the starting signal is given, there's only one rule: destroy every other robot in sight. A robot that stops for 30 seconds or is flung out of the fight zone is eliminated. A few champions have already emerged, including Chaos2 with its swinging arm or Razor, that slices its opponents in two once they climb on top of it. In order to stimulate the creativity of robot builders,

winners of annual competitions can make millions in prize money, a fair reward for months or years of work and an average investment of $6,500.

Could this be the show of the future? Probably. An unmistakable sign is the increasing variety of competitions: sumo, running, obstacle races, firefighting, and so on. This new variation on an old theme is part of a new perception of popular entertainment and participates in its way in the changes in the arts.

Robot Wars

A combat-robot competition created by Marc Thorpe
1. Fight between Snake and Scorpion robots
2. Will Wright (left) famous creator of the Sims
3. The arena

© Courtesy of Marc Thorpe

Interview
Marc Thorpe

Marc Thorpe worked for George Lucas's Industrial Light and Magic special effects company as an animatronics (see chapter 2) designer for many years. In the 1990s, he created Robot Wars, *the first combat-robot show, later to become a television program.*

Rumor has it you got the idea for *Robot Wars* in 1992 after trying to build a robot vacuum cleaner that didn't work. Is that true, and did you ever build a combat robot yourself?
This invention that never quite made it was a weird combination: a tank-mounted vacuum cleaner. The idea was to turn a chore into play to make vacuuming fun. I built a radio-controlled tank and modified it by putting a battery-powered vacuum cleaner on it. It was fun but not very useful, especially with the batteries available at the time so eventually I gave up on it, but continued tinkering with the prototype.

In 1992, I was working as a senior designer for LucasToys, a new division of LucasFilm. Early that year, I pitched a concept for

Marc Thorpe, inventor of **Robot** Wars
© Courtesy of Marc Thorpe

fighting vehicles to a major toy company. The response was strangely indifferent. All that was said was "some day someone is going to figure out how to do this." I was perplexed, but let it go at that. One day I took the vacuum off the tank and as I looked at it, I envisioned its potential as a dangerous toy with battery powered tools mounted on it. That reminded me of my fighting-vehicle toy concept, which brought forth the entrepreneur part of me as it was instantly clear that this was how to make the idea work: I could stage events and invite competitors to build their own vehicles to compete in them.

Was it hard to get such a concept accepted at the time?
I remember New Years Day, 1994. Lying on my couch with the flu, beginning to have serious doubts about the future of the project, I put together a package with images, text, flyers, ads, letterhead— everything I had done to create *Robot Wars*—and sent it to *Wired* magazine. Within a few days *Wired* called me and wanted to do an article with a photo. I was overjoyed— and panicked. A photo? My robot was a Photoshop-enhanced captured image of my radio-controlled tank with various stuff on it—cool looking stuff—but none of it worked! Then I realized that the robot didn't have to work, but it clearly needed to appear capable of serious destruction. So I went to the hardware store and bought a chain saw, and when the photographer arrived I showed him the saw and the tank and told him that the saw sits on the tank; and he said that was cool and that it would look great. So we set up the shot and the photographer shook the camera a bit while taking the pictures, giving a really sinister look to the photos. Incidentally, I took the saw back to the store after the photos.

When were the first combats organized?

Although I first had the idea in 1992, the first event didn't take place until 1994. There were four events in all, and they were a raging success.

Why did they not continue in the US?

It's a complicated story. A *Robot Wars* license was sold to an English TV company in 1995. They changed the whole philosophy of my concept, stressing entertainment rather than sport, just the opposite of what I intended. Still, *Robot Wars* was very popular in the UK and they sold sublicenses to dozens of countries around the world.

What kind of individual takes part in *Robot Wars*?

Many of the early participants had done amazing things in a number of different fields. There was Caleb Cheung, who was later to invent the very successful Furby. Jeff Raskin, who wrote the article in *Wired* was the real inventor of the Apple Mac. Will Wright, legendary for creating Sim City and now the Sims, took part in the first four competitions. John Knoll invented Photoshop. There was also Charlie Tilford, a mechanical engineer with a wild, joyous flamboyance and a winning robot called the Mauler. The team behind the Thor robot came from a company that made leading-edge robots for tasks such as underwater exploration. JD Street made gadgets and accessories for the *Home Improvement* TV show. And there was the very young Scott LaValley with his exquisitely beautiful robot called DooLittle, which was prophetically named. Later incarnations by Scott were appropriately named DooMore and DooAll.

... Interview cont.

What are the most amazing combat robots you've seen so far?

There's Thor, the robot jackhammer. With its titanium body, it's as impressive to look at as it's powerful in action. Likewise, the Master, with its gas powered circular saw was a pleasure to see from the moment it stepped into the ring. The Machine was the Cinderella of the event. Not to look at. It was very rapid and powerful, and its hull design inspired many later models. Blendo was the first to spin like a top, and after being disqualified each year as too dangerous, gained a kind of cult status as a robot that could probably carve its way out of the arena and destroy San Francisco. These spinning robots are really dangerous since you can attach sharp accessories to them.

What are your favorite stories concerning *Robot Wars*?

A funny thing happened during a German tour. I was being interviewed on the theme, Do sports encourage violence? when behind us two robot builders started to fight. They asked me to go and calm them down so we could continue recording.

The robots' weapons each have their advantages and disadvantages. Does that tell us anything new about human and animal behavior?

Yes, they seem to follow adaptive principles. A successful aggressor soon meets defensive resistance on a similar level, leading to new attack and defense strategies.

What kind of thought does the development of these robots provoke?

Death and destruction are fundamental parts of life. They are inevitable, but still cause anguish. *Robot Wars* gives you the opportunity to experience these aspects in a theatrical way via mechanical substitutes. It's a healthy celebration of destruction and survival. Nobody gets hurt, no blood is shed. The atmosphere is that of a supportive community.

Marc Thorpe, with the robot vacuum invention.
© Courtesy of Marc Thorpe

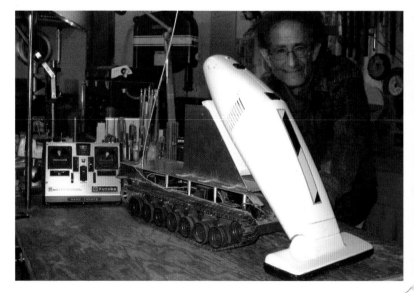

The robot toy bridges the generation gap

Children have always been fascinated by movement. Cartoons and video games use this attraction for fast-paced action, riotous colors, falling objects and wild zigzags. The most gifted children often become programmers or image creators themselves.

Today, the action has stepped out of the screen and has become a dynamic material that reacts to stimuli in the environment. You can talk to your robot dog, make it stand on its hind legs, beg, give a paw or whatever. Later, the infant will grow into a teenager giving orders to the machine, telling it to dance, climb a wall, or play the plumber, depending on the lines of programming code.

As for other disciplines, meeting other designers boosts productivity. Competitions hone talent, channel effort towards designated paths, and encourage increased rigor, dedication and technical competence. RoboCup, Eurobot and the hundreds of matches throughout the year all spur development. Competition conjures up images of grandiose extravaganzas, with modern-day metal and electronic Spartacuses

exploring the ins and outs of cunning strategies to destabilize, slice or shred. A generation of robot programmers is emerging. Their domain forces them to examine numerous subjects, since competition winners come from a clever melting pot of competen-

cies, ranging from suspension systems to predictive programming for unexpected situations. The playful nature of robotics can even help to bridge the generation gap, with parents and children contributing scientific know-how or the will to rise to a challenge.

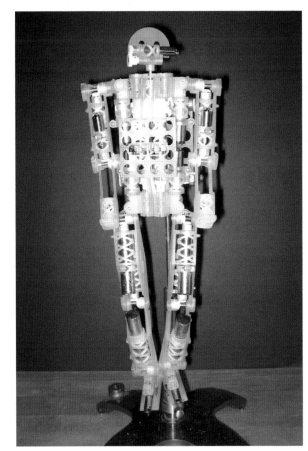

The research prototype of a play robot from Electronic Arts.

© Courtesy of
Luc Barthelet

Interview
Luc Barthelet

Frenchman Luc Barthelet has been working for Electronic Arts (EA), the world's number one gaming publisher, since 1988. In 1997 he was appointed director of EA's Maxis subsidiary, and stimulated the production of video gaming's greatest success: the Sims. He is now researching a promising new market—robotics.

Luc Barthelet
© Courtesy of Luc Barthelet

What makes the world's number one video games publisher interested in robots?
EA's core competence is the development of complex programs for leisure-oriented applications and robots. Interactive toys incorporate a large software component, which makes it an interesting area for us. It's clear that a robot craze is on the way, and we're looking at how we could accompany or boost this new market.

Why were you chosen to carry out this research at EA?
I've always carried out technology research at EA, even in the early days of consoles. Since we were finishing Sim 2 I wanted to look a bit further ahead, see what was interesting on the horizon. I've been interested in what's "outside the computer" for a long time.

Robots are a step in this direction; they are integrated into the world beyond the keyboard and screen.

What robots have most impressed you up till now?
For the time being, the two robots that interest me most are Sony's Qrio, one of the few to have solved the problem of walking, and Silph 2 as well, since it was made by one person in Tokyo, Ito. Silph 2 is a very promising humanoid, and represents the best that's been done at this size till now. Ito should be very proud of it.

Given the nature of EA, any robots you sell would probably be for gaming. Can you give us some clues as to the paths you're exploring?

We're closely following humanoid developments in Japan. Two or three avenues seem interesting. The first is combat robots, since they reproduce what's already done in video games. The second is dancing; robots could become quite good dancers. This could be as a teacher, but the fact that they could dance to music would be quite novel. The third point concerns personalization, the possibility for the user to transform the robot, in a similar way to what you can do with the Sims. That could include programming, as Lego does. If a creative platform is accessible, people will start making things. To put it another way, we'd offer an "open" toy rather than a closed one.

Do you foresee some kind of robot taking over from the video games boom of the last twenty years?
It's difficult to say if it will arrive sooner rather than later. There are already some indications, but you'd have to make a robot that could interest users and interact with them for a long period, say twenty or even a hundred hours. For the time being, it's hard to be interested in Robosapiens for more than twenty minutes. After that you get the impression you know all there is to know. You have to be able to justify buying an electronic toy. The Pong video

game showed the way in this respect—you could play for hours and develop your personal skills. In my opinion, the best robot toy on the market now is Sony's dog, since they supply it "unfinished" and you have to spend fifteen hours or so training it. So Sony was able to artificially extend the discovery time of Aibo.

Does EA intend to sell construction kits like Lego?
We're still doing the research and don't have any product on offer for the time being. Our competence is in programming and we'd prefer to stick to that. There's nothing in the pipeline.

But the photo you gave us shows an android.
It's only a research prototype. We need platforms to work on when developing the programs. The ones you can buy are too limited, they make many tradesoffs to keep the price down. We wanted to start from a platform with a greater capacity.

Are you going to link robots to existing video games, e.g., the FIFA soccer game with robot players, or life-size robot Sims?
A Sims that was situated outside the game itself would certainly be interesting. It's what you see in some sci-fi movies - video games bursting into life. Toys will

certainly become smart, and integrate programs and processors. If we manage to make gaming robots, we could adapt our software to them (Sims, Command&Conquer, etc.) Teenagers could have a character that led an attack with miniature tanks in their bedrooms.

In your opinion, how will robots change human activities, and in particular leisure and games?
What's interesting is that the robot will become an extension of the computer that's far more interesting for children. Computers have limited perception and are not easily transportable. If the computer is integrated into a doll or action figure, the interface changes: the robot could talk, have a playful side. A whole virtual world whose access was limited by the screen and the keyboard would become accessible from new entry points. You'll notice that in any toy shop, there are more dolls and action figures than any other kind of item. As with stories, this allows a projection onto humanlike characters, and projecting a story onto a character is a distinguishing trait of human beings. Afterward, you have to ask yourself if you'd prefer a doll you could ask to find a map or what's on at the movies to typing a search into Google!

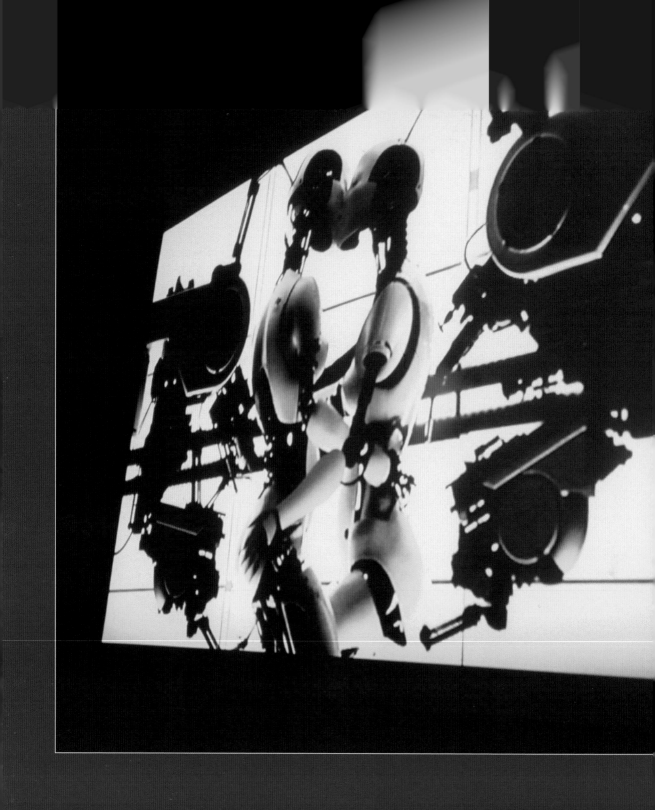

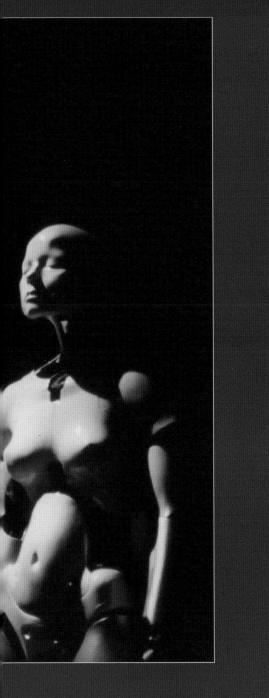

ROBOTS IN THE ARTS

The showtime of the future is here today!

The curtain has risen on a show that will go on...and on, and from strength to strength, the deeper into the 21st century we get. Robots are moving into the arts as actors, as strutting stages, and as masters of ceremonies. They are mobile, moving, and magical. They are among us, of us, express us. They have evolved from the creations of Méliès, the magician and movie-maker, and Giacometti the great sculptor of fleshless beings.

Whereas some thought that everything had been said, thought, played, shaped and painted, a new and boundless sphere of expression is opening up. The automaton as artist and its whole robotic repertoire of motion, empathy and unpredictability have brought a new dimension and powerful new breath of freedom to the performing and visual arts.

Björk is widely considered to be among the most inventive practitioners of musical video currently at work. She hired Chris Cunningham for her song All Is Full of Love. *The video clip and robot were presented at the exhibition* Robots !, *held in the northern French city of Lille, European Culture Capital 2004, and curated by Richard Castelli.*

© Chris Cunningham (UK).

Photo: Courtesy of E. Valette - Production Lille 2004

Chris Cunningham
All Is Full of Love

"When I first heard the track, I wrote down the words 'sexual,' 'milk,' 'white porcelain,' 'surgery.' My immediate association with sex was vindicated when Björk arrived at my London office with a book of Chinese Kama Sutra prints as her only guiding reference. I knew them and liked them, but I couldn't figure out how to keep the explicit sexuality and still make it broadcastable.
It's a combination of several fetishes: industrial robotics, female anatomy, and fluorescent light, in that order.
It was perfect, I got to play around with the two things I was into as a teenager: robots and porn."

Chris Cunningham

From catalogue to Robot !, *a Lille 2004 European Culture Capital exhibition.*

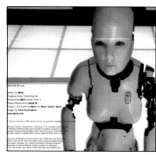

© Chris Cunningham.

Far left: Sketch of a robot designed by Chris Cunningham. The robots were built by Paul Catling, who taught Cunningham the craft of model making.

Left: Cover design for Björk's maxi-single CD, All Is Full of Love, *published by Warner in 1999 and produced by Chris Cunningham.*

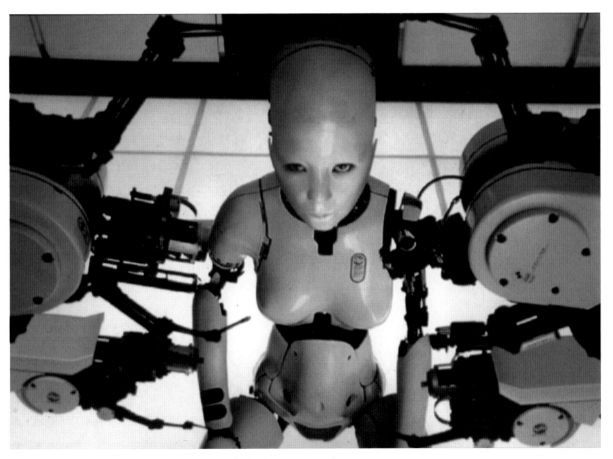

The video is an elegant, moving description of two Björk robots enraptured.
As they are pieced and wired into existence, they sing to each other and fall in love.
The video reaches its harmonious climax as the robots join in embrace,
still being detailed by the robotic machines beside them.
© Chris Cunningham.

Robots into art

Artists reflect their times. Yet for centuries it was as if they were overwhelmed by the sheer grandeur of the world around them. They strove to reproduce the visible world as it was, painting its plant and animal life, human bodies and customs, loath to render them through the prism of their own vision. In historical terms it is only very recently that they have broken free of mimetic conventions to give shape and color to their own emotions and perceptions; that they have dared to venture into a world of time travel and artificial beings that was once the exclusive province of literature. The Surrealists first bent, fashioned and invented words; sculptors forged new shapes and form, and saxophonists unleashed pure sensation as they unshackled music and took it soaring into interstellar space. Gradually, avant-garde schools developed notions of interactivity between spectators and mobile works.

**Cartons transformables
(Transformable cardboard boxes)**
© Jeremy Heringuez - Photos: Courtesy of E. Valette - Production Lille 2004

The beauty of kinetics

In 1919 a German architect by the name of Walter Gropius founded a pioneering school of design called the Bauhaus. What made its teaching distinctive was that it sought to pluck the arts from their isolation and bring them together with crafts and engineering. In 1923 a teacher at the Bauhaus School, Lászlo Moholy-Nagy, began work of an unprecedented kind in that it was not static but shifting. Moholy-Nagy projected light onto a white wall through the moving metal and glass parts of a machine. He called the piece *Light-Space Modulator*, and when he completed it in 1929 it came to be regarded as the first kinetic sculpture. The endlessly shifting effects of light reflected by glass and metal produced mobile shapes of startling beauty—all produced by a machine in motion! More and more artists grew fascinated by kinetics and created objects and assemblies that moved. In 1932 Alexander Calder began building his early motorized sculptures, a form that the influential painter and sculptor Marcel Duchamp called "mobiles". Other artists followed in Calder's wake. They included Swiss sculptor Jean Tinguely who, like others, used industrial scrap metal to create kinetic sculptures of all kinds.

Jeremy Heringuez
Cartons transformables
France

Some robots nestle snugly inside each other like Russian dolls, gradually revealing their innermost desires. Others are static and silent—architectural models of places of worship or renditions of functional machines. All, though, are part of a gigantic puzzle, like the ruins of the past or a world still be built to arouse the imagination.

"I studied for five years at the Tourcoing School of Fine Arts, where I discovered arte povera, Walter Gropius's Bauhaus and modular architecture, and his philosophy of art for all. And there were the figures and real-life materials of constructivism and the speed and movement of futurism."

From the catalog to Robot !, a Lille 2004 European Culture Capital exhibition.
© Jeremy Heringuez - Photos: Courtesy of E. Valette - Production Lille 2004

Cybernetic sculpture

Cybernetic art was pioneered by Hungarian-born Nicolas Schöffer, who moved to Paris in his mid-twenties. In 1951 he began work on his "spatio-dynamic" sculptures, designed to evolve over time. He completed his first cybernetic sculpture, *Cysp 1*, in 1956, an autonomous mobile work that responded to sounds and colors. Schöffer subsequently drew on research by Moholy-Nagy to design and build another spatio-dynamic work, a light-emitting tower (1961). The tower contained projectors whose beams of light were refracted and reflected by arrangements of mirrors and prisms. Perhaps its most distinctive quality, though, was interactivity. It responded to sounds and colors, but also to people moving around.

The development of robotic art

According to Eduardo Kac, installation artist and teacher at the School of the Art Institute of Chicago, three seminal works from the 1960s stand as landmarks in the development of robotic art. The first was *Robot K-456* (1964) by Korean artist Nam June Paik and video engineer Shuya Abe. With *Robot K-456*,

a remote-controlled robot that went towards people to greet them, Nam June Paik explored issues of remote control and free mobility in a wider social, political context. The second was *Squat* (1966) by Tom Shannon, who was only 19 at the time. *Squat* was a big robotic sculpture cybernetically wired up to the leaves of a plant. It undulated when anyone touched a leaf. Kac describes *Squat* as "the first interactive artwork that is an organic and inorganic hybrid". *The Senster* (1969-1970) by Polish artist Edward Ihnatowicz was the first biomorphic, computer-controlled sculpture. "We find [with the *Senster*] the first instance of behavioral autonomy in art," writes Kac, "in which a given personality is assigned to the robot, which then responds to humans and changing situations on its own." Nicolas Schöffer, meanwhile, was continuing his exploration of light and space, to which he now applied music. In 1973 he produced *Kyldex 1*, a work of "total art", staged by the Hamburg Opera. It consisted of fifteen sequences designed to elicit active responses from the spectators. They could demand that sequences be repeated or played faster or more slowly, and even call a halt to the play. One sequence used cybernetic sculptures that danced with the opera house's ballet corps as

the lighting changed and screens rose and fell with projectors playing on them. In 1978 Harold Cohen exhibited a robot in Amsterdam. It was computer controlled and found its way around using ultrasound, a system that also enabled it to make great abstract drawings on paper spread on the floor. The smart piano-playing robot had gone beyond practicing its scales by 1984 as Wabot-2, the android developed by Tokyo's Waseda University demonstrated. It could read and play an easy musical score on an electric organ. One year later its successor, Wasubot, played a Bach piece to the accompaniment of a symphony orchestra. There is no art involved, argue some, when a robotic musician merely plays a piece with no personal feeling for the music—though the same could be said of some humans.

Nevertheless the inception of works of art using robots marked a new stage in their history. They were part of attempts to explore the present time, while the sheen from their metal bodies lit the way to the future. They bore no resemblance to the tin men of 1950s B movies, nor were they forerunners of the high-tech Terminator. Robots were now an integral component of inquiries into the nature of man and society, having taken their place in a whole new kind of experimentation with art forms.

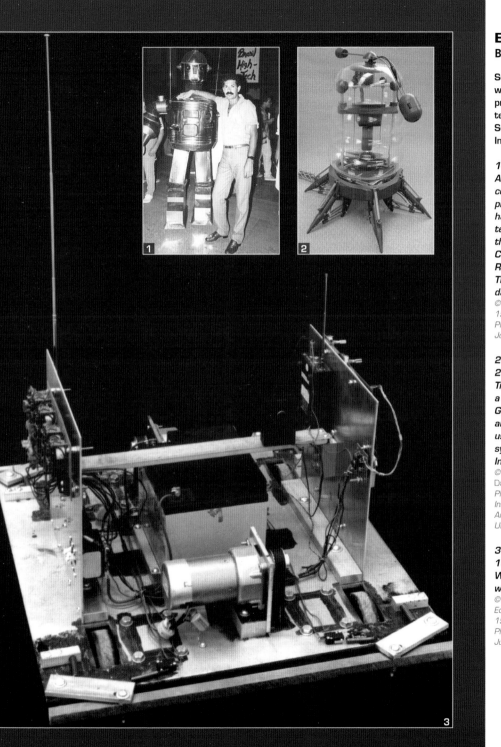

Eduardo Kac
Brazil

Some examples of the work of Eduardo Kac, professor of art and technology at the School of the Art Institute of Chicago.

1- RC ROBOT, 1986
A two-way radio-controlled telerobotics piece, where the robot's handler was only telepresent. Shown at the Galeria de Arte do Centro Empresarial Rio, Rio de Janeiro, 1986. The robot was built by da Silva de Cristóvão.
© Eduardo Kac, RC Robot, 1986, wireless.
Photo: courtesy of Julia Friedman Gallery.

2- THE EIGHTH DAY, 2001
Transgenic piece with a biobot, GFP plants, GFP amoebae, GFP fish and GFP mice, and using acoustic and video systems and the Internet.
© Eduardo Kac, The Eighth Day, 2001 (detail).
Photo: courtesy of Institute for Studies in the Arts, Arizona State University, Tempe.

3- ORNITORRINCO, 1989
Wireless telerobotics work.
© Eduardo Kac and Ed Bennett, Ornitorrinco, 1989 (detail).
Photo: courtesy of Julia Friedman Gallery.

Robots, a material spirit

Gradually the robot has carved out a place for itself in the visual arts. Movies like *Star Wars*, *Terminator* and others have, of course, played a key role in clearing the way for a line of inquiry to which only writers had dared to give thought on a large scale. But other artistic media are increasingly making the robot a part of their exploration of new forms.

One particularly interesting experiment is the *Exoskeleton*, a 1998 work by Australian artist Stelarc, short for Stelios Arcadiou. Like some fantastical hydra, *Exoskelton* is a six-legged pneumatically-powered walking robot. Stelarc explores walking machines, or locomotors, which have long, jointed spiderlike legs and are controlled by a third, prosthetic arm. He has experimented extensively with often spectacular virtual reality performances that take place in the open air. Stelarc argues that the human body is no longer in phase with technology and seeks to show how it should evolve in response to its artificial environment.

Björk, that most other-worldly of pop singers, chose to metamorphose into a robot in the 1998 video clip for her song *All Is Full of Love*. Chris Cunningham, who did special effects for the film *A.I.*, produced the clip, which shows two Björk-like robots gradually fusing into emotional embrace as they are assembled. They sing their song of love before finally coming together in reciprocal rapture. The robots have Björk's eyes and mouth and she seems to reveal a part of herself in newly built, bionically erotic bodies. Both are dazzlingly white under the metal encasing them, with their eerie serenity and female forms exerting a disquieting effect.

Some artists use robots as models, creating works that attest to newly invigorated inspiration. In the 1980s sculptor Fabian Sanchez paved the way with his sewing-machine robots. At the time of the Gulf War, Vincent Brodin sculpted an articulated metal android powered by electric motors, which offered an odd blend of heroism and despair.

French sculptor Jeremy Heringuez makes cardboard robots, nesting pieces within each other to produce strange creatures with modular limbs like huge 3D puzzles. Similarly, Thierry Deroche collects all sorts of scrap metal and appliances, like toasters, piping, and umbrella stands, which he assembles to form androids that look as if they have stepped out of a 1950s B movie. Their maker is the first to admit that he is as surprised as anyone by the beings that emerge from his piles of paraphernalia.

The Robotics and IT research laboratory LIRMM, in the southwestern French city of Montpellier, has designed a weird and wonderful teleoperated sculpting system. Using a haptic arm (which produces the sensation of touch) an artist sculpts a virtual material while a remote robot executes the sculpture. The Louvre Museum is said to have expressed interest.

Waseda University
Wabot-1 / WL-10RD
Japan

Two historically important robots: WABOT-1, the first bipedal robot, developed in the 1970s, in the company of the younger WL-10RD. The latter incorporates advanced mechanisms that give it a human gait and enable it to climb stairs.

© Waseda University - Photos: Jean-Pierre Duplan - Production Lille 2004

LIRMM
Creating a 3D model
France

In the background the 3D data acquisition device is being calibrated. It scrutinizes a real-life object from all sides before modeling it. In the foreground, a sculptor works on the virtual object, using a haptic arm with force feedback that gives the sculptor the sensation of touch. Although the sculpture is virtual, the original object is real.
© CNRS photo library - Photo: Emmanuel Perrin

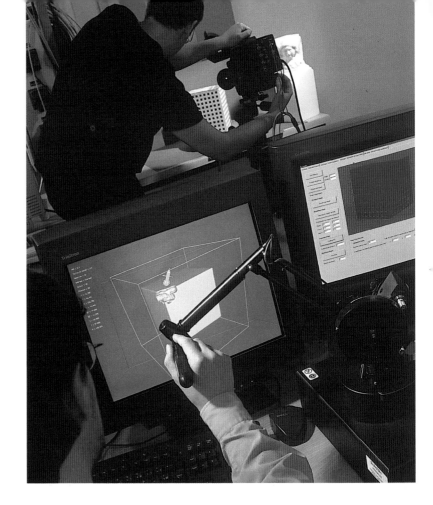

A robotic art new wave

Nothing can resist the inspiration that fires artists. At the touch of their skillful hands the ugly old PC has let itself go, grown curves, donned translucent bodies with textures and color that are a wonder to touch and to behold. Screens that were once dark and glum are now bright and merry festivals of pixel-swarming *son et lumière*.

After the PCs, it was the robots' turn. Artists dressed them, cross-dressed them, primped and powdered them so they could take their star-struck place among the tragedians and comedians of stage and silver screen. Robots are unwitting performers, acting out the ideas of creative artists who envision them as movie actors, stage villains or…in roles in some entirely new art form. Like the electric guitar and the synthesizer, the robot is perceived as an instrument for exploring uncharted territory.

On the cusp of the 21ˢᵗ century, the robot is something new, offering artists of all sorts a material, a medium unlike anything that has gone before, something that moves, interacts, is three-dimensional, sentient, tactile and unpredictable. Like a tango dancer, the artist is teacher and pupil, listener and listened to. The robot is like a mirror whose surface is ruffled by crisscrossing currents, like the surface of water.

As in the great days of Fauvism and the psychedelic sixties, a wind of freedom is blowing across the land of creation and a new art movement is taking shape. It blends the dramatic sweep of great film, the absurdity of experimental performance art, and the mind-blowing sensations afforded by technology and its power to change dreams into palpable reality.

Vincent Boureau
Huis clos
(No Exit)
France

Huis clos *is an organic installation made from numerous treelike elements.*

© Vincent Boureau - Photos: Courtesy of
E. Valette - Production Lille 2004

Reactive landscapes and stages

Vincent Boureau's *Huis clos* (No Exit) is an artificial field full of reactive trees. Should you approach one, it immediately responds to the movement it has sensed either by vibrating, making a noise or swiveling. Some trees even respond by forming random groups. There is also something branchlike to the hanging arms, made from Cabernet-Sauvignon grapevines and plastic, in *Autopoiesis* by

American artist, Kenneth Rinaldo. The hanging objects are in fact moving sound sculptures that sense and re-spond to a visitor's movements without seeking to touch him or her, as if at once repelled and attracted.

Orchisoid 03 by Masaki Fujihata and Yûji Dôgane adopts a similar approach. Orchids in plant pots are connected to each other by electrodes. A human operator triggers an electric charge to which the robot plants respond, varying their behavior according to the voltage they receive.

Robots also have architectural potential. A Dutch architecture' firm, NOX, has developed a sort of mutant substance that mixes traditional building materials with electronic technology. It was used to build *Fresh-Water House* at the Fresh H2O Expo exhibition, held between 1994 and 1997 on the Dutch island of Neeltje Hans. *Fresh-Water House* is an installation made with the hybrid material and embodies the "liquid" architecture of which American artist Marcos Novak is a proponent. When a visitor strolls around inside, the house changes shape and curves like a wave, while lights, too, respond to the movements perceived.

Decoi's *Aegis Hyposurface* is a 3D tactile mural screen that responds to digital input and human movement and noise by changing shape. It is called *Aegis* after Athena's breastplate, which could also change shape, turning into a cloak, as it absorbed blows—i.e., stimuli—from its surroundings.

The *Happy Robo Room* by Toshio Iwai and Babakayo is peopled with smart toylike devices that respond to visitors' touch. What looks like a mundane cardboard box, for example, changes into a waking robot. "We'd like the machines and tools of the future to be more than practical, so that people can grow fond of them," reveal the two artists.

An *art nouveau*, a *nouvelle vague* is gathering pace. It could be as big and radical as the Impressionists were in their day.

Vincent Boureau
Huis clos (No Exit)
France

When a tree senses the presence of a visitor it responds by vibrating, swiveling, making a noise or a combination of all three reactions. Some even react as a group, which raises issues linked to social cohesion and communication systems within a group. Conversely, trees react differently and to different degrees, which underscores individual differences of perception within a single community.

© Vincent Boureau - Production Lille 2004

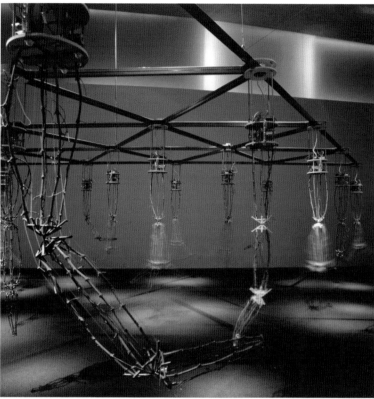

Kenneth Rinaldo
Autopoiesis
USA

Fifteen moving sound sculptures hang like arms from the ceiling. Equipped with infrared sensors, they sense and respond to the presence of visitors at distances of up to three metros and communicate amongst themselves. Made of Cabernet-Sauvignon grapevines, electronic elements and urethane plastic, the robotic sculptures embody the group consciousness of machines.

© Kenneth Rinaldo
Photo left: Jean-Pierre Duplan
Photo below: Production Lille 2004

Kenneth Rinaldo
Augmented Fish Reality
USA

Augmented Fish Reality *is a biocybernetic installation of "rolling robotic fish-bowl sculptures" in which Siamese fighting fish actuate software that moves their bowls. When fish see each other or visitors approaching, they swim towards what they see. Infrared sensors around each bowl sense a fish's movements. When it comes to the edge of its bowl feedback from the sensors turns motorized wheels and the robotic sculpture moves. This is "transspecies interaction" in that the fish respond to the presence of humans and move the robots towards them. Visitors can imagine fish perception from the screen projections mounted around the installation.*

© Kenneth Rinaldo
Photo left: Courtesy of E. Valette -
Production Lille 2004
Photo below: Production Lille 2004

Living sculptures

Performances and installations involving robots is like nothing that has come before. They can be compared to life-size 3D video games. But that would not do them justice. Anyone who has strolled through a space where robots and interactive sculptures move will confirm that it is an experience difficult to put into words. The monsters of the Berlin-based collective Dead Chickens make up a menagerie of weird creatures that are so blatantly crazy that the metal bars around them might not be such a bad thing!

One of the leading lights in the unheralded robotic-art movement is New York artist Chico MacMurtrie (*see interview, pages 472–473*). Like some strange founding father, he has installed robot villages whose inhabitants go tranquilly about their business. The performance installation *The Amorphic Landscape* unfolds over one hour during which robots are born into a world that they then build, establishing social and geographical structures while interacting with the spectators.

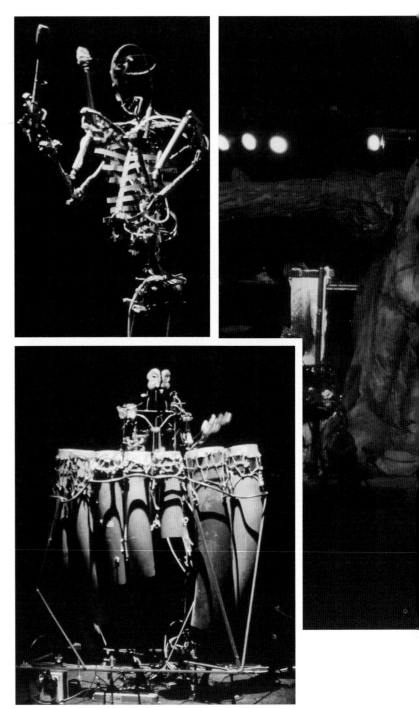

Chico MacMurtrie
The Amorphic Landscape
USA

The Amorphic Landscape *is the original home and venue of more than 60 musical and kinetic works
by the Amorphous Robotic Workshop (ARW). In its one-hour performance* The Amorphic Landscape
*depicts the formation of Earth and its inhabitants, the rise of communication, and the eventual erosion
of the environment as its creatures carve roads and structures from their original surroundings.
This being an interactive installation, visitors can experience one-on-one exchanges with the Amorphic
Landscape's denizens as they evolve from, and with, their environment throughout the day.*

© Chico MacMurtrie – Photos: Courtesy of E. Valette – Production Lille 2004

Too Big Dog-Monkey is Mac-Murtrie's nine-meter-tall (29 feet 6 inches) hydraulically driven robot. "She" is an artificial mother whose sheer size symbolically awes her dog-monkey children. It is as if she exercised an implicit authority over them through her illusory ability to carry and give birth to them.

Skeletal Reflections is a robotic sculpture that resembles a human skeleton, topped by a face made from plastic. Visitors can teach the robot movements and it can register emotions like joy, anger, surprise and impatience.

Chico MacMurtrie
Skeletal Reflections
USA

Of all the robotic sculptures that ARW has designed, Skeletal Reflections *is the one the bears the closest resemblance to the human body. It is built from aluminum and plastic and driven by compressed air. Spectators can teach it movements and attitudes. It can express emotions that range widely from fear, grief and anger to impatience, surprise and joy.*

© Chico MacMurtrie – Photos: Courtesy of
E. Valette – Production Lille 2004

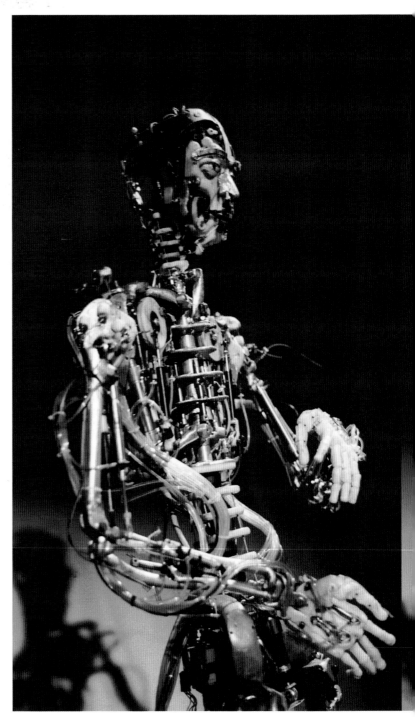

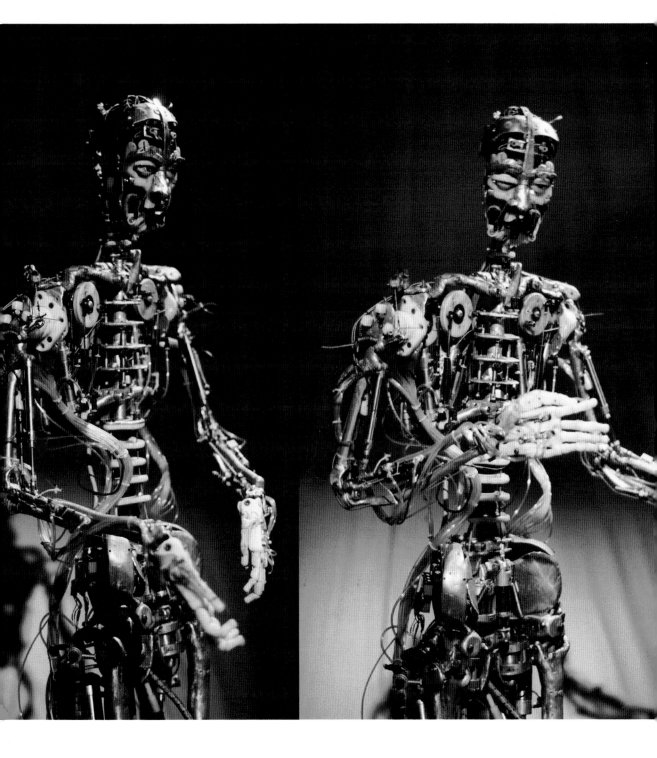

Interview
Chico MacMurtrie

The futuristic work of New York artist Chico MacMurtrie includes science-fictional landscapes that are home to robot settlements. There is interaction between visitors and robots as MacMurtrie explores the nature of the evolving human condition.

In what spirit do you create your works?

I create places where machines can live. I call them landscapes. The robots in them have a purpose. The dog-monkeys, for example, are shepherds that lead others back to the village. There is interaction with the public and each spectator can trigger a machine. So when people come to the Cave of the Subconscious, where there are 24 machines, the more people there are, the more the robots move and noise levels become overwhelming. Visitors experience a whole range of feelings from fear to laughter, and the experience is different according to whether you're alone or come with friends. Most of my machines use percussion or emit sounds.

What is the underlying message of your landscapes?

They symbolize the beginning of life. I wanted to depict the primitive side of the human condition. So I show how language developed from percussion and copying outside noises. The landscapes prompt people to draw conclusions about the survival of man today where only the rich can afford machines to purify the air. The robots' behavior also shows how humans use gods to control the people. One robot swells and swells, which is a comment on consumerism and the mind-set whereby everything is take, take and there's less and less giving. The robot is in a bubble that swells until it's stretched to the limit. It's a metaphor for a world that is turning obese.

What about *Skeletal Reflections*, which reacts to postures that you take?

We have given it a vision system, but most important, we taught it the history of art— Michaelangelo, David, etc. When someone strikes a pose, the robots tries to mimic

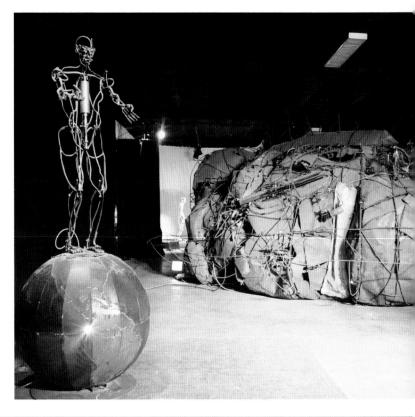

it while factoring in all it's learned about art. There's no more private experience.

Could you tell us briefly about the robot that tries to do somersaults?
It's a powerful machine that behaves like a child, like man-made machines that then destroy the earth. The robot is like a child. And what's really interesting is to see how the public reacts. They want it to succeed even though they don't really understand what it's trying to do. Not far from this child

robot there's a huge sculpture that represents the mother. She's huge, but she moves very slowly.

Do spectators feel emotions like the emotions they feel when they go to see humans performing?
No. I can't offer what humans can. I won't be able to for some years. In themselves, machines have no emotions.

Is it true that you don't like to be called a roboticist?
I'm an artist who creates in order to show the history of our society.

I draw the robots that I create sitting down at a table. Technology intervenes only at the end.

Chico MacMurtrie
Cave of the Subconscious
USA

All those who enter the Cave of the Subconscious *are players in an interactive experience. The* Cave of the Subconscious *and its denizens are made up of 30 archetypes through which the visitor embarks on a psychological and cultural exploration of subconscious terrain. The completed* Cave of the Subconscious *was on show at a performance held at the Lille 2004 European Culture Capital.*

© Chico MacMurtrie – Photos: Courtesy of E. Valette – Production Lille 2004

Kafka's *The Trial* with a robot jury

The Trial by Canadian artist Bill Vorn (*see interview, pages 478–479*), is an interactive multimedia installation that delivers a powerful emotional charge. Bill Vorn and Louis-Philippe Demers have recreated Kafka's novel, with robots playing the part of a merciless jury that makes more noise than a class of children when teacher's out. Kinetic, audio, and luminous robots encircle the visitors who are installed in an arena. The unbearable mechanical hullabaloo is like the evil, cackling, dogmatic, degenerate judges who recall some of the most iniquitous trials in the darkest hours of the 20th century.

Bill Vorn and Louis-Philippe Demers
The Trial
– part of the Zulu Time performance produced by Robert Lepage.

Canada

Artists Louis-Philippe Demers and Bill Vorn have worked together on a number of performances and installations. They have once again teamed up, as they memorably did for Zulu Time produced by stage director Robert Lepage. This time, Louis-Philippe Demers's assembly installation, L'Assemblée, is the venue for The Trial, a performance adapted from Kafka's famous novel and featuring an all-robot cast.

© Bill Vorn - Louis-Philippe Demers - Robert Lepage
Photos: Courtesy of
E. Valette – Production Lille 2004

Bill Vorn
Hysterical Machines—Stele
Canada

My Hysterical Machines *are just machines that move more or less depending on the strength of the stimuli they sense, says Bill Vorn. He adds that they get hysterical when people get too near, so they usually keep their distance.*

© Bill Vorn

Louis-Philippe Demers
L'Assemblée
(The Assembly)
Canada

Louis-Philippe Demers is one of the few artists who create and choreograph luminous singing, dancing robots. He partnered with stage and movie director Robert Lepage for a project called L'Assemblée, a strange arrangement where 48 robots encircle spectators in an arena, watching them and sizing them up.

© Louis-Philippe Demers
Photos: Courtesy of E. Valette – Production Lille 2004

Interview
Bill Vorn

The subject of Bill Vorn's doctoral thesis in communications was "Artificial Life as Media". He has been working in the field of robotic art since 1992. His freakish, unsettling creations leave nobody cold.

How does one become a "robotic artist"?
Hmm…There's no straight answer to that question. I think that the passion comes through interest in so many disciplines— scientific, technical, artistic and creative. There's also the desire to bring together sound, music, light and the image of mechanical motion, by which I mean tangible structures moving in space. What interests me most is to create machine shows, that's what drives me, even if it involves working as much on musical composition as programming microcontrollers, arranging lighting, putting together electronic circuitry and assembling mechanical parts.

Would you say that robots have become a subject that raises issues related to the human condition in the same ways as the fine arts?

I believe that automata and robots have always raised those kind of questions. They're part of the timeless quest to create our double that takes us back to the dawn of humankind. It's hardly surprising that robots are now a medium in their own right, a vehicle of meaning and expression that can convey a message and even aesthetic values. But even though people talk of robotic art, it's still a long way from being considered one of the noble arts, like painting or sculpture. You have to be careful about comparing them.

Much of your work revolves around man's relationship with machines. What is it that you are trying to say about that relationship?
My aim is to design and build interactive robotic environments that question, restate and subvert the notions of behavior, self-projection and empathy that characterize relations between humans and machines. I want to create worlds where a machine's human qualities intersect and blend with man's machinelike nature.

Many of your pieces use light bulbs to symbolize the life, death and regeneration of cells. What, ultimately, are you trying to portray?

What matters is not the life or death of each element, but how the whole evolves. The cellular automation installation *Evil/Live* used variations in brightness to impact first on visual perception. Spectators had to come to terms with continually changing shapes that they couldn't get away from because the light was so bright. The basic cellular structure is secondary. What mattered was to overpower the spectators' gaze and to put them in a trance by overloading the senses.

In 1999 you installed Kafka's *The Trial* in a big room where robots passed judgment. The effect was both theatrical and oppressive. What was the intention behind staging the installation in that way?
The Trial is vaguely inspired by Kafka's book. It's more a working canvas, a possible interpretation that's much more abstract than Kafka. It's an unusual work because it's staged as a performance; in other words, it has a beginning and an end and spectators are meant to attend. It works much more as choreography than on a behavioral level, because the machines were programmed to make a certain movement at a set time, rather than to react to the spectators approaching them.

What is the idea underlying the Hysterical Machines that hang from the ceiling and twitch and flail desperately?
As with the *Court of Miracles*, which is a freak show that shows the poverty of the machine's condition, this project is designed to deconstruct dysfunctional, absurd and deviant behavior through functional machines.

How does the public perceive these works?
The most interesting facet of my work is how the spectator interprets it. Obviously there are always as many different interpretations as there are spectators, but I think that as a rule what I do works. In fact, people's fascination with automata is what enables this type of public performance to work, in addition to the immersive experience.

Have you ever heard from spectators, if that's the right word, telling you that the work has impacted on their lives?
I think that some people have been affected. I even saw somebody crying at the *Court of Miracles*. That suggests that in a way we've done what we set out to do, because our prime aim was to project feelings and empathy.

Are you aware that much of your work has a dark, unsettling side as well as a playful side? If so, is it intentional?
In the first place it's an aesthetic choice. Since machines are at work in a particular environment, the environment needs personality. We've chosen an atmosphere that is dark and disquieting because it springs from the creation of a new world, one that is different from the one we know. This environment further stresses the mysterious side of machines, and overrides the fact that they are in reality assemblies of wires, motors and aluminum. That's the personality I tried to give my automata, which makes them very different from the robots on the Disneyland conveyor belt.

If Leonardo da Vinci were alive today, do you think he would have experimented with robotics?
Maybe. But maybe, also, he would have preferred virtual robots with 3D animation. Who knows? They're a lot less dirty!

Artificial animals

It looks like a fantastical beachcombing insect, some giant mutant creepy-crawly born of a genetic slip or a DNA remnant from Jurassic times. But it is, in fact, *Animaris currens ventosa*, a very "now" creation and the work of Dutch scientist-turned-artist Theo Jansen, who endeavors to resolve a problem that still dogs robotics: energy self-sufficiency.

Jansen designs his machines on a computer, endowing them with legs filled with compressed air that pumps them along with minimum effort. His weird and wonderful creatures proceed at a gait that is half stagger, half prance, carried by their own momentum. They have saillike wings that catch the gust of wind that give them the impetus they need to get moving. Their structure and their numerous piston legs then keep them creeping along irrespective of the weather. A slap on the rear sometimes suffices to get them going.

Jansen's research has paved the way for robotic locomotion that is independent of any energy supply.

Visitors to the *Robots!* exhibition, an event held as part of Lille 2004 European Culture Capital celebrations, gasped at the sight of the *Animaris currens ventosa*. Lille 2004 artistic advisor and *Robots!* exhibition curator, Richard Castelli says "people were particularly impressed by their organic appearance, which eschewed all high-tech feel".

Jansen has created a company called Epidemic which organizes avant-garde exhibitions and festivals. He hopes one day to see herds of his creatures wandering across the land under their own steam. In September 2004 one machine, *Animaris rhinoceros*, took a stroll on Valkenburg Airport in a force-six wind—almost a gale.

Theo Jansen
Animaris currens ventosa
Netherlands

Animaris currens ventosa *is the first generation of Jansen's creatures. When they were built their joints were heated with hot air. Animaris currens ventosa is 5.5 meters (18 feet) long, 4 meters (13 feet) wide and weighs 160 kilograms (352 pounds). Ten years ago Dutch scientist-turned-artist Theo Jansen had a vision in which he saw art in motion. On a computer he now designs strange machines that, once actuated by a gust of wind, can walk on their own.*

© *Theo Jansen*
Photos page right and pages 482-3, courtesy of Lourens & Van Der Klis – Production Lille 2004

Photos page right and pages 482–483, work by Theo Jansen.

A new artistic dimension

The world of visual and performance art can now welcome a new, left-field arrival—robotic art. In their day the French sculptor César, of compressed-car fame, and the early Charlie Chaplin surprised and fascinated spectators. Robots are doing the same today, revealing a new form of expression that is still finding its bearings.

Robotic artists advance gingerly, as if across a terra incognita riddled with pitfalls. Artificial intelligence has written itself into the script, introducing chance, unpredictability and improvisation that defy all existing rules. Robotic sculptures are not fashioned from static matter, but one that seems animated and sentient.

Mind-boggling are the prospects of conducting orchestras of robot musicians, teaching robots to paint then handing them their own brushes and canvas, or fitting them into mobile, constantly shifting installations. A new aesthetic that draws its force from the element of chance in robotic relationships is taking form. Art has a new dimension.

Carlos Corpa
Machina Artis 3.0
Spain
Carlos Corpa invites us into his robots' painting studio and music room for an astonishing artistic performance.
© Carlos Corpa. Photos: Courtesy of E. Valette – Production Lille 2004

Jürg Lehni
Le Che
Switzerland
Hektor test number 7, Che, 2002.
© Courtesy of Jürg Lehni

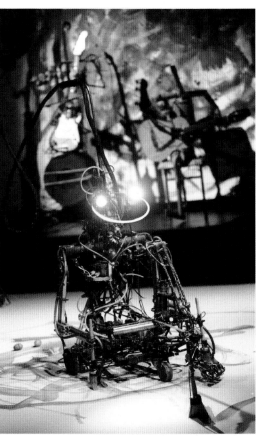

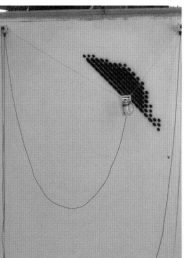

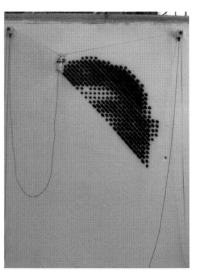

Humanoid Robotics Laboratory
Bar Bot
Austria

The Bar Bot enters bars, asks people for money and, when it has collected enough, buys itself a beer. The Bar Bot is of no use to humans. It is useful only to its selfish self, getting beer money out of people purely to slake its thirst. The way the Bar Bot uses people to serve its own egotistical ends makes it probably the most human robot ever built.

© Courtesy of Humanoid Robotics Laboratory - Production Lille 2004

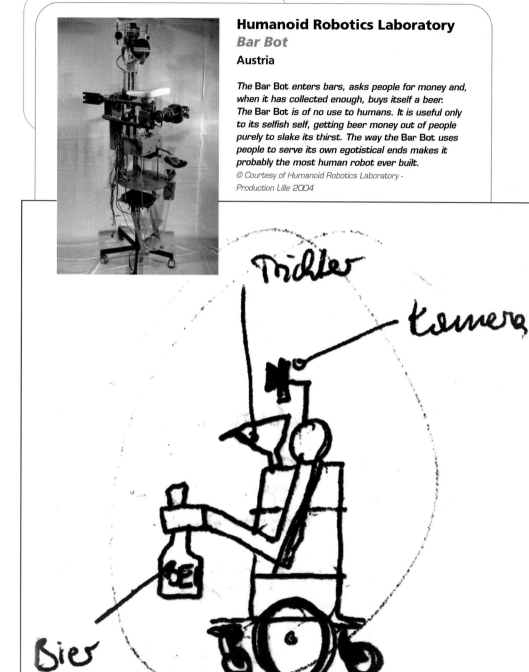

Dead Chickens
Monster clock-jack
Germany

Hannes Heiner, founding member of the Berlin collective
Dead Chickens, has designed a mobile folding clock-jack that is
five meters in width and almost the same in height. It is peopled
by frightening, gesticulating monsters all day long, from when
it is turned on in the morning to when it stops at night.
© Dead Chickens

Photos: Courtesy of Henryk Weiffenbach – Production Lille 2004

Michell & Jean-Pierre Hartmann
France
Parallel worlds

1- Sculpture: Lem.
The doll driver swivels 280° in its bubble, while the
whole structure is intermittently animated by
a vertical movement.
2- Sculpture: seven life-size clones. A swinging
pendulum motion drives them as they utter strange
and random noises.

1

3 - Le Réveil du Troisième Homme (The Awakening of the Third Man). *Steel and bronze sculpture with 18 programmed movements and 350 fiberoptical lights. It is triggered by an infrared sensor. The android is inside a six-door capsule that has traveled from planet to planet and has arrived on Earth. When a human passes by, the doors open, the being wakes, breathes, opens its blue sulfur eyes and tries to sally forth in the strangest fashion. Then it changes into a light machine, looks right and left, yawns and decides the earth is not worth the visit. The door to the capsule closes behind it.*

© Michell & Jean-Pierre Hartmann
Photos: Courtesy of Christophe Recoura

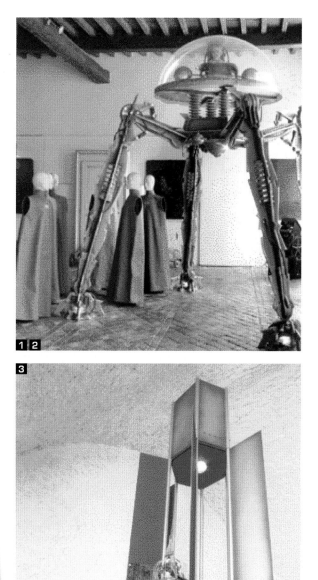

Michell & Jean-Pierre Hartmann
France
Galerie des clones

Installation consisting of seven clones, seven portraits of clones (the ancestors),
a box of clones, a box of clone eyes, the great Ordodinateur (Compoputer),
which is the clone-making robot, and the Lem, the device driven by a baby clone
(see previous page).
Michell and Jean-Pierre Hartmann imagine that biotechnology,
IT and micromachines will accomplish what social engineers have failed to do.
Their thinking is obviously on an interplanetary scale. The robots will
manufacture, inspect, judge, entertain, dream, create and think for us.
And they might even replace us by far more evolved life forms or mechanisms.
They will be post-human beings that will ensure the survival of the planet far
more effectively than we have done. Theirs will be the brave new world.

ROBOTS IN THE FUTURE

Whether we find robots fascinating, fearsome or freaky, we have to come to terms with the fact that they are here to stay.

By the end of the decade, the in-crowd will have one or more at home to do the cooking, mind the children or park the car in the garage. But in the longer term, they will be able to do a great deal more than that. They will deliver, or even be, the microscopic medication for curing sickly cells; they will be used as construction materials that can change appearance and shape; and they will be indefatigable workers, building and assembling at will. Drawing on the behavior of ants or bees, some will devise strategies for group tasks, moving in the hundreds or thousands into exploration zones. Androids will chat with their owners and become an integral part of society. At the same time, technology will create crossovers between man and machine as human bodies are augmented with robotic accessories.

Where will this frenzied wealth of ideas and experimentation lead? Which avenues of research will bring practical application? Which will remain utopian visions? It is still too early to say, but the sheer depth and scope of research reflects a creative energy that recalls the early days of cinema, gaming and the Internet.

The European Space Agency's Ariadna Project includes biomimetic robots.
© Courtesy of ESA

Glossary

Future Terms

Self-replication
Self-replication is a process in nanotechnology *(see below)*, whereby devices can create copies (running into the thousands) of themselves.

Biomimetics
Also called biomimicry, biomimetics refers to the art and science of building devices, materials and processes that mimic nature.

Bionics
The application of biological principles to the study and design of electromechanical engineering systems.

Cyborg
A human whose capabilities have been enhanced by the implant of mechanical and/or electronic devices.

Swarm Robots
Like swarms of ants or bees, a swarm of robots is a large group of networked devices, numbering in the hundreds or thousands, programmed to perform individual tasks in order to complete an overall task.

Nanoassembler
A programming language in nanotechnology that, researchers hope, will enable nanoassemblers to manipulate atoms at will and fabricate materials from them.

Nanotechnology
A branch of engineering that deals with the design and manufacture of molecular-size electronic circuits and mechanical devices, measured in nanometers. A nanometer is one billionth of a meter. The underlying hope is that it may be possible to create new materials by mastering the physical properties of microscopic entities.

Augmented reality
The enhancement of a person's experience of the real environment by augmenting it with computer-generated information and images.

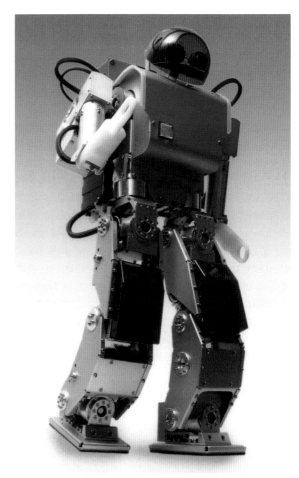

HOAP.
© Courtesy of Fujitsu

What does the future of robotics hold?

Just as the appearance and design of towns and cities changed in the 20th century and life in the home was made easier by an influx of electric and electronic appliances, so a major new sea change is afoot in the world around us. The driving force is the robot, in forms ranging from the home-help android and unmanned vehicle to serpentine bots that can slide into cracks, and smart materials incorporated into buildings.

But what is the purpose of all these robots? Are they part of humankind's innate urge to invent new forms so as to rise above natural constraints? Or will they contribute to a new, more insidious kind of enslavement than any that have gone before?

For the time being, optimism prevails. The organizing committee of the Robot Fair, held in Fukuoka, Japan, in February 2004, issued the World Robot Declaration that made three defining statements about next-generation robots:

1- Next-generation robots will be partners that coexist with human beings

2- Next-generation robots will assist human beings both physically and psychologically

3- Next-generation robots will contribute to the realization of a safe and peaceful society.

The World Robot Declaration reflects the marvellous mindset of Japanese robotics researchers who want robots to improve society. We can but laud such intentions, backed as they are by the strength of industry. But it would be illusory to believe that there will be no blots on the future robotic landscape. Human society is all too human. Accordingly, there exist robotic projects for grim androids that guard embassies and shoot on sight and others for preemptively settling newly discovered planets, just in case someone else gets there first. All of which raises other questions, given that the human race has survived for thousands of years without the aid of preprogrammed artificial beings. What is the right response to the rise of the robot, already felt by some to be all-pervasive? Should there not be a red line that delineates research and experiments, particularly regarding the development of crossovers between the living and the "undead"?

Hardly has the third millennium dawned than already the sheer tangle of avenues that research is exploring has grown inextricable. "There is probably no company in the consumer electronics and automobile industries that doesn't have its robot project," says Dan Kara of the industry analyst firm Robotics Trends. Not to mention ongoing work by university researchers, space agencies and the military. Carnegie Mellon University has inventoried over 120 projects. MIT has eight robotics laboratories, each of which employs dozens of researchers. Similarly, Tokyo's pioneering Waseda University is also a hive of activity. In France, too, a dozen projects are under development as part of the ROBEA Program, in which the national scientific reserach agency, the CNRS, and its systems architecture research laboratory, LAAS, are at work.

It is important to keep a level head faced with such myriad projects. Some work is advancing slowly, which does not prevent researchers from drawing on limited experiments to predict a carefree future. Others temper that optimism and advise caution, stressing that robots have not yet fulfilled their promise of 50 years ago. "It'll be a long time before robots can make beds as well as people can," quips Chuck Thorpe of Carnegie Mellon. "The manipulation skills required to make a bed—flipping up the corner of the mattress and tucking in a sheet so it looks straight—is complicated." Yes, indeed, what is second nature for us can outwit android butlers.

Still, the horizon is crowded with prospects for robotic applications that were once confined to science fiction. Even if they are not with us until 2050, with us they eventually will be. A few examples will suffice.

Nanotechnology will introduce intelligent materials that can change shape and color. At the press of a button, the walls in your house will switch from a rustic timber look to a smooth and soothing blue. Devices will be able to assemble themselves and robots to change their conformation.

Biomimetics will give robots powers drawn from the living world, enabling them to cling spiderlike to skyscrapers and wash their windows or to wiggle into holes like earthworms.

Swarms of workerbots will, like ants, single-mindedly undertake rescue operations or replace roadies to set up and dismantle sound systems for rock concerts.

Hybrid beings – part biological, part-electronic – could revolutionize medicine by acting as antibodies for failing or impaired bodily organs. Research has given rise to bionics and the real prospect of augmenting human capabilities.

Androids with the gift of gab are already limbering up to slip seamlessly into everyday life and keep their owners company.

In the future, which daily draws nearer, the span of robotics applications will embrace all walks of life.

Mobile robotic cell in an electronic microscanner. Such robotic cells can be equipped with probes and microtweezers.
© Courtesy of AMIR - University of Oldenburg

Nanorobots

Much of mankind's history and prehistory was identified with the minerals from that humans made their tools and which lent their names to periods in the distant past: the Stone Age, the Bronze Age and Iron Age. With the much-hyped advent of nanotechnology we will enter the Materials Age, so called because nanotechnology will enable us to control the creation of new materials from molecules.

Nanotechnology can work to one billionth of a meter—a molecular scale. Like some all-pervasive software, it will rearrange molecules to form wood or steel, a seashell or a cucumber. Ultimately, there will be nothing it cannot create.

Let us travel ahead in time to 2030 to a household that enjoys the services of a super-strength, nanoassembling robot that can provide solutions to a limitless range of everyday situations. The mother of the household is a budding writer who drops her hundred-page manuscript all over the floor.

Super Nanobot picks up the sheets of paper, sorts them, and makes paper clips to hold them together. Later on a dinner party is in full swing, when the doorbell rings and an (uninvited) friend drops in unexpectedly.

There are not enough chairs, so the bot builds one. And when the mozzarella runs out for the pasta dish, it makes some more. Where, though, does it find its raw material? From its sandy soil carbon reserve or household waste—it merely rearranges the molecules!

Such a scenario would be too way-out even for science fiction, were it not for Richard Feynman, winner of the Nobel Prize for Physics in 1965, who originated the concept of nanotechnology. It was not until the mid-1980s, however, that another researcher, Eric Drexler, propounded theories for practical use of nanotechnology in his book *Engines of Creation*. Experts now agree that within thirty years the emerging technology will bring changes that are more far-reaching than all the technologies of the previous 100 years put together.

As always in cutting-edge technology, Japan is very active. Its Ministry of International Trade and Industry (MITI) has invested $150 million in nanotechnology over the next 10 years. In the US the National Nanotechnology Initiative received an endowment of $679 million for 2003, while Canada's National Institute for Nanotechnology is to receive $120 million over five years.

The principle of a microgenerator actuated by muscles. Left: a microscopic image of piezoelectric plates. Right: schematic diagram.

© Courtesy of Carlo D. Montemagno, UCLA

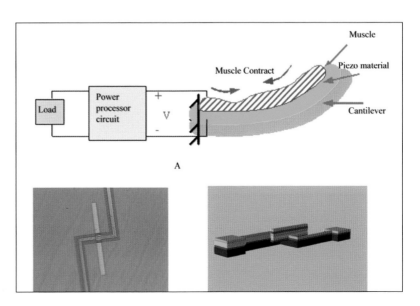

Nanoassemblers

Nanotechnology is set to develop two types of fabrication processes, both of which defy the imagination. The first is nanoassembly, also called positional assembly, in which controlled nanoscopic robots work to orders to assemble atoms. The second is self-replication, whereby minute robots duplicate themselves.

The first company dedicated to nanotechnology fabrication processes was founded in 1997 by Texan millionnaire James Von Ehr. It is called Zyvex. On its Website it makes this claim: "If we rearrange the atoms in coal, we get diamonds. If we rearrange the atoms in sand (and add a pinch of impurities) we get computer chips. If we rearrange the atoms in dirt, water and air we get grass."

Research, meanwhile, is progressing slowly but surely and aims to produce the first nanoassembler robots within ten to twenty years. Professor Hod Lipson of Cornell University is working on the automated construction of systems that can self-replicate.

IBM has achieved important breakthroughs in the field. In August 2001 a team headed by Dr. Phaeodon Avouris produced the first carbon atom computer chip. It was 100,000 times finer than a hairsbreadth.

Intelligent materials

Nanotechnology will lead to the development of intelligent materials that can interact with their environment, such as photochromic glass, which darkens when exposed to light. A smart structure could also repair itself, a capacity once associated exclusively with living creatures. A bridge, for example, could use a store of spare materials to fix a point where there is too much strain, applying a material until the force exerted is restored to the correct level.

Carolyn Dry of the University of Illinois has developed smart concrete. It holds an adhesive substance in hollow pores that it releases when it senses fissures forming. Ideally, smart structures should be able to draw on external storage points

Nanorobots

In May 2004 two chemists from New York University, Nadrian Seeman and William Sherman, built the world's first bipedal nanorobot made from fragments of DNA and able to walk on a track also made from DNA. The 10-nanometer-long legs of the minute stroller assemble to form a sort of helix. Each leg has a footlike strand that works in conjunction with the DNA in the track to lift the robot's feet and then set them down, so producing the walking motion.

that hold materials made from particles caught in the blowing wind.

Experts paint pictures of tomorrows that surpass the imaginations of today. Ralph C. Merkle has predicted nanorobots that are smaller than a human cell and able to destroy cancers, infections, blocked arteries and even delay old age. Others look forward to a brave new world that has overcome hunger by producing synthetic food to meet needs, while the wildly sanguine entertain the possibility of reintroducing extinct plant species.

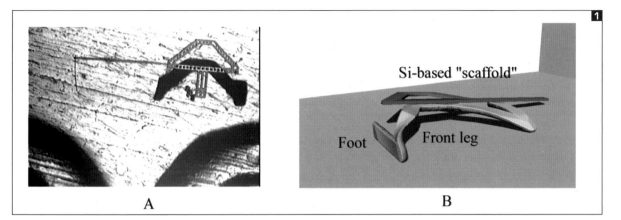

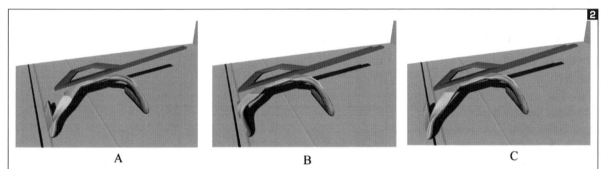

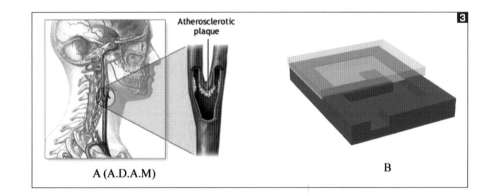

1- Microscopic view of a microrobot (A)
and schematic diagram of
a microrobot (B).
The leg is *160* micrometers long.
© Courtesy of Carlo D. Montemagno - UCLA

2- Schematic diagrams show
the sequential movement of the
microwalker taking a step.
A: It prepares to tense its leg.
B: It tenses its leg.
C: It relaxes its leg.
The blue, green, and red lines delineate
the foot's positions.
© Courtesy of Carlo D. Montemagno - UCLA

3- Two potential microrobot applications.
A: Microsurgery to remove
atherosclerotic plaque.
B: Delivering a drug or antibody
through microchannels.
© Courtesy of A.D.A.M. Inc.

The field has its detractors, like Dr. Brad Cox of George Mason University, who believes that problems will arise from limited human knowledge and the "brute force" of chemistry. Most, however, like Merkle, are nanotech believers, arguing that the laws of physics are impervious to our hopes and fears and that nanometric manipulation and fabrication are not just possible, but inevitable.

The Shape of Robotic Things to Come

Some robots, which, though tiny, are big guys compared to nanobots, rely on a basic concept that is similar to nanoassembly. One example is the polymorphic, or shape-changing, robot. Hop Lipson and Jordan Pollack of Brandeis University in Massachussets have produced a prototype. Its skeleton is made up of tubes and pistons, and its muscles of electric motors. The integrated computer creates a model of the shape required by the situation and environment that it identifies. In simple terms, the robot "reinvents" itself as circumstances dictate.

The shape-changing robot has inspired the work of British scientist Joseph Michael, whose polymorphic creation can flow through holes, using a system of sliding rods. Other projects, too,

are ongoing, though none has yet materialized. That in itself is an added spur to researchers.

Nadrian Seeman and William Sherman designed the first bipedal walking nanorobot made from fragments of DNA. It moves along a track also made from DNA. Each of its two legs measures 10 nanometers in length and consists of two strands of DNA that form a helix. To begin with, the nanowalker's feet make up a single DNA strand, but each can pair up with a complementary DNA strand on the special track. To make the foot take hold of the track, Seeman and Sherman add a strand that attaches the foot, then one for detaching it. Then the released foot grabs another attach strand to gain a foothold. In this way it walks! The two researchers are now working on getting the nanowalker to carry a load.

© Courtesy of Nadrian C. Seeman

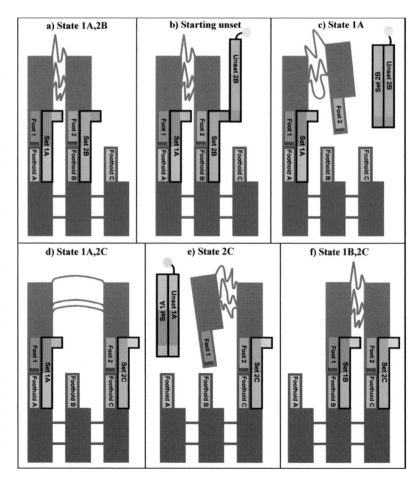

Biomimetics: toward the biobot

To build robots with superior capabilities (*see Chapter 6*) scientists are increasingly turning to biomimetics, which draws on the properties and behavior of the living world to produce new materials. As A. Beukers and E.V. Hinte put it: "Animals and plants, in order to survive in competition with each other, have evolved ways of living and reproducing using the least amount of resources. This involves efficiency both in metabolism and optimal apportionment of energy among the various functions of life."[1]

Robots, then, stand to gain from copying nature. The sense of smell, for example, intrigues numerous robotics reasearchers worldwide, including Dominique Martinez from INRIA who is working on a project called NOSE[2]. His idea is to develop electronic odor sensors that copy insects' olfactory systems and can home in on specific smells. He believes it is both feasible and necessary, for unless robots are equipped to sense their environment in ways other than by vision alone, they will not be able to assess it accurately.

HOAP-1 is an android marketed by Fujitsu, which plans to design its successor with a faster, simpler system for learning how to move in a coordinated motion. The company's engineers have studied the rippling movement of earthworms and lampreys as examplars.

The jovial Timothy Wolf Bretl has made a robot that looks like some primitive insect, except that it has only four legs. It uses them to scale walls like a real climber, hooking on to any handholds it spots. Bretl is also working on a smaller version of a climbing robot prototype developed by the Jet Propulsion Laboratory and Stanford University. It is the Lemur II. He has modeled Lemur on himself—he enjoys rock-climbing—and hopes that the robot will be able to hoist itself up exclusively by getting a purchase on holds and hollows on the climbing surface. Another wish he harbors is getting the robot to right itself should it slip. Much of the focus of his work is on motion planning with the aim, he says, of extending the mobility range of existing robots. Applications could include rescuing stranded mountaineers and people trapped in collapsed buildings and underground exploration.

Headed by Professor Jean-Arcady Meyer, a French research team at Animatlab has been building a self-adaptive robot, whose real-life model is the rat. It goes by the odd name of Psikharpax. Laboratories worldwide are engaged in the project, which began in September 2001 and brings together biologists, roboticists and IT engineers. Animatlab's Psikharpax will not only have a ratlike physiology and physionomy, but—so hope its creators—the cognitive powers of the ubiquitous rodent.

Swarms of cooperative robot workers

Swarms of robots working together like insects on a single task is an area of research that has similarities with biomimetics. It is, however, based on another theory, that of collective intelligence. The idea was first aired by the celebrated roboticist Rodney Brooks in the early 1990s when he posited that bees in a hive and ants in an anthill were like so many robots. No single one, said Brooks, knew what overarching task—e.g., building a new nest—they were working towards. Yet once they came together, their interaction revealed the grand design.

This understanding leads to the bottom-up approach, whereby robots are given straightforward tasks that accrue to gradually increase the complexity of their capabilities. Brooks argued that it should be possible to create robots that learn from their experience, much as primitive living organisms evolved to perform highly complex tasks over billions of years. They could not, he felt, be programmed to perform what they were able to learn to do empirically.

The Toulouse-based animal cognition research center CRCA, in southwest France has also predicated that no individual ant in an anthill knows what the overall purpose is behind its task. A kick in an anthill sends ants scurrying frenziedly all over the place. It is impossible to make out what each member of the colony is doing, yet eventually eggs and larvae have been rescued and rehoused and collapsed galleries rebuilt. CRCA plans to introduce robotic ants into a real colony so that they interact with living insects. The objective behind the experiment will be twofold. First, by coming to a better understanding of the real ants' group intelligence, the researchers hope to apply it to robots. Second, they will investigate the fascinating possibility of using the preprogrammed intruders to influence the ant colony's collective decision making.

Rodney Brooks, meanwhile, continues to work in the direction that he adumbrated in the 1990s. His company, iRobot, has secured a contract to create swarm robots for military applications like reconnoitering high-risk buildings or defusing explosive devices. The project draws on the findings of a young prodigy by the name of James McLurkin.

Currently studying at MIT for a doctorate, which he hopes to obtain in 2006, McLurkin built his first robots when he was still in high school. On admission to MIT in 1991, he sought to apply the concept of an "anthill of robots" to tiny machines equipped with sensors and microprocessors that could communicate with each other, a supposed feature of group intelligence. He has claimed that he drew his ideas from watching real ant colonies. He has since offered his services as consultant to iRobot, Walt Disney Imagineering, and SensAble Technologies.

McLurkin's doctoral thesis focuses on programming swarms of robots so that they mimic the behavior of bees—how they assemble, disperse, follow each other, and fly in circles. McLurkin believes that the communication system of bees is richly instructive because it enables a great many individual tasks to be performed for the overall benefit of the community. He envisions developing thousand-strong groups of robots, which could carry out security missions like mine-detection patrols and searching the wreckage left by earthquakes. In 2003 he won the Lemelson-MIT Award.

On a simpler level, the IT and robotics research center LIRMM,

1- Alice robot swarms.
© Courtesy of CNRS, CRCA of Toulouse

2- Alice robots are the result of research by the autonomous systems laboratory of the Swiss engineering university EPFL. They can avoid obstacles, track stationary and mobile objects, and collaborate through wireless and other connections. Although very small, they can be fitted with modules such as video cameras, sensors and communication devices. They are frequently used in robotic group work experiments.
© Courtesy of ASL/EPFL

3- The micromechanical flying insect (MFI) project aims to develop a 25-millimeter device that can fly autonomously in the manner of a fly.
© Courtesy of Ron Fearing - Berkeley University

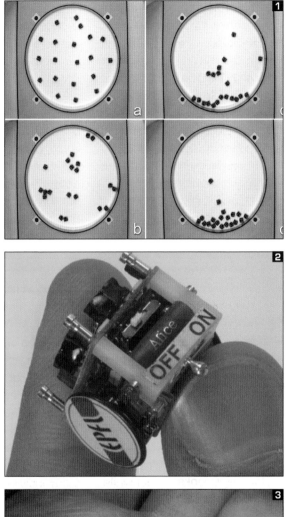

in the southwestern French city of Montpellier, is working on a project called MP2 that involves coordinating robots with two parallel arms to work simultaneously on lifting heavy loads. The load is distributed across each robot's parallel arms. The result is spectacular gains in accelerative force of over 20 to 30G, compared to 1G or 2G for single-armed robots. These findings could be applied to ultrarealistic flight and yaw simulators and even high-precision surgical tools.

Robots working together in space is a hope that NASA nurses. Such interplanetary labor collectives would be dispatched into outer space to build structures of over 10 kilometers (6 miles) in length. The University of Southern California is researching weightlessness by experimenting with robots that float on air cushions.

Hybrids

Hybrid robots lie in an area where nanotechnology and biomimetics bisect and that some scientists would like to see them merge into a separate discipline, with a dash of swarm robotics thrown in for good measure. The technology of hybrid robots brings together the organic and the mechanical, the artificial and the living.

In practice, the human muscle system resembles a swarm of cooperative robots and, from a certain point of view, can be likened to what is known as massively parallel processing. It is coordinated by complex molecular machinery honed by millions of years of natural engineering. Why not harness muscle cells to electronic microchips to make micromachines directly fed by glucose?

Researchers Jianzhong Xi, Eric Dy and Carlos Montemagno from UCLA's bioengineering department have tackled the problem through two experiments. In the first experiment the team used a polymer, PNI[3], to coat a silicon microchip and anchor a heart muscle cell culture. Once PNI has been steeped in water, its phase transition behavior responds to the ambient temperature. At the temperature of the cell culture medium it was hard, allowing the muscles to grow. At room temperature it dissolved and released the muscles, which immediately contracted. With each contraction they moved, driving the silicon chip in what was the first muscle-powered microdevice.

The second experiment used piezoelectric[4] microgenerators and rat cardiac muscle fibers grown and mounted on a silicon arc. The muscles contracted then relaxed and in this way propelled the silicon arc. Carlos Montemagno thus succeeded in creating a hybrid microbot prototype. By converting glucose into electrical power it generated enough energy to stay in motion on a microscopic scale for four hours. Possible applications could include the controlled delivery of antibodies to sick cells through microchannels, and diaphragm nerve stimulation to actuate breathing.

Projects for robotic prosthetics are another example of the union of the organic with the mechanical. A bionic hand that obeys instructions from the brain could become reality if the work of biorobotics professor Paulo Dario of the Sant'Anna engineering school in Pisa, Italy, comes to fruition. Elsewhere, Quebec-based researcher Stéphane Bédard set up a company in early 2004 to commercialize the first motorized smart leg. It efficiently and effectively mimics natural motion, particularly for getting up and down stairs.

Research has also taken the direction of a robot that is controlled by an animal. A team at Duke University ran an experiment in 2000 with monkeys. They manipulated a joystick to control the image of a robotic arm on a computer display, making it perform simple gestures like grabbing food on a tray. A robotic arm some 1,000 kilometers away reproduced the monkey's movements. Using electrodes implanted in the monkeys' brains, the research team analyzed and modeled signals.

In Chicago, Professor Sandro Mussa-Ivaldi carried out a similar test in which a master fish guided a slave robot towards a light.

Cyborgs

One of the next major robotics evolutions will be electro-human fusion leading to the emergence of cyborgs à la Robocop. That certainly seems to be a dream nursed by avant-garde researchers like cybernetics professor Kevin Warwick. He actually became a cyborg when he had an electronic chip grafted under his skin so that his computer could recognize him. He claimed afterwards that his ability to interact with his environment was enhanced *(see interview, pages 522–523).*

Transhumanist artist Natasha Vita-More has plans for an "optimized human" called Primo, which will bristle with high-tech add-ons like sonar, a fiberoptic cable along the spine, and a head crammed with nanotech data stor-age. The cyborg will also be able to change shape, height and color, and regenerate itself.

For Better Humanoids

Technology is still far from able to recreate the fabulously fine-tuned workings of the human body. Pierre Rabischong, professor at Montpellier Medical School and a leader of the European project, "Lève-toi et marche" (Arise and walk), goes some way toward explaining why in an interview on the French-language Website *Automates Intelligents*[5]: "Man is a highly complex machine that is usually driven by an ignoramus.

The awareness of the individual controlling a machine is confined to the simplest decision making—take that thing, go here, go there. He or she doesn't care about the 600 engines in the body." How will it ever be possible to build into a robot human capabilities that draw on such a multitude of underlying high-precision mechanisms?

V. Mattoli '02

The Cyberhand system is a comprehensive functional substitute for the human hand. It is a robotic prosthetic attachment that feels and can be used like a real hand.
© Courtesy of Lucia Beccai, Scuola Superiore Sant'Anna, Pisa, Italy.

Self-adaptive robots

Numerous research teams have founded their work on another important idea propounded by Rodney Brooks in the early 1990s—that robots could be made to program themselves.

One approach is to let them randomly find solutions by submitting them to tests one by one. Those that fail are deprogrammed, while those that find answers transmit their knowledge to the next generation of software. Artificial DNA thus develops along the lines of Darwinian theory, but on an incomparably shorter time scale.

It is a line of thought that is considered enlightened by Hans Moravec, who taught at the Carnegie Mellon University's Robotics Institute for thirty years before setting up his own company. "Nature proceeded by trial and error, keeping solutions that worked as it built complex structures. Robots will follow suit. Increasingly they will no longer have to be taught, they will exper-

iment by themselves incalculably faster than the biological world," says Moravec. He believes that the strong point of robots like those he builds is their ability to perceive and operate in three dimensions. With the advent of very high-speed computing power, robotic models can be developed following a pattern similar to biological intelligence and evolution.

Yet there are thinkers and scientists who argue that the evolutionary genetic model yields results too slowly. Stephen Thaler is one of them, theorizing that robots should be forced to find answers themselves by being confronted with chaotic situations. He draws on his work into neuronal networks, a branch of computer science that explores how a system reacts when a factor is introduced into its logical processes to disturb them.

To market the results of this research, Thaler created his own company, Imagination Engines, and is working on developing creativity machines. He says that creativity machines draw on his important scientific discovery (protected by several copyrights) that a neuronal network that is "internally irritated, in a specially prescribed way, tends to generate coherent ideas derived from its absorbed wisdom."

The ability to focus on a particular detail in its surroundings is,

claims Thaler, a characteristic of his creativity machines. He explains that when we find ourselves in an unfamiliar environment, our gaze moves around randomly until it lands on something deserving of attention. It scans what it sees and moves in on it. Imagination Engines has developed machine vision that enables a chaotic neural network driven by internal noise to scan a scene up and down. When the second neuronal network detects items of interest in the camera view, it emits a signal that reduces the amplitude of the noise applied to the network controlling camera movement. From this dialogue, concludes Thaler, we get something that seems identical to the way in which the human eye works. The camera is automatically drawn to items of interest and it all happens in just a few seconds.

Thaler cites an example to illustrate his point. When a creativity machine is shown bitmaps of twelve different faces for one minute, it is then able to produce new faces, each one of which is distinct from the initial exemplars. Thaler claims that a conventional genetic algorithm would have taken months of programming on the world's most powerful computers running for years at a time to arrive at an equivalent result. The creativity machine could help self-learning robots to find solutions by themselves at very high speeds.

Learning by imitation

Robots can also learn by imitating what they perceive, an approach taken by researchers from Norway's technology and industry research agency, SIN-TEF, in a project called PPM. When a human operator performs a task, the robot painstakingly deconstructs and analyzes his or her every movement, then imitates it. The PPM system is already much in demand from industry.

Reasoning robots

Intelligent robots will eventually have to venture into the world of humans and communicate with them. But first, they must overcome numerous obstacles.

At Sony, Aibo's creator, Toshitada Doi, has taken charge of the new research unit, Life Dynamics Laboratory, whose task is to develop an enhanced version of the humanoid, Qrio, which will be able to communicate with people. French researcher Frédéric Kaplan is working on related cognitive capabilities at Sony CSL in Paris. He says that the experiments he has conducted have revealed that it is possible to teach Aibos to draw on their past.

At the LAAS systems architecture research laboratory in Toulouse, Raja Chatila and his 40-strong team are pursuing a similar tack, designing machines that can perceive and reason.

Nevertheless robots will not be able to take their place in human society for a long time to come. A DARPA linguistics expert Ronald J. Brachman heads the Information Processing Technology Office, whose purpose is to teach machines learning and reasoning processes, a line of work that has received funding of $29 million. The DARPA consensus is that robots will not be able to chew the fat with or relate intelligently to their human counterparts until 2030.

The era of the willing slave

Hans Moravec knows more than most about mechanical minds and mores after working so long with robots. His view of their, and our, future is surprising. Even if, in several centuries' time, robots gain the upper hand over their creator, there will first be a long period of placid transition when they will be entirely at the beck and call of humans, from whom they will learn everything they eventually know. "It will be perfectly plausible to build machines madly in love with the idea of being totally servile," Moravec says pithily.

From washing dishes to cleaning sewers and from peril-fraught missions to the drudgery of the assembly line, robots will embrace their tasks cheerfully and even—who knows—with a hint of gusto? Humans will then be able to devote themselves to rewarding activities, which will include building, programming, fixing and servicing robots, all likely to create much employment. The era of the willing slave devoted hardware and software to its master will come.

(1) Quotation from *Lightness, the Inevitable Renaissance of Minimum Energy Structures,* by A. Beukers and E.V. Hinte. Published by 010 Publishers, Rotterdam
(2) NOSE: Neuromimetic Olfactory Sensing.
(3) PNI: poly-N-isopropylacrylamid.
(4) Piezoelectricity: Electricity produced by mechanical pressure on crystals such as quartz. Electrostatic voltage deforms the crystal.
(5) Website: automatesintelligents.com

Other imaginary worlds

The future is built by those who are bold enough to lay the foundations of new realities. Neither the theory of relativity nor the discovery of gravity were the work of minds that respected the established order. They blossomed from hybrid shoots, grafted inspiration and chance combinations. From penicillin to quantum theory, great discoveries have taken shape in laboratories where conformism was not a term that entered into the equations. Inventions rest on the creed of power to the ideas of people, for conservatives have no place in assemblies of innovators.

The predictions of the specialists interviewed on the following pages might seem fantastical, but the very fact that they have sharpened their minds on the future and stated their cases is in itself worthy of merit. Future reality is rooted in the imaginings of today, a fertile, untamed soil that is prone to extravagance and experimentation.

What will a future peopled by robots be like? How will they fit into daily life, whether they take the shape of androids, household appliances, molecular mechanisms or smart prosthetics? We can only conjecture at a time when robotics is still in its infancy. It is as if we were back in the days of the first photographs, silent movies, or when computers were nothing more than alphanumeric characters shimmering on a hulking screen. In these pioneering times theories abound and impress. Yet caution should be exercised. The audacity of some predictions prompts mixed feelings. There is something immature about the way in which some innovators fixate on their findings, building castles on flimsy foundations and experiments that have not been fully tried and tested. They seldom have the breadth of vision needed to see their own work as part of a wider thrust in which other disciplines and sciences play a crucial part.

Those caveats aside, it is important to devote active thought to the role of robots in our future. Be they imposing or unobtrusive, life-size or barely perceptible, robots are intelligent machines that will walk next to us, fly over us and nestle in the materials that make up our everyday environments. Their prototypes are already part of today's world.

The interviews that follow spring from our determination to hear directly from some of the boldest robotics thinkers, who have broken away from the pack to reconnoiter the uncharted land of the future. Their ideas are always challenging, even moving, though some may think them shocking. They deserve credit for standing back and envisioning original applications for robotics that lie off the beaten track.

It is up to the reader to separate the wheat from the chaff and spot any kernel of truth; to strip away overblown assertions and understand that the interviewees are not telling the future, merely expressing their hopes.

Anyway, the future is elusive. It has a habit of thwarting prediction and prophecy and taking unsuspected directions. No crystal ball can factor in the human parameter, with all its aspirations, emotions, doubts, and fundamental independence. Who predicted the Internet, networked gaming, the rise of MP3 downloads? They took off by themselves, triggered by sudden, unaccountable popularity. The human refusal to be catalogued, pigeonholed or pinned down will probably be a decisive factor in shaping the robot's place and role in the future.

The concepts expounded in the following pages should be considered as messages in a bottle: only very few will wash up on shore.

What will robotic software consist of? What will their destinies be? Should they necessarily be part of the human adventure? Who will benefit and who will suffer? How will they shape society? Should their impact and influence be limited? Should they always be slaves to man or should their capabilities be allowed to enhance the human body?

These are just some of the questions we wanted to ask of some committed avant-garde practitioners.

Interview
Takao Someya

Takao Someya is more of an expert in electronics than robotics. But in his capacity as professor at the University of Tokyo's school of engineering, Someya has been involved in developing surfaces that are sensitive to pressure exerted upon them. The surfaces are designed to be the skin of next-generation robots.

How long have you been interested in robots?
As with most Japanese, including scientists and engineers, my affection for robots goes back to Astro Boy, a character in one of Japan's most popular comic strips and cartoon series. They were created by Osamu Tezuka, who lived from 1928 to 1989. Astro Boy's original Japanese name is Tetsuwan Atom. Some market analysts say that Astro Boy is one of the reasons why Japan is such a force in researching and developing humanoid robots. But in more practical terms, we use them as domestic helpers.

Why do you think that artificial skin are important in robotics?
Personal robots that work in the home, doing the housework and looking after people need to be touch sensitive. Unlike industrial robots they don't perform routine, repetitive tasks. If a robot has to move an invalid, it must lift him or her up gently. That requires a sophisticated sense of touch. But today's robots have trouble even picking up an egg. Another thing is that if a robot doesn't have a sense of touch, then it could tread on a baby crawling on the floor and not even notice. So we're hard at work on making electronic skin whose properties are as much like those of human skin as possible. They must stretch and sense other stimuli in addition to mere pressure.

Would you say that artificial skin is the next stage to be mastered in robotics?
There are thousands of problems still to be resolved before robots can take their place in our daily lives. But artificial skin is obviously a major hurdle to be overcome. Quite frankly, robots that don't have sentient skin will be hazards in everyday life.

Interview
Frédéric Kaplan

On completing his doctoral thesis on artificial intelligence, Frédéric Kaplan accepted a proposal from researcher Luc Steels to join him at the Sony CSL laboratory in Paris, shortly after it was set up in 1997. It was a chance not to be missed for the young engineer, who considers that building intelligent machines is first and foremost a way of coming to a better understanding of human beings.

You've said that you are more drawn to creating likable robots than utilitarian ones. How does that view inform your work?
When Sony first introduced entertainment robotics towards the end of the 20th century, it cleared the way for some very exciting avenues of research. The idea is to build machines that continue to be interesting in the long term, but with which you can interact just out of enjoyment. In that light it is crucial that a robot should evolve and grow as it gets to know its environment and enters into relationships with the people around it. It has to build its own story. That's why new artificial intelligence technologies have to be developed—to help robots to

learn and to want to learn. So our prime objective is to construct robots endowed with a form of artificial curiosity.

What do you mean by artificial curiosity?
Pierre-Yves Oudeyer and I have been working on the project for several years. Our robots choose to perform actions that place

them in situations from which they can learn. To get them to behave in that way we build into them a mechanism that enables them to measure the accuracy with which they predict the consequences of their actions. Say, for example, a robot is playing with a red ball. It will then try to predict the visual effect it will produce by handling

Frédéric Kaplan
© Courtesy of F. Kaplan

the ball in a certain way. Once it has executed the movement, it can then measure its prediction's margin of error. That enables it to make a more accurate prediction the next time and to gauge how easy or difficult the task is to learn. When it sees that it's making steady progress, it practices more and more until it has mastered it. It then grows

tired of the trick and moves on to something new.

Couldn't such a robot be tempted to try out tasks that might be dangerous both for itself and people around it?

Absolutely. There are some forms of behavior that are interesting from the point of view of the effect that they produce. But they can damage the robot. That's why we program a pain system into robots to prevent them from self-destructing just out of sheer inquisitiveness. Pain probably plays the same role in the human makeup. As for any damage or harm robots could cause to those around it, there's only one possible solution—to make robots small enough not to cause too much damage whatever they do.

You describe your role as regards robots as that of a mediator rather than a teacher. Could you explain what you mean by that?

In the models that we build, the role of the experimenter—and, later, the owner—is not to teach robots knowledge and skills, but to coax them towards learning by themselves. That involves placing

a robot in an environment that is conducive to learning and directing its attention towards one object rather than another. That is why I prefer to describe myself as a mediator who helps robots to structure their world. In the end, though, it's always the machine that decides what it's going to try to learn.

Will such robots one day be able to talk?

The prototypes that we build have a system of motivation, which drives them to autonomous self-development. But the road that leads to a form of intelligence is a long one and the dynamics of robots' behavior among themselves and their interaction with humans play a key role. Before robots can learn to speak, they must, like children, learn to be attentive to different aspects of their surroundings and the people around them. They have to learn not only to read and understand looks and gestures designed to catch their attention, but also to sense the intentions of humans and other robots. To read the intention behind a certain kind of behavior, a robot has to be able to interpret it in terms of its underlying ends and means. A robot must be able to see that another robot is *trying* to hit a ball or *wants* to show it something.

... Interview cont.

That is a crucial step toward language and learning social skills. And building robots with that capability is my biggest challenge as a researcher.

In your first book, *La Naissance d'une langue chez les robots* (The Birth of Robot Language), you describe how some of your robots have together invented their own language. What have you learned from that experience?
Well, in 1999 we showed Luc Steels how, if you gave robots certain "preadaptive" capabilities, like the ability to divide their attention among several focus points, a group of them could build a new vocabulary without any need for a central controller. That finding was the result of experiments that we ran for over a year. We put couples of "robotic bodies" into museums and laboratories. Software agents could take over the bodies and play at word games to try to come to an agreement on the meanings of certain words. By interacting with each other in that way, thousands of software agents gradually agreed on a shared vocabulary, even though at no time did they have the opportunity to read each others' minds, as it were. They used the word "wapaku"

for example to denote the color red and "bozopite" for things that were big or wide. They ascribed shared meaning to these words through collective bargaining.

As long as a robot has only two of the basic senses, sight and hearing, isn't its perception of its environment severely restricted?
The idea is not really to mimic every last aspect of human development. We try to understand some of the basic fundamentals of individual and collective learning processes. We show how robots can develop through those same dynamics by carrying out experiments that wouldn't be feasible with human beings.

Douglas Lenat spent twenty years trying to create his intelligent computer, Cyc, and the end result wasn't exactly a great success. To an outsider it looks as if artificial intelligence has tried to master something it isn't cut out for.
Douglas Lenat's work underestimates the importance of bodily processes in the acquisition of knowledge. We are exploring avenues of research that, on the contrary, place the body at the core of learning. Robots learn through their bodies initially. The gradual process over time is missing from

Cyc. But the bottom line is that artificial intelligence may never achieve its end, not for technical reasons, but because its objectives are constantly being redefined as research advances. All artificial intelligence's successes have shown, more than anything, is exactly what artificial intelligence isn't. In the last fifty years, the progress we have made in robotics has allowed us to change our definition of intelligent behavior, so that only we humans have intelligence as our defining feature. And by forcing us to pinpoint exactly what sets us apart from machines, robots have helped us to understand human beings better.

Some distinguished roboticists have said that robots will eventually end up enslaving humans. What do you think?
Science fiction is always describing armies of robot soldiers seizing power on earth and reducing humans to slavery. It's important not to take those books and movies at face value. What is much more interesting is to see them in terms of the role of machines in our culture. Man is dependent on the machines that he builds in many ways. In addition to the practical applications that it creates, technology changes society and the way in which we see the

world. There is a true symbiotic relationship between the cultural and the technological. Culture steers technology in certain directions and in return technology modifies culture. We'll soon see new kinds of machines emerging. They will change our ways of life and, more than that, they will alter the way we see ourselves. This last aspect is, I believe, the most important.

Among the objectives that Frédéric Kaplan and CSL Sony have set themselves is to build robots endowed with a form of artificial curiosity.
© Courtesy of F. Kaplan

Interview
Haipeng Xie

Dr. Haipeng Xie is the chief technology officer of Dr. Robot Inc., a company that plans to commercialize a multifunctional personal robot controlled from a PC. He is also adjunct professor at the University of Guelph in Canada and has a Ph.D. in robotics from the University of Western Ontario.

Dr. Haipeng Xie.
© Courtesy of Haipeng
Xie - Dr Robot.com

Do you think that the day will come when we will just have to select a program for a robot to perform a certain task?
That is the long-term vision, and the dream of most robotics engineers. But we won't achieve a universal robot for decades rather than years. In the near-term, it might be possible to produce a robot capable of a range of actions and functions that would be sold in modules, allowing consumers to tailor a suite of tasks to best suit their needs. For example, if you wanted a robot to patrol your home for intruders while you're out, but you don't really care whether it feeds your pet cat, then you could just select the right application. It's entirely analogous to selecting the applications you'll install on your PC.

Why did you decide to control Dr. Robot's HR-7 from a PC or Mac? What was the thinking behind the configuration?
Everything will be networked eventually, so why should the robot carry a heavy, high-power PC, when the PC could be sitting on the floor and plugged into a wall power outlet? Our approach is different from many companies developing next-generation robotics. A number of Japanese developers are pouring money into creating truly remarkable robots with impressive functionality, but whose technology is too expensive to be marketable to consumers at this point. Their investment horizon is very long-term. Our goal is to maximize

functionality and minimize cost by leveraging as much as possible technologies that are widely available today. HR-6 has an onboard digital signal processor, but it offloads heavy computation to a remote source. Primarily this reduces the robot's cost and power consumption, two challenges to producing an affordable humanoid robot. Also, whether a local or a remote PC supports the robot, the benefits of any upgrade to the PC are also enjoyed by the robot.

What do consumers most want from a robot according to the market surveys that you have conducted?

Asking this question without specifying that current technology is still limited yields very ambitious orders. Most people are hoping for a robotic personal assistant or a robot that will do everything from cooking and serving a meal, to washing the dishes afterward. Put in the near-term context, most people would prefer robots that assist with domestic chores, followed by security, education and entertainment.

Above:
Prototypes of the HR7, the multitask personal robot.

Left:
DRK-8000, an intelligent personal robot.

© Courtesy of Haipeng Xie - Dr Robot.com

Interview
Natasha Vita-More

Founder and president of the Extropy Institute[1], Natasha Vita-More calls for bodily faculties to be enhanced by incorporating components developed in robotics and electronics technology.

Natasha Vita-More.
© Courtesy of Natasha Vita-More

Your creed is that the human body can be improved and augmented by fusion with machines. What benefits does such a mutation bring?

The benefits are numerous, but not without recognizing some of the potential problems that lie ahead. Each time humanity steps ahead and away from its past, it must consider at what cost. What are the psychological, emotional and financial costs to humanity of progress? But these costs are minimal when we consider the advantages of progress. Today, we have solved the problems of many diseases and eradicated life-threatening ones. We have been able to give eyesight to the blind, hearing to the deaf, mobility to the paralyzed, voice to the mute, and peace of mind to the mentally disabled. We have saved millions of births that would have perished without medical advances. We have made life better for the elderly and the

unfortunate. Many of us today have a full mouth of teeth, rather than decay. Many of us today have healthy hearts, minds, and bodies, thanks to technological innovations. The benefits of such human–machine mutation are right before us—in our present circumstance and written in our history. Each new machine, be it one's eyeglasses or one's prosthetic leg, has been beneficial to the very person who was in need of it.
From outside the situation it is quite easy to say, "They don't need it" or "We have gone too far with technology." But inside the situation—if you, your child, or your parent, is in need of a heart replacement valve, or cancer treatment—there is no way that you would say that such technology is not needed or has gone too far. It is all relative.

But what is the point of augmenting the capabilities of the human body even when there's no problem to begin with?

The benefits of incorporating more machinery onto or into our bodies are to protect ourselves and keep us out of harm's way. Bodily augmentations have been enormously helpful to people throughout the world for many, many years. If we consider what machinery actually is, the idea of meshing with machines becomes more sensible and natural. For example, eyeglasses are a piece of technology. Wheelchairs are machines, and prosthetic limbs are machines. These two design products have supported the mobility of humans who have been unable to move about on their own, whereas otherwise they would be immobile. What I am proposing is to design a full body prototype that functions like a human body, but is not 100 percent biological. Rather, it is a whole-body prosthetic that acts as either an alternative body or a spare body. This body, "Primo", would house the brain and whatever organs and essential parts would not be replaced. The other parts would be prosthetic, synthetic models working together; forming a system, that acts to transport us just like our human body does today.

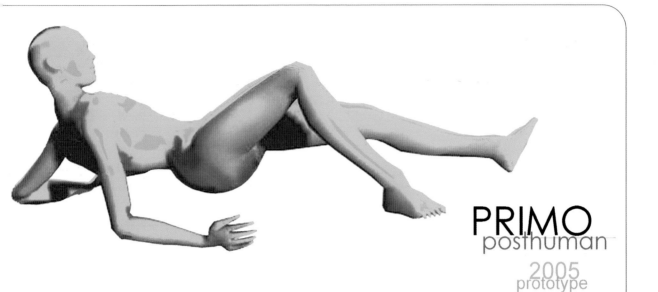

PRIMO
posthuman
2005
prototype

© Courtesy of Natasha Vita-More

Could you cite some examples of the benefits of having a posthuman body, where most of its capabilities and functions would be enhanced by technology?

The human body is undergoing change. Plastic surgery, prosthetics, robotics, electronic and digitized vocal chords, implants for hearing, chemicals to adjust and fine-tune brain functioning, genetics and genetic engineering, and cloning organs are ways to augment and upgrade our physique.

Let me backtrack a moment: The first prosthetics were way back in ancient times.

In India in 600 BC nasal reconstruction was elaborated by using a forehead flap over a nasal defect. Much later, in the late 1800s, improvements in anesthesia techniques encouraged surgeons to augment the nose with such materials as paraffin, gold, silver, aluminum, platinum, porcelain, celluloid, ivory, cork and some stones from the Black Sea.

The magnifying lens was designed by Roger Bacon in 1267 and eyeglasses by Salvino D'Armate in 1291. Speaking tubes were introduced as an add-on to the body in the 1700s, ear trumpets in the 1800s, hearing aids in the 1900s, cochlea bionic implants in the 1960s and digital signal processing in 1984. Humans have been augmented machines for eons; we just don't think of it as mutation. The idea of a prosthesis seems distant, but today we can modify a person's face with a prosthetic nose, chin, or jaw. Prosthetic robotic limbs have advanced radically over the past decades. It is moving so rapidly that patients are often asked to beta-test new technologies and components at the leading edge of new developments. And today prosthetic limbs can be controlled by microchips.

Robotic prosthetics are becoming so sophisticated that they often look better and function better than the biological body part. With this in mind, it is plausible that one day in the future there will be a "whole-body prosthetic" for people who need "bodies" and these bodies will be a combination of robotic computerized microchips, nanorobots, artificial intelligence and cosmetics.

What do you say to those people who refuse transhuman evolution because they believe it's unnatural?

The idea of "natural" changes over time. What we once considered to be natural is no longer in vogue and visa versa. Here is an example. At one time, the idea of lipstick was "unnatural" as was hair dye, teeth veneers, and the telephone, not to mention the airplane. Then, at a later date, what was considered unnatural was breast augmentations, face lifts, and the microwave oven, not to mention polyester, computers, and space tourism. People individually adapt and adjust to change according to society's standards, or what society accepts as acceptable. Society adapts and adjusts to

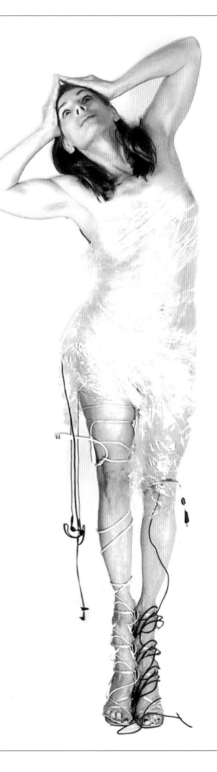

change according to economics. There are two levels within this scenario—the needs level and the wants level. If we need to have something, we will do all that we can to have it. At the most vital level, if we need food and shelter, we will get it. If we need love, we will find it. Further, if we want something we will do our best to make it happen—if we want a Mercedes-Benz we will get more work to trade in the Honda to get it. Or better, if we want an energy efficient, low-mileage hybrid car, we will trade in the Benz to get it. Applying this motivation to needing to find a cure for disease and wanting to live, I think that most people, given the understanding and assessment of their options and given the ability of society to learn to critically analyze information for its benefits versus its disadvantages, I think that most people would opt for Primo.

(1) Extropy Institute: A not-for-profit, educational organization that acts as a networking and information center for those seeking to foster our continuing evolutionary advance by using technology to extend healthy life, augment intelligence, optimize psychology, and improve social systems.

(2) Cochlea: The coillike tube in the inner ear. A cochlear implant is a medical device that bypasses damaged structures in the inner ear and restores partial hearing to the deaf.

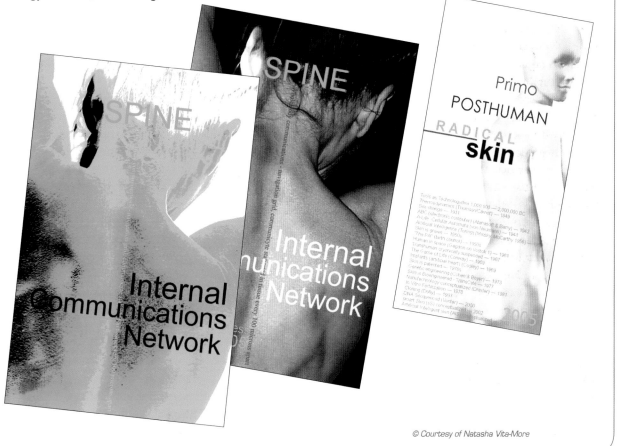

© Courtesy of Natasha Vita-More

Interview
Hod Lipson

Professor Hod Lipson led the Golem Project at Brandeis University to create the first forms of artificial life. Since 2001 he has been working on self-designing robotics at Cornell University.

Hod Lipson.
© Courtesy of Hod Lipson

As far as I understand, your works could lead to self assembling robots / machines. What are the principles underlying it?

Self-assembling machines are likely to be composed of many similar modules, with common interface connections and communications protocols—like the cells in animal bodies. Self-assembly could be done stochastically[1], by having these modules float around and connect where appropriate, or deterministically—where a machine directly assembles each and every component into the correct place.

You also envision self-designing systems. How would it work?

We are working on automatically designed systems—where there is an algorithm that explores the space of a possible machine (both body and brain—morphology and control). The algorithm seeks the design that fits a given task. A self-designing system would be a one that can execute this design-search algorithm autonomously, and then change its own shape to meet the new design.

Have you already had working prototypes based on such research?

We have prototypes of design automation algorithms working using simulated evolutionary processes. We also have shown some machines where this algorithm is executed on board, so that the robot's controller is automatically designed in response to a task given to a robot.

Do you envision the making sooner or later of "shape-changing robotics"?

Automatically changing the morphology (body shape) of the robot is difficult. One of the concepts suitable for this task is known as "reconfigurable robotics"—a concept that has been pioneered by Daniela Rus (MIT), Mark Yim (University of Pennsylvania) and Satoshi Murata (Tokyo University). We also have robots built on these principles capable of changing their morphology, self-repairing, and self-replicating. The more modules these robots have the more versatile their shape can be. Most robots today have a dozen or so units. This will probably expand in the future.

Could you give practical examples of applications of self-assembling objects?

Self-assembly is used already today in some microscale fabrication applications (MEMS). Large-scale robotic self-assembly, such as the type we are considering here, would be useful for long-term, self-sustainable robotic systems that need to repair themselves and adapt to new tasks without human assistance. Space applications (lunar, Mars, Space Station), or operation in hazardous conditions (in reactors, undersea, in rescue) are examples of conditions where such properties would be needed.

When do you think your works may lead to a workable system that anyone could buy?

Lets make a distinction between

Can computers enhance or replace human inventiveness and genius? Hod Lipson thinks they can and is working on machines that can design other machines.
The pictures show examples of self-assembly.

© Courtesy of Hod Lipson

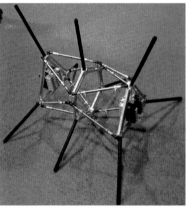

"self-designing" or design-automation systems, and "self-reconfiguring" or shape-changing systems. Design automation is already being used commercially in a number of application areas. Complete robotic systems that can both automatically design their new shape and physically change their shape have been investigated for about a decade in the lab. I would expect that they would be used in commercial applications about a decade from now.

---

(1) Stochastics: That which is based on probability.

Interview
Kevin Warwick

Kevin Warwick is professor of cybernetics at the University of Reading, England. He enjoys having the media spotlight trained on his work and ideas. As a result he attracts his fair share of criticism. In the mid-1990s he issued a stark warning that robots would become too powerful and called for legislation to curb the development of intelligent machines. In 1998 he again hit the headlines when he became (one of) the world's first cyborgs by having an active computer chip implanted in his arm.

Kevin Warwick is convinced that cyborgs will soon be among us.
© Courtesy of Kevin Warwick

How do you foresee the future of robots ?

More intelligence will be put into networked robotic systems, and they will be given more decision-making roles, replacing many positions where today human-thinking is involved. The military domain is a good example. It is already difficult for human-piloted fighter planes to compete with computer-piloted versions—one reason is that humans think too slowly. By 2020 it is difficult to imagine warfare not being mainly technical.

Could robots also help to build a better world ?

In the short term I feel this is what will happen. We will witness more domestic robots helping around the home, more entertainment robots, robots helping in hospitals, and robots performing valuable tasks in many other walks of life.

On two occasions you have had chip implants so that you could become the "world's first cyborg." What was the reasoning behind the move?

In 1998 some scientists, like Peter Cochrane, were predicting that in the future people would have an implant instead of a passport or credit card. But no one had

actually done anything like that at the time—so I did. The computer in my building knew where I was at any time, so my lab door opened for me, lights came on, and the computer welcomed me with a "hello"— all because it could recognize me through my implant. The second one, 2002, was much more—this experiment really was the basis for my book, *I, Cyborg.* Then my nervous system was linked to a computer so that we could send signals back and forth. I wanted to investigate a number of things—in particular the possibilities for extrasensory input and the possibilities ultimately of communicating by thought signals alone.

Could you give practical, everyday examples of the consequences of being a cyborg?

I was able to successfully use an ultrasonic sense, which is literally an extra sense. This could of course be useful for someone who is visually impaired. I could switch on lights and operate a wheelchair directly from neural signals. I could operate a robot hand directly from neural signals and feel how much force the hand was applying. I communicated with my wife (who also had electrodes inserted in her nervous system)—the first direct nervous-system-to-nervous-system communication experiment.

Do you feel superior to what you were before and in what way?

Not now because the 2002 implant was removed after just over three months. Now I am just an ordinary chap again. But when the implant was in place—of course. I had an extra sense and could control technology on the other side of the world directly from my neural signals—I didn't need to press a button or speak.

Your wife took part in a similar experience. How did this change the relationship between the two of you ?

The experiment we conducted was far more successful than we could have imagined. Neural signals were sent from nervous system to nervous system. Since that time we have felt much closer, more intimate.

Do you think most humans might take the same "cyborg route" in the future ?

This is a big question. Some will want to, of course, and some will, I guess, choose to stay as they are. But just who can get upgraded and under what conditions raises enormous ethical questions. Will it all be financially or politically driven. In fact, the technology necessary is not particularly expensive. However, commercial interests haven't yet taken control. So, in theory at least, we could get upgraded en masse, but in reality I think that is extremely unlikely.

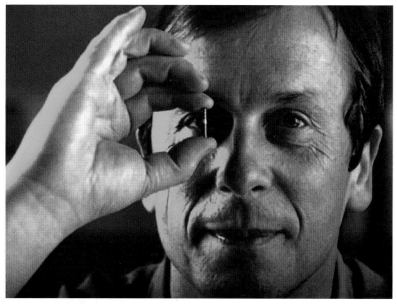

In 1998 Kevin Warwick had a chip implanted in his arm that enabled him to control a robotic hand directly by neural signals.
© Courtesy of Kevin Warwick

Poster for the world premiere
of R.U.R. *by Karel Capek*
(Prague, 1921).

BRIEF GUIDE FOR

THE ROBOTICS ENTHUSIAST

> **Helena:** No, you don't understand me.
> What we really want is to free the robots.
> **Helman:** How do you plan to do that?
> **Helena:** By treating them like human beings.

From *R.U.R.* (Rossum's Universal Robots) by Karel Capek.

Whether presented as fact or fiction, the robot has long been a figure on the stage, in exhibitions, museums, specialist publications, novels and short stories, movies, and DVDs. These different media have, over time, revealed the multifaceted nature of robotic beings, prompting questions as to their place among us humans. We would like to close this book, which we have tried to make as comprehensive as possible, with a brief guide to human perceptions of the robot and passionate attempts to understand it.

Visit

Robots, toys, musical boxes, self-playing instruments and other curiosities in museums and exhibitions.

■ **Musée d'Art et d'Histoire (Art and History Museum)**
Esplanade Léopold-Robert
BP 2001 Neuchâtel, Switzerland
Here you can admire Jaquet-Droz's famous androids.

■ **Keith Harding's World of Mechanical Music**
The Oak House - High Street
Northleach - Glos. GL543, UK
An old factory in the heart of England houses a museum and workshop, where fine craftsmen restore and repair clocks, musical boxes and self-playing instruments. Through the workshop is a museum that exhibits an extraordinary variety of clockwork pieces.

■ **La Maison de la magie (House of Magic)**
1, place du Château
41000 Blois, France
With its 2,000 square meters of exhibition space, sound and visual effects, and performances, Robert-Houdin's "Maison de la magie" is unique in Europe. This voyage through the world of magic is an amusing, spectacular experience.

Hellbrunn Park or "The Devil's Fountain"

Hellbrunn Castle, with its park, peerless trick fountains, and water-driven automata, is located nine kilometers south of Salzburg. It was in 1612 that whimsical archbishop Marcus Siticus had this elegant Italianate residence built in a park where nothing is quite what it seems. Today visitors can enjoy the same mechanical trickery and illusion that delighted the master of Hellbrunn. Nearly 400 years later, the machines are still in perfect working order.

■ **Hellbrunn Park**
Schlossverwaltung Hellbrunn
Fürstenweg 37
5020 Salzburg, Austria

Kyoto Arashiyama Orgel Museum

Antique Orgel Hall - Akaikutu
1-38 Tateishi Tenryuji Saga
Ukyo-ku Kyoto, Japan
This museum houses exceptional pieces, including some rare items from the Reuge collection. Visitors can feast their eyes on the mechanical music box created in 1796 by Antoine Favre and acquired by Reuge, and on Napoleon's musical snuff box, crafted in 1809 by Piguet and Meylan.
www.cjn.or.jp/automta/

Water-driven automata at Hellbrunn Park
© Schlossverwaltung Hellbrunn - Photo: Foto Sulzer

Hellbrunn's mechanical theater
© Schlossverwaltung Hellbrunn - Photo: Foto Sulzer

www.hellbrunn.at

Worth discovering:

■ **www.automates-anciens.com**
Philippe Sayous is a passionate automata enthusiast.
He created this website dedicated to old automata as a tribute to the gifted craftsmen who fashioned music boxes and automata of such lyrical charm and superb finish.

Universal Exhibition, EXPO 2005, Aichi, Japan
March 24 to September 2, 2005
In partnership with the Japanese Ministry of Economy, Trade and Industry and the New Energy and Industrial Technology Development Organization, the Robot Project at EXPO 2005 offers a grand overview of prototypes.
www.expo2005.or.jp/

Roboethics :

An international symposium on "roboethics" will be held in June 2005. It will be chaired by Gianmarco Veruggio, the president of Italy's Scuola di Robotica (Robotics School). It will address the ethical, social, humanitarian and ecological aspects of robotics, bringing together philosophers, legal

© Emanuele Luzzati - Courtesy of Fiorella Operto

thinkers, sociologists, anthropologists, and other scientists to explore the social and economic implications of robotics. The venue will be the Villa Nobel, at San Remo, in Italy. Not to be missed!

www.roboethics.org

Musée des Automates (Museum of Automata)
12, rue des Arts - 38000 Grenoble, France
This museum features an exhibition of the "Mechanical Dreamworld," which mirrors the inner passion of its founder, Francis Lara. Visitors will discover puppets, automata, moving figurines, and musical boxes.

Little Pitch, an automaton that can lean at a sharp angle without falling. It is based on a circus performer who really lived.

© Courtesy of Francis Lara - Musée des Automates, Rêves mécaniques de Grenoble

Fairs

Hanover Fair 2005,
Showcase for innovation in industrial automation technology
April 11 to 15, 2005
Hanover, Germany
www.hannovermesse.de

IEEE/IFR Innovation Award
April 20 to 22, 2005
Barcelona, Spain
www.ncsu.edu/IEEE-RAS/

The Fifth European Dependable Computing Conference (ADCC-5)
April 20 to 22, 2005
Budapest, Hungary
sauron.inf.mit.bme.hu/EDCC5.nsf

ROBOBusiness
Hyatt Regency Cambridge
Cambridge, MA
www.roboevent.com

IPMM'05, 5th International Conference on Intelligent Processing and Manufacturing of Materials
July 19 to 23, 2005
Monterey, CA
www.mining.ubc.ca/ipmm

ISR 2005, The 36th International Symposium on Robotics
November 29 to December 3, 2005
Keidanren Kaikan - Tokyo, Japan

ROBOTICS
International Robotics and Mechatronics Seminar
November 23 to 26, 2005
Russian Exhibition Center
Moscow, Russia
www.expo-design.ru

International Robots and Vision Show
September 27 to 29, 2005
Rosemont (Chicago), Illinois
info@reuterexpo.com

Visit

Karel Capek Museum
Pamatnik - Karla Capka
Stará Hut 125 - 262 02 - Czech Republic

© Courtesy of Kristina Vánová, Památník Karla Capka

This house was a wedding present to Karel Capek and his wife, Olga Scheinpflug, in 1935. It is called "Strz" (ravine), and was the birthplace of numerous works of literature. It now houses a permanent exhibition of the work of Capek, still best known for his play, *R.U.R.* The house is located near Stará Hut, 40 kilometers (30 miles) south of Prague.
www.capek-karel-pamatnik.cz

Take part

▧ PASS

Scientific Adventure Park, Belgium
The PASS is an unusual museum that offers visitors a fun, hands-on approach to discovering the place of science and technology in society. It is housed in a listed colliery in Frameries, a village not far from the city of Mons.
www.pass.be/en/home.shtml

The PASS stages a Robotics Contest.
© PASS - Serge Rovenne

▧ Eurobot
The nonprofit organization Planète Sciences, production company VM Group, the municipality of La Ferté-Bernard, and their sponsors have created Eurobot, an international amateur robotics contest. Winners of national tournaments go on to compete in the International Eurobot Final. Eurobot is a highly technical, scientific, but fun challenge, in which entirely autonomous robots are pitted against each other. Eurobot also organizes numerous conferences.
All the information (rules, how and when to enter) are available on the Eurobot site:
www.eurobot.org

◾ ARTEC
International Festival of Art and Technology
13, rue Viet - 72400 La Ferté Bernard, France
Every year 80,000 visitors throng to ARTEC. They include
4,500 engineering students, which makes it Europe's largest
gathering of budding roboticists. ARTEC staged the first
French Robotics Cup in 1993, followed by the **European Cup**
in 1998. The tournaments are ARTEC's flagship events.
For four days competing teams try to outwit each other in
a unique atmosphere. The contests are organized by
the educational nonprofit organization, Planète Sciences.
www.robotique-artec.com

© Courtesy of ARTEC - Photo: Franck Badaire

◾ ROBOlympics
March 24 to 27, 2005
Ft. Mason's Herbst Pavilion
San Francisco, CA
At the ROBOlympics, the US chapter of the
International Robot Olympiad Committee,
there are numerous team and individual
events, and demonstrations of robotics
technology.
www.robolympics.net

Applaud

Robots,
Roses for Jusinka
Christian Denisart

A play by the theater company
Les Voyages Extraordinaires,
founded by Christian Denisart.
A man, a woman and three
machines are the intriguing
cast for this silent tragedy in
three acts. The story is narrated
by music from eight musicians.
You will find further information
and the dates of the world tour
on this website:
www.robots-theatre.ch

fichier de travail projet ROBOTS scénographie silbert

Watch

Chronology of robot movies and DVDs

1916
Homunculus
Otto Rippert
A deeply lonely android tries to conceal its true nature from humans.

1920
Der Golem, wie er in die Welt kam
The Golem
Paul Wegener
Gotham Distribution
In the heart of the Prague ghetto, a rabbi instills life into a clay sculpture according to the ritual prescriptions of the Kabbala. But the creature revolts against its creator.

1931
Frankenstein
James Whale
Universal Pictures
Henry Frankenstein brings to life a being made from corpses that he and his servant have robbed from graves.
The creature's brain is unfortunately a criminal's. Disaster looms when it accidentally kills little Maria.

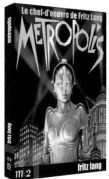

1927
Metropolis
Fritz Lang
MK2
Futura is a robot modeled on Maria, a young woman who works in a subterranean factory. The automated Doppelgänger charms the workers, creating rivalries that drive them to destructive acts in peril of their lives.

1935
Gibel sensatsii
Loss of Sensation
Aleksandr Andrievski
The bourgeoisie caste manipulates a people of mechanical workers in order to pit them against the toiling human proletariat.

1940
Doctor Satan's Robot
William Witney & John English
A crazy scientists makes an invincible robot.

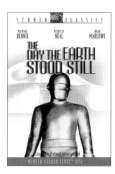

1951
The Day the Earth Stood Still
Robert Wise
Twentieth Century Fox
Klaatu, an extraterrestrial in human likeness, lands on Earth together with a robot called Gorty. Klaatu has come to stop the madness of the human race before it has terrible interstellar consequences. Gort saves his life several times.

1953
The Invaders from Mars
William Cameron Menzies
Image Entertainment
Invaders from Mars capture a city's dignitaries, including its chief of police and the young hero's parents. They become robots remote controlled by the Martians.

1953
Robot Monster
Phil Tucker
Image Entertainment
Ro-Man is a robot

from Mars who looks like a scuba-diving gorilla. He wipes out Earth's entire population, all except for one family. Although ordered to kill them, he is reluctant to do so because he has fallen for Alice, one of the daughters. His leader, the Great Guidance, finishes off the grim work himself.

1954
Tobor the Great
Lee Sholem
Republic Pictures Corporation
Doctor Harrison and Professor Nordstrom create a robot, Tobor, to take over space exploration in place of astronauts. Just when they are about to announce their invention to the press, Nordstrom and his grandson are kidnapped by a spy. Tobor, the faithful robot, sets out to bring them back to safety.

1954
Gog
Herbert L. Strock
Warner Vision
Security agent Sheppard arrives at the secret underground space research base to investigate possible sabotage. He finds that the whole base is coordinated by supercomputer NOVAC and its robots Gog and Magog.

1954
Devil Girl from Mars
David MacDonald
Image Entertainment
An evil female Martian lands on Earth. Her mission is to abduct Earthlings in order to repopulate the red planet. She is accompanied by a terrifying robot, Chanti, who totes a devastating disintegrator gun.

1956
Forbidden Planet
Warner Home Vidéo
Fred MacLeod Wilcox
On planet Altair 4, Robby, a walking-talking robot of superior intelligence, is on the (none too) welcoming committee, waiting for the crew of a spacecraft from Earth. It lands on Altair 4 on a mission to discover what happened to a previous expedition.

1957
The Mysterians
Iroshiro Honda
Beings from outer space who have destroyed their own planet attempt to invade Earth. They are aided in the wicked quest by a gigantic robot.

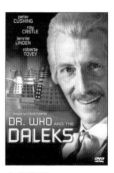

1957
The Invisible Boy
Warner Home Vidéo
Hermann Hoffman
Robby the Robot, who had endeared himself to millions in *Forbidden Planet*, is back again. In this movie he must try to escape from the clutches of a dastardly supercomputer that tries to turn him against humans.

1965
Alphaville
Jean-Luc Godard
Studio Canal Vidéo
Secret agent Lemmy

Caution takes on a totalitarian state ruled by a dictatorial computer, Alpha 60. In Alphaville, the state capital, live the brainwashed Alphabètes, who behave like sedated robots.

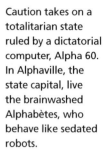

1965
Dr. Who and the Daleks
Gordon Flemyng
Anchor Bay
An eccentric time-traveling scientist continues his lone fight against the Daleks. They are malevolent robots that resemble pepperpots studded with flashing lights.

1965
Docteur Goldfoot and the Bikini Machine
Norman Taurog
MGM/UA

The evil scientist Dr. Goldfoot creates a set of robots that assume the guise of young bikini-clad girls. The plan is to trap the world's richest millionaires by marrying them.

1965
Human Duplicators
Hugo Grimaldi

An alien from a distant galaxy lands on Earth. To accomplish its iniquitous mission of taking total control of our planet, it uses its secret power of duplicating human beings with artificial replicas that do its bidding.

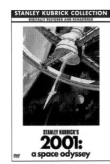

1968
2001 : A Space Odyssey
Stanley Kubrick
Warner Home Vidéo

It is the dawn of humanity. Our primeval, apelike ancestors discover a strange, black monolith and become endowed with intelligence. Four million years later, Dr. Heywood Floyd goes to the Moon to investigate the presence of a strange, black, slablike stone emitting signals to Jupiter. Eighteen months later, the spaceship *Discovery* takes off on a voyage to Jupiter. But the supercomputer HAL, which controls all the craft's onboard functions, causes the death of one of the two astronauts and of the hibernating crew.

1971
THX 1138
George Lucas
Warner Bros

In a totalitarian dystopia, law and order is the task of desexed human beings clad all in white. They, too, are constantly under surveillance from CCTV and android policemen.

1972
Silent Running
Douglas Trumbull
Universal Studio

All vegetation has disappeared from the face of the Earth. Lowell Freeman, aided by three robots, decides that he must try to save plant life from extinction.

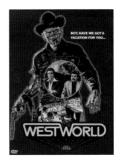

1973
Westworld
Michael Crichton
Warner Home Vidéo

Delos is a Disneyworld where adults can indulge their fantasies. In Westworld, the section that recreates the Wild West, robots play the part of stock characters from westerns. Customers enjoy shoot-outs with them...until the robots rebel and start killing the customers.

1973
Sleeper
Woody Allen
MGM/UA

In the early 1970s the owner of a health-food store is put into cryogenic storage when he undergoes a routine operation that goes wrong. He emerges to find himself in the USA in 2173. The country is now a police state served by robots.

1977
Futureworld
Richard T. Helffron

The follow-up to Michael Crichton's *Westworld*. Two journalists, covering the reopening of the Delos amusement park, discover that a plot is afoot to replace the world's major leaders with robots.

1977
Demon Seed
Donald Cammel

An artificial intelligence engineer designs a supercomputer, Proteus, which is so powerful that it takes over the home PC and traps its creator's wife. Then, using a robot, Proteus rapes her. Later she gives birth to a hybrid baby girl.

1977
Star Wars
George Lucas
Fox Pathé Europa

Princess Leia gives the robot R2D2 a message of warning for the Jedi. It is to tell them that Darth Vader is planning to take over her planet. Together with the robotic diplomat-cum-interpreter, C3PO, R2D2 accompanies and helps Luke Skywalker in his adventures.

1979
Alien
Ridley Scott
Fox Pathé Europa

Lieutenant Ripley looks on aghast as an evil alien, unknowingly shipped onboard the *Nostromo*, kills off its crew members one by one. Too late does she discover that the crew's scientific advisor is an android secretly drafted onboard to bring the alien back to Earth.

1980
Galaxina
William Sachs
Rhino

Galaxina is a beautiful female android. She remains impervious to the advances made by Sergeant Thor on the spaceship *Infinity*. The ship then receives orders to make a 27-year detour. The ship's crew go into cryogenic hibernation, and Galaxina starts feeling something for Thor.

1980
Saturn 3
Stanley Donen
Pioneer Studios

This movie is set in space and features a large robot with an organic brain. An evil psychopath can control it with his own thoughts when he wishes to.

1982
Blade Runner
Ridley Scott
Warner Bros

In a futuristic Los Angeles a former police officer, now a bounty hunter, by the name of Rick Deckard, is hired to kill four replicants who have become dangerous.

1983
Androïde
Aaron Lipstadt

On a space station in 2036 a mad scientist and his assistant undertake illegal research into androids. Later three outlaws on the run, one of whom is a woman, break into the station to hole up there.

1984
Runaway
Michael Crichton
Columbia/Tristar

Police officer Jack Ramsay is given the job of tracking down the evil geek who is reprogramming domestic robots to become killers.

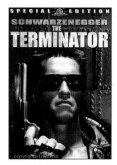

1984
Terminator
James Cameron
MGM/U.A

Terminator is a robot programmed to kill. It is sent back in time on a mission to find and kill all the women who could be the mothers of the man who will lead humans to victory over the robots in 2029.

1985
Daryl
Simon Wincer
Paramount Pictures

A couple adopt a little boy who is just too well behaved. Gradually they discover that he is in fact a robotic prototype that has escaped from a military research laboratory.

1986
Deadly Friend
Wes Craven
A teenager grafts the electronic brain of a robot onto his girlfriend's brain after she has had an accident.
She recovers, then rapidly runs amok.

In this, the seventh *Star Trek* series, a new character joins the crew of the starship *Enterprise*. It is lieutenant-commander Data, a highly refined android who dances, sculpts, paints and acts.

takes refuge in an underground shelter. For company he has the regime's former dignitaries who have fallen from grace. Outside, they are surrounded by dysfunctional androids.

1986
Short Circuit
John Badham
After being hit by a bolt of lightning, robot Johnny 5 apparently becomes endowed with self-awareness.

1987
Robocop
Paul Verhoeven
MGM/U.A
After being beaten to within an inch of his life by a gang of killers, a police officer undergoes major surgery. His damaged limbs and organs are all replaced by electronic parts that augment his powers. He becomes the cyborg Robocop.

1989
Roboforce
David Chung
A gang of bank-robbers use a military robot to try to overcome the police.

1989
Cyborg
Albert Pyun
MGM/U.A
Pearl Prophet is half woman, half robot. She alone has the knowledge that can save the world from certain doom. She is captured by evil cannibals. The only one who can save her is Jean-Claude Vandamme.

1995
Screamers
Christian Duguay
In 2078, on a planet mined for its mineral wealth, scientists have created a race of machines equipped with electric saws. They can change shape and reproduce, and their only aim is to kill.

1999
Bicentennial Man
Chris Columbus
Disney/Buena Vista
A roboticist helps a household robot to become a human being by implanting human organs in it.

1998
Lost in Space (The movie)
Stephen Hopkins
New Line Home Video
In 2050 Earth has become overpopulated and is facing ecological disaster. The eminent scientist John Robinson is given the mission of opening a hypergate—a tunnel in space and time—through which the human race can evacuate to the planet Alpha Prime. The Robinson family blasts off on board *Jupiter 2*, piloted by Major Don West. But Dr. Smith tries everything in an attempt to sabotage the mission.

Lost in Space (The 1965 series)
Twentieth Century Fox

1987
Star Trek, The Next Generation
Gene Roddenberry
Paramount Pictures

1989
Bunker Palace Hotel
Enki Bilal
Sometime, somewhere in a futuristic world, Holm, a businessman,

2001
A.I. Artificial Intelligence
Steven Spielberg
DreamWorks Home Ent.

In the 21st century, overpopulation and global warming have forced governments to institute strict birth control policies. Each couple is allowed to have only one natural child. However, couples can adopt android children. When their daughter falls seriously ill, the Swintons adopt David, an android programmed to feel love for its human parents. But its mother, Monica, cannot bring herself to feel affection for David. When the Swintons' daughter recovers, the mother asks him to leave home.

2004
i, Robot
Alex Proyas

In 2035, robots go peacefully and busily about their daily duties. Human beings have grown used to them and have complete confidence in them. After all, they are programmed to abide by the Laws of Robotics. But when a detective investigates the murder of a roboticist, he begins to suspect a robot as the prime culprit. Then his suspicions gradually widen to include all robots.

Education films, documentaries, shows in VHS
Available in multilingual versions.
For further information, check out:
www.automates-anciens.com

Les Androïdes de Jaquet-Droz
(Jacquet-Droz's Androids)
Til Productions / Talia Films
www.automates-anciens.com
An official selection for the 25th UNESCO International Art and Educational Film Festival.

Between 1770 and 1773, after the death of his young wife and daughter, craftsman and clockmaker Pierre Jaquet-Droz founded a new, "little clockwork family." The automata he built included a writer, a draughtsman and a musician. They were milestones in medical research, for Jaquet-Droz fabricated artificial organs and mechanical prosthetic limbs, based on his extensive work on the human anatomy. The end of the 18th century was marked by a veritable android craze.

La Musicienne
(The Musician)
Olivier Roux,
Jean Cadran
Til Productions / Talia Films
www.automates-anciens.com

The aim of this film is to capture by image and sound a quite exceptional event—the restoration of *La Musicienne*, an android in the likeness of a young woman playing the organ. "She" was built in 1774 by Pierre and Henry Jaquet-Droz, both clockmakers and craftsmen of genius, and both natives of the Swiss canton of Neuchâtel. This documentary recounts what was a great historical, artistic, technological and musical enterprise. It is also a chance for serious movie-lovers to enter a world where dream and reality closely intermingle.

Read

1796, La Grande Histoire de la Boîte à musique
(1796, History of the Music Box)

This film takes a deeply original look at the enchanting world of mechanically driven music. It uses still shots never before shown publicly and rare, period documents to illustrate its fresh approach. It traces the evolution of mechanical music, placing special emphasis on the wonders of musical boxes and automata.

www.automates-anciens.com

Robert-Houdin, une vie de magicien
(Robert-Houdin, a Magician's Life)

Jean-Luc Muller

The aptly named "Theater of Fantastical Evenings" was where Robert-Houdin, the great illusionist, performed. This masterful conjurer's magic tricks were so bewitching that they set a standard that has rarely been equaled, even by today's artists, with all the sophisticated technology at their disposal. The film draws only authenticated facts and historically accurate documents.

www.automates-anciens.com

L'Âge d'or des automates 1848–1914
(The Golden Age of Automata, 1848 to 1914)

Christian Bailly
www.automates-anciens.com

A reference work for collectors and specialists, this book will also fascinate anyone interested in automata. It is the first publication devoted exclusively to 19th-century automata.

Le Monde des automates
(The World of Automata)

Alfred Chapuis et Éduard Gélis
www.automates-anciens.com

This two-volume reference work was written in 1928 by two specialists in the field of automata. It paints pictures of the great craftsmen who made automata and of the mechanical figures they created from antiquity to the beginning of the 20th century.

Waikiki : Le Palais enchanté
(Waikiki: The Enchanted Palace)

Talia Film / Musée Grévin / Musée Baud

A spectacular music-hall show for children aged three and older, with 19th-century automata making up part of the cast. This wonderful idea has given rise to an entertaining "show-in-a-movie" with varied performances from clowns, puppets, illusionists, android musicians, automata and robots.

www.automates-anciens.com

Histoires d'automates
La grande anthologie de la science-fiction
(Stories of automata. The Grand Anthology of Science Fiction)

www.automates-anciens.com

This work is a veritable encyclopedia on the theme of science fiction, exploring a wide spectrum of facets and providing an exhaustive overview of authors and styles from the 1930s to the present day.

Les Univers de Liberatore
(The Worlds of Liberatore)
Preface by Pierre Lescure
Published by Editions Albin Michel

Tanino Liberatore, the creator of Ranx the Robot, is not only a comic strip artist. His talent also extends to painting, illustration, drawing, and work in the movies and music. This book offers a selection of his most outstanding graphic work. It is worth checking out the comic strips of Druillet in the same collection, *Les Univers de Druillet*.

Wired
Wired is indisputably the benchmark for magazines dealing with new technologies and cyberculture. It offers news and in-depth analyses of current developments and future directions. *Wired* regularly runs pieces on robotics, written in its own inimitable style.

www.wired.com

Sciences et Avenir
This French-language magazine frequently publishes pieces on all aspects of robotics. Here are a few examples of issues featuring articles on robots:

▨ 636, February 2000: "Nous sommes les robots" (We Are the Robots).

▨ 669, November 2002: "Les robots coupent le cordon" (Robots Cut the Umbilical Cord).

▨ 688, June 2004: "Les exploits des robots sous-marins" (The Exploits of Underwater Robots).

sciences.nouvelobs.com

Vie artificielle
(Artificial life)
This Website, created by Jérôme Damelincourt, is constantly updated and provides wide-ranging overviews of important developments in the world of robotics, as well as interviews. It also offers a selection of scripts, software, development kits and a forum on the Mindstorms robot.

www.vieartificielle.com

Pour la Science
Pour la science is the French edition of *Scientific American*. It runs regular, comprehensive pieces on robotics. The following issues provide some examples:

▨ 267, January 2000: "L'ère des robots" (The Robot Era) by Hans Moravec.

▨ 300, October 2002: "Vers une robotique animale" (Toward Animal Robotics) by Jean-Arcady Meyer and Agnès Guillot

▨ 315, January 2004: "Une armée de petits robots" (An Army of Little Robots) by Robert Grabowski, Luis Navarro-Serment and Pradeep Khosla.

www. pourlascience.com

La Recherche
Special issue: 350, February 2002
"Les nouveaux robots"
("The New Robots")
Addressed through the three themes of designing, building, socializing. This special issue explores work in robotics research, ranging from the most hands-on to the most abstract. It features some remarkable articles by:

Jean-Jacques Slotine, Rodney Brooks, Dario Floreano, Antoine Danchin et Daniel Mange, Pierre Bessière et Emmanuel Mazer, Jessica Riskin, Thierry Viéville et Olivier Faugeras, Simon Lacroix et Raja Chatila, Christian Laugier et César Mendoza, Pierre Vandeginste, Francis Rocard, Agnès Guillot et Jean-Arcady Meyer, Luc Steels, Krestin Dautenham, Olivier BLond, Nicolas Chevassus-au-Louis, Frédéric Kaplan, Masahiro Fujita et Toshi T. Doi, Robert Triendl, Alexis Drogoul et Jean-Daniel Zucker.

www.larecherche.fr

Read

Flesh and Machines: How Robots Will Change Us
Rodney Brooks
Pantheon Books, 2002
Campus Verlag, 2002
The breadth of his imagination and work enables Australian-born roboticist Rodney Brooks to take a critical overview of developments and directions in robotics. Although it focuses almost exclusively on American research and output, his work is a fine introduction to the history of robotics, particularly behavioral robotics.

Modéliser et concevoir une machine pensante
Approche de la conscience artificielle
(Modeling and Designing Thinking Machines. An Approach to Artificial Awareness)
Alain Cardon
Published by Editions Vuibert, collection Automates Intelligents
Is our knowledge of systems enough to find an answer to the question, "What is thinking?" This book explains how to design and build an artificial system that will enable an autonomous robot to experience sensation, emotion, and even to produce (artificial) thought. It is enriched with forty-two diagrams that are accessible to most readers.

World Robotics Report 2004
IFR
The *World Robotics Report* is an annual statistical report that supplies economic and industrial data on robots in the world. The IFR organizes the International Symposium on Robotics, an annual event that takes place in a different country and on a different continent every year. The 2005 symposium is scheduled to take place in Japan.
www.ofr.org

La naissance d'une langue chez les robots
(The Birth of Language in Robots)
Frédéric Kaplan
Published by Editions Hermes, 2001
Using simple models, this work shows how collective dynamics can produce patterns of convergence towards shared ideas, so giving rise to the gradual emergence of systems that are more regular and easier to learn and to pass on. Kaplan shows how robots go about attempting to imitate human language.

Designing Sociable Robots
Cynthia L. Breazeal
The MIT Press, 2004
In this work Cynthia Breazeal introduces the sociable robot of the future. More than a sophisticated tool, it is a synthetic being, a robot that is socially intelligent in the same way as a human being. The author uses the results of her experiments with the robot Kismet, and explains what is at stake in nonverbal interaction between humans and robots.

Automates intelligents
"Automates intelligents" is an online review on artificial intelligence, virtual reality and robotics that also features interviews with researchers. Christophe Jacquemin and Jean-Paul Basquiat, who designed the site and contribute to its content, do not merely report new developments. They provide in-depth analyses of the state of the robotic art and its implications for our lives and social development. This site is not to be missed:
www.automatesintelligents.com

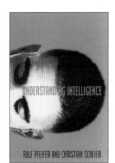

The MIT Encyclopedia of the Cognitive Sciences (MITECS)

Edited by Robert A. Wilson and Frank C. Keil

Since the 1970s cognitive sciences have offered multidisciplinary methods of trying to understand the mind and knowledge. The *MIT Encyclopedia of the Cognitive Sciences* is a reference work that mirrors the diversity of methodologies and theories in the field.

Understanding Intelligence

Rolf Pfeifer and Christian Scheier

The MIT Press, 2001

This book is a clearly structured introduction to new developments in the study of artificial intelligence. Case studies illustrating some of the concepts behind cognitive science help readers steer a path through notions of intelligence.

Robot Shaping

An Experiment in Behavior Engineering

Marco Dorigo and Marco Colombetti

The MIT Press, 1997

The term "shaping" comes from experimental psychology, and denotes the process whereby animals learn by building behavior through successive approximations. The authors propose a new discipline, "behavior technology", which supplies the tools and methodologies for creating autonomous robots.

Des moutons et des robots

(Of Sheep and Robots)

Pierre Arnaud

Published by Presses Polytechniques et Universitaires Romandes

This work discusses the problems of group robotics. The author proposes a new reactive control architecture for mobile robotics, which builds on a simple, powerful, unified model. The book prompts the reader to ponder the very notion of intelligence, from sheep to robots and human beings.

Vie artificielle

Où la biologie rencontre l'informatique

(Artificial life. Where biology meets information technology)

Jean-Philippe Rennard

Published by Editions Vuibert, 2002

Artificial life is a still-mysterious discipline that brings together computer scientists, philosophers, and biologists. This work takes the reader on a journey to the heart of the discipline. Jean-Philippe Rennard, an economist and computer scientist, introduces the topic clearly, accessibly and logically.

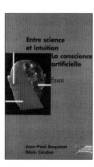

Entre science et intuition La conscience artificielle

(Between Science and Intuition. Artificial awareness)

Jean-Paul Basquiat et Alain Cardon

Published by Editions Automates Intelligents, 2003

The authors aim, in this work, to show that machines with artificial intelligence will be able to help humans solve concrete problems that are often beyond their capacity.

www.automatesintelligents.com

Approche dynamique de la cognition artificielle

Traité des Sciences Cognitives

(Dynamic Approach to Artificial Cognition—a Treatise on the Cognitive Sciences).

Agnès Guillot et Emmanuel Daucé

Published by Editions Hermes, 2001

A veritable reference work that describes in detail dynamic systems and how they fit into the cognitive sciences.

Acknowledgments

Paul Johannes & Derek Birkett (One Little Indian), Bjork, Chris Cunningham, John Payne (RSA / Black Dog Films), Richard Castelli (Epidemic), Stéphanie Vukovic, Jeanne Castoriano, Jean-Pierre Stephan (Minerva), Brian Carlisle (Precise Automation), Michell & Jean-Pierre Hartmann, Matt Denton (Micromagic Systems), Philippe Druillet, Siobhan Hall (Nimba Creations), Tanino Liberatore, Chico MacMurtrie, Bill Vorn (billvorn.com), Rachel Loya (iRobot), Christian Denisart, Frédéric Kaplan (Sony Computer Science Labs Paris), Tom LaPuzza (Space and Naval Warfare Systems Center, San Diego), Luc Barthelet (Electronic Arts), Lucia Beccai (CRIM Lab), Lieutenant Colonel Cyril Carcy (Escadron d'Expérimentation Drones de la base aérienne de Mont de Marsan), Raja Chatila (LAAS-CNRS, www.laas.fr/~raja), Jérôme Damelincourt (Robopolis), Thierry Deroche, Dominique Duhaut (Laboratoire de Recherche en Informatique et ses Applications de Vannes et Lorient), Diane Foley (Nec), Jean-François Germain (Kuka France), Erika Louis-Roy (PSA Peugeot Citroën), Fumio Kanehiro (Humanoid Research Group of AIST), Hiroaki Kitano (Sony Computer Science Laboratories), Hod Lipson (Cornell University, www.mae.cornell.edu/lipson), Frédéric Marchand, Jacques Marescaux (IRCAD/EITS), Christophe Masclet, Francesco Mondada (EPFL www.mondada.net/francesco/), Hans Moravec (Carnegie Mellon University, www.frc.ri.cmu.edu/~hpm/), Marie Obringer (IRCAD), Philippe Sayous (TIL Production), Nadrian C. Seeman (International Society for Nanoscale Science) Math. Des Noes, Luc Soler (IRCAD, www.virtual-surg.com), Takao Someya (University of Tokyo, www.ntech.t.u-tokyo.ac.jp), Tom Suto (Kawasaki Robotics), Stephen Thaler (Imagination Engines), Marc Thorpe, Nicolas Tomatis (BlueBotics), Herman Verbrugge (Internation Federation of Robotics), Régis Vincent (SRI), Natasha Vita-More (Extropy Institute), Kevin Warwick (University of Reading), Walter Weisel (Robotic Workspace Technologies), Chris Willis (Android World), Will Wright (Maxis / Electronic Arts), Haipeng Xie (Dr. Robot), George York (YFX Studios, www.yfxstudio.com), Aude Billard (EPFL), Hisako Ohta (Humnaoid Robotics Insitute, Waseda Univeristy), Hiroyasu Iwata, Sumiko Katayama (National Institute of Advanced Industrial Science and Technology), Wolfram Burgard (Institut für Informatik, Albert-Ludwigs-Universität Freiburg), Chie Ushiwata (Kitano Symbiotic Systems), Takashi Uehara (Sony Corporation), INRIA Photothèque, Tomotaka Takahashi (Robo-Garage, VBL Kyoto University), Rich Walker (Shadow Robot Company), Takao Someya (University of Tokyo), Céline Berger (LIRMM), Jürg Lehni, Emmanuel Chevreul (ARTEC), Hideo Ikuno (Mitsubishi), Tamim Asfour (University of Karlsruhe, http://i61www.ira.uka.de/users/asfour), Faye Martin (Floridarobotics), Aurélia Gance et Anne-Marie Wattellier (Electrolux), Martine Matthys (Kärcher), Yasuhisa Hirata (Tohoku University), CNRS Images – Photothèque, Manami Inoue (Secom Co), Danièle Lemercier (www.ifremer.fr), Ayano Kubo (Japan Agency for Marine-Earth Science and Technology), Michael Jenkin (York University, Canada), Simon Lacroix (LAAS/CNRS), Volker Karpen (International University Bremen), Alastair Bourne (Seiko Epson Corp.), Amir Shapira, Roger D. Quinn (Case Western Reserve University), Nikki Lin (University of California Los Angeles), Céline Noulin (Maison de la magie Robert-Houdin), Kristina Vanova (www.capek-karel-pamatnik.cz), Olivier de Géa, Bob Mottram, Colcord Webster, Val Cureton (Adept), Cécile Diaz (Cybernetix), Yael Edan, François Hirigoyen (Robosoft), Walter Weisel (Robotic Workspace Technologies), Jack J. Conie II (Ca-Botics Fiber System), Tim Bretl, Dominique Martinez, Jin-Wook Kim (FIRA), Aline Petit (www.planete-sciences.org), Veronique Raoul (VM GROUP), Changjiu Zhou (ARICC), Pierre Bureau (K-Team), Trine Nissen (Lego), Florence Bruyère (Le Pass), Thomas Bräunl, The University of Western Australia, Stéphane Nicot (Galaxies), Rich LeGrand (Charmed Labs), Takahiro Wada (Kagawa University), Giulio Sandini (University of Genova), James Trevelyan, Alexander Vogler, Kristin Capece (Evolution Robotics), Alexander H. Slocum (MIT), Joelle Graver (Sony), Rie Sudo (Tmsuk), Josef Syfrig (Tecan Group), Chrystal Phipps (Cardinal Health), NASA/Jpl, Nadia Imbert-Vier et Stéphane Corvaja (European Space Agency), Hydro Quebec, Jean-Baptiste Pean (MK2), Les Cahiers du Cinéma, Ken Kincaid, Sandra Brossard.